terra australis 28

Terra Australis reports the results of archaeological and related research within the south and east of Asia, though mainly Australia, New Guinea and island Melanesia — lands that remained *terra australis incognita* to generations of prehistorians. Its subject is the settlement of the diverse environments in this isolated quarter of the globe by peoples who have maintained their discrete and traditional ways of life into the recent recorded or remembered past and at times into the observable present.

Since the beginning of the series, the basic colour on the spine and cover has distinguished the regional distribution of topics as follows: ochre for Australia, green for New Guinea, red for South-East Asia and blue for the Pacific Islands. From 2001, issues with a gold spine will include conference proceedings, edited papers and monographs which in topic or desired format do not fit easily within the original arrangements. All volumes are numbered within the same series.

List of volumes in *Terra Australis*

Volume 1: Burrill Lake and Currarong: Coastal Sites in Southern New South Wales. R.J. Lampert (1971)

Volume 2: Ol Tumbuna: Archaeological Excavations in the Eastern Central Highlands, Papua New Guinea. J.P. White (1972)

Volume 3: New Guinea Stone Age Trade: The Geography and Ecology of Traffic in the Interior. I. Hughes (1977)

Volume 4: Recent Prehistory in Southeast Papua. B. Egloff (1979)

Volume 5: The Great Kartan Mystery. R. Lampert (1981)

Volume 6: Early Man in North Queensland: Art and Archaeology in the Laura Area. A. Rosenfeld, D. Horton and J. Winter (1981)

Volume 7: The Alligator Rivers: Prehistory and Ecology in Western Arnhem Land. C. Schrire (1982)

Volume 8: Hunter Hill, Hunter Island: Archaeological Investigations of a Prehistoric Tasmanian Site. S. Bowdler (1984)

Volume 9: Coastal South-West Tasmania: The Prehistory of Louisa Bay and Maatsuyker Island. R. Vanderwal and D. Horton (1984)

Volume 10: The Emergence of Mailu. G. Irwin (1985)

Volume 11: Archaeology in Eastern Timor, 1966–67. I. Glover (1986)

Volume 12: Early Tongan Prehistory: The Lapita Period on Tongatapu and its Relationships. J. Poulsen (1987)

Volume 13: Coobool Creek. P. Brown (1989)

Volume 14: 30,000 Years of Aboriginal Occupation: Kimberley, North-West Australia. S. O'Connor (1999)

Volume 15: Lapita Interaction. G. Summerhayes (2000)

Volume 16: The Prehistory of Buka: A Stepping Stone Island in the Northern Solomons. S. Wickler (2001)

Volume 17: The Archaeology of Lapita Dispersal in Oceania. G.R. Clark, A.J. Anderson and T. Vunidilo (2001)

Volume 18: An Archaeology of West Polynesian Prehistory. A. Smith (2002)

Volume 19: Phytolith and Starch Research in the Australian-Pacific-Asian Regions: The State of the Art.
    D. Hart and L. Wallis (2003)

Volume 20: The Sea People: Late-Holocene Maritime Specialisation in the Whitsunday Islands, Central Queensland. B. Barker (2004)

Volume 21: What's Changing: Population Size or Land-Use Patterns? The Archaeology of Upper Mangrove Creek, Sydney Basin.
    V. Attenbrow (2004)

Volume 22: The Archaeology of the Aru Islands, Eastern Indonesia. S. O'Connor, M. Spriggs and P. Veth (2005)

Volume 23: Pieces of the Vanuatu Puzzle: Archaeology of the North, South and Centre. S. Bedford (2006)

Volume 24: Coastal Themes: An Archaeology of the Southern Curtis Coast, Queensland. S. Ulm (2006)

Volume 25: Lithics in the Land of the Lightning Brothers: The Archaeology of Wardaman Country, Northern Territory.
    C. Clarkson (2007)

Volume 26: Oceanic Explorations: Lapita and Western Pacific Settlement. Stuart Bedford, Christophe Sand and Sean P.
    Connaughton (2007)

Volume 27: Dreamtime Superhighway: Sydney Basin Rock Art and Prehistoric Information Exchange, J. McDonald (2008)

Volume 28: New Directions in Archaeological Science. A. Fairbairn, S. O'Connor and B. Marwick (2009)

Volume 29: Islands of Inquiry: Colonisation, seafaring and the archaeology of maritime landscapes. G. Clark, F. Leach and
    S. O'Connor (2008)

terra australis 28

New Directions in Archaeological Science

edited by Andrew Fairbairn, Sue O'Connor
and Ben Marwick

ANU
THE AUSTRALIAN NATIONAL UNIVERSITY

E PRESS

ANU
E PRESS

© 2009 ANU E Press

Published by ANU E Press
The Australian National University
Canberra ACT 0200 Australia
Email: anuepress@anu.edu.au
Web: http://epress.anu.edu.au

National Library of Australia Cataloguing-in-Publication entry

Author: Australasian Archaeometry Conference (8th : 2005 : Canberra, A.C.T.)

Title: New directions in archaeological science / editors, Andrew Fairbairn, Sue O'Connor and Ben Marwick.

ISBN: 9781921536489 (pbk.)   9781921536496 (pdf)

Series: Terra Australis ; 28.

Notes: Bibliography.

Subjects: Archaeometry--Congresses.
               Archaeology--Congresses.

Other Authors/Contributors:
               Fairbairn, Andrew S.
               O'Connor, Sue.
               Marwick, Benjamin.

Dewey Number: 930.1028

Series Editor: Sue O'Connor

Typesetting and design: Bunyanut Suwannakul

Cover photograph: Pierre Proske
Back cover map: *Hollandia Nova.* Thevenot 1663 by courtesy of the National Library of Australia.
Reprinted with permission of the National Library of Australia.

Terra Australis Editorial Board: Sue O'Connor, Jack Golson, Simon Haberle, Sally Brockwell, Geoffrey Clark

Contents

New directions in archaeological science

Foreword

The papers in this volume were presented at the 8th Australasian Archaeometry Conference (AAC) hosted from the 12-15th December 2005 at the Department of Archaeology and Natural History in the Research School of Pacific and Asian Studies of the Australian National University (ANU), Canberra, Australia. The Australasian Archaeometry Conference was initiated in 1982 by ANU archaeologist Wal Ambrose in collaboration with the then Australian Atomic Energy Commission at Lucas Heights. Since its inception, the conference has been an important meeting place for the widely dispersed and diverse community of Australasian archaeological scientists, including both those resident in the region, primarily in Australia and New Zealand, as well as Australasians working around the world. The 2005 meeting at ANU was the first to be held there since 1991 and was the venue for the presentation of 79 papers in 10 sessions covering all aspects of archaeological science from analytical chemistry to GIS, fire histories and ancient technology. The linking concept with all of the sessions was the integration of archaeological science research within broader archaeological projects – in other words forwarding interpretations alongside consideration of method. Over 120 delegates from 6 countries attended the lecture sessions which were accompanied by a poster session, plenary lecture and several workshops. Student participation was especially encouraged and, partly as a result of industry and University sponsorship (see below), subsidised places and travel bursaries helped to boost student delegates to almost one third of the overall total attendance.

One of the key roles of the AAC has been to promote the work of the region's archaeological scientists by producing refereed publications of conference papers. As academic publication has radically changed in recent years so has the publication format and venues for conference publications. Rather than producing a single conference proceedings, papers from the 2005 AAC can be found in a variety of international journals, including *Lithic Technology, Internet Archaeology, Archaeometry* and the *Journal of Archaeological Science*. This volume presents original research papers from three key subject areas covered in the conference, all of key research interest to Australasian archaeologists, namely geoarchaeology, archaeobotany and archaeometry. In keeping with the aims of the AAC, student papers feature heavily among those published here and we have included papers studying a range of geographical regions. All provide new insights into the past and several herald significant shifts in analytical priorities for their component disciplines.

Geoarchaeology is a key focus of research interest for Australasian researchers often dealing with ancient and eroded landscapes and where dating cultural materials in surface and even stratified contexts can prove extremely challenging. Several of the papers focus on dating and interpretation of surface exposures of artefacts and provide novel approaches integrating studies of geoarchaeology with geochronology to this end. One group of papers tackles these problems in western NSW, an arid region where archaeological sites such as surface stone artefact scatters and hearths have in the past proved difficult to date and to compare with the archaeological record in other regions of Australia. Many of these studies are not just about new ways of doing or dating archaeology: some of the results presented have significant implications for the inferences archaeologists draw from their data.

A good example of this is the paper by Holdaway, Fanning and Littleton which provides a salutary example of how time and taphonomic processes impact on the sedimentary and archaeological record; a factor often overlooked by archaeologists who use radiocarbon dates as proxies for numbers of people in the landscape to generate inferences about human behaviour in the past. They compare the frequency distributions derived from two different data sets for Holocene western NSW, heat retainer hearths and dated human burials. The heat retainer dates show a dramatic decrease in frequency with age, whereas the burial dates have a more even distribution. Their study demonstrates the impact of differential preservation, not just of cultural materials but also of sedimentary contexts, accessible to archaeologists to study. They conclude that the "accumulative hearth record in northwestern NSW, and, by association, the accumulated record of stone artefacts, increases during the last 2000 years because of geomorphic surface preservation rather than greater numbers of Aboriginal people and/or increases in occupation span".

Fanning, Holdaway and Phillips contribute a new method of identifying heat-retainer hearths which have been dispersed. These important sites are ubiquitous in the arid landscape of western NSW but as they are exposed on the surface they are highly subject to erosion and bioturbation. The eventually loose the charcoal and spatial patterning is lost and identification becomes problematic. Fanning *et al.* recommend a technique based on a combination of condition, typology and use of a fluxgate gradiometer to detect remnant thermo-magnetism in the heat retainers. If adopted by other researchers this technique will assist with standarisation of descriptions of dispersed hearths and monitoring of changes in their condition over time. Shiner's paper focuses on stone artefact scatters in the same region, dated by association with the hearth sites to the late Holocene. His paper examines patterning in the lithic artefacts and raises the issue of 'scale' in interpretation of data which in effect represents a palimpsest of diverse human activities over several thousand of years. This paper should be of interest to all archaeologists working in Australia where surface stone artefacts form the major component of the archaeological record.

Similar problems of recording cultural materials dispersed over a vast geographic area present themselves to archaeologists dealing with the historic time frame. Artefacts of the 19th and 20th centuries can often be dated with some precision using typology and manufacturing dates, but when spatially dispersed this data can appear unpatterned, is difficult to record and is often neglected by archaeologists. Bolton's paper describes an effective method of recording archaeological data from historic period sites *in situ*.

Prendergast, Bowler and Cupper employ a multidisciplinary approach to study the environmental evolution and human occupation of the Victorian Mallee region. They show the while humans were present in this landscape from at least 15,000 years ago the fluvial regime and climate of the region were substantially different to that of today. They emphasise that the recognition of changes in fluvial and aeolian environments though successive humid-arid cycles is necessary for interpretation of the sedimentary and archaeological record and human adaptation in the Murray Basin.

Six papers focused on archaeobotany and related studies in Australasia and elsewhere. Four detailed research into plant microfossil taphonomy and analysis, which continues to be an important focus for Australasian researchers. Lamb's paper builds on recent results confirming the utility of Congo Red to identify cooked starch, with details of contamination experiments. Haslam's paper details an experimental observation of starch movement in soils as a result of water percolation and provides unsettling evidence that even large starch grains may be subject to lateral and vertical post-depositional movement. As with the subsequent paper by Crowther on raphide identification, the paper's conclusion demands the routine investigation of site formation processes when evaluating microfossil occurrences in archaeological strata and thus closer integration between archaeobotanists and geoarchaeologists. Crowther's paper is a significant contribution to the Pacific microfossil debate, critically evaluating the reliability and methods of raphide identification, which has become an important indicator of prehistoric aroid cultivation. Raphides are calcium oxalate crystals, and Crowther's work shows that light microscopy may confuse commonly occurring calcite crystals with raphides. Globally, archaeological starch analysis has produced significant new results in recent decades and the papers here, along with recent publications (e.g. R. Torrence and H. Barton 2006 *Starch Analysis*), suggest that Australasian practitioners, who were among the first in the field, are entering a stage of consolidation and necessary critical evaluation of methods and results.

Longford, Drinnan and Sagona's paper provides a good example of integrated macrobotanical research, utilising both seeds and wood charcoals, providing important environmental, economic and behavioural information about Sos Höyük in northeastern Turkey. The paper presents preliminary data from the Late Chalcolithic to Iron Age, periods whose archaeobotany is poorly understood in Eastern Anatolia and under-represented in southwest Asia as a whole, especially when compared to the Neolithic. The paper serves to illustrate how wood charcoals can serve to provide much more information than simply radiocarbon dates and presents results from an important Australian field project in Turkey. Macrofossil analysis is part of an integrated suite of techniques critically evaluated by Denham, Haberle and Pierret as a source of information for understanding a key topic in the region: the emergence of tropical agriculture. Once again, the Kuk site provides the focus for an engaging and well-supported exploration of how integrated archaeological science can provide novel and vital information about past agricultural and land-use practices. Concurring with a theme repeated through this volume, the paper argues strongly for even greater integration of tailored geoarchaeological and archaeobotanical methods if archaeological studies of Australasian plant use, and perhaps those of tropical regions across the Asia-Pacific as a whole, are to overcome taphonomic and classificatory problems to provide the depth of data that has revolutionised understanding of agricultural origins in the Americas and the Old World. As with all of the papers in this section, this final study hints at a methodologically and theoretically healthy future for this area of archaeological science.

The third and final section includes papers covering a variety of archaeometry disciplines beginning with Petchey's review of marine shell dating in Oceania. The paper provides an up-to-date review of the subject, with guidance on date selection and a call for the construction of more shell-terrestrial (charcoal) paired date sequences to help improve Oceanic marine calibration, which is highly variable in time and space. Bourke and Hua's paper provides such a calibration series from the Beagle Gulf in the Northern Territory of Australia, providing a terrestrial-marine sequence that concurs well with near by data.

Rhodes *et al.* present the preliminary results of a novel application of the Optically Stimulated Luminescene (OSL) dating technique for estimating the age of heat-retainer hearths exposed in surface contexts in arid Australia. The comparisons with radiocarbon ages derived from charcoal from the same hearths suggest that the technique holds promise. This technique had the advantage that it does not require excavation of the hearth so provides a good substitute for radiocarbon dating in situations where non-destructive dating measures are called for.

The two following papers by Grave and Craig and Grave provide important new insights into the potential of chemical characterisation techniques to investigate both the production and use of pottery. Graves' paper presents a novel approach to statistically and mathematical analysis of ICP-OES results from a 'closed group' of shipwreck ceramics from South and Southeast Asia, which allows the distinction of firing technologies and provenance, providing a significant new approach for disentangling the complexity of ancient technology. Craig and Graves' paper identifies the underused method of HPLC-MS as a means of helping to identify some food residues, in their case using experimental samples and pots from ancient Gordion, Turkey, home of the famous King Midas. Moving to yet another geographical region, Herries' paper uses examples from South Africa to illustrate the potential of archaeomagnetism to investigate issues as diverse as fire histories and palaeoclimate. The techniques have clear potential for further understanding aspects of Australasian prehistory and we hope to see more local studies in the future. The final paper by Macgregor brings us closer to home with a discussion of conservation of megafaunal bone remains from the famous Cuddie Springs spite in New South Wales, Australia. The paper details the planning, field and laboratory methods applied to the megafaunal bone from this important site, ensuring that are available for detailed future study and curation. Conservation is a key discipline in global archaeology and is alive in Australasia still, though now without the University of Canberra's internationally renowned conservation program, which we note with dismay was closed several years ago.

For the conference we acknowledge and thank the major sponsors of the conference, namely the Department of Archaeology and Natural History, The Graduate School, the Research School of Pacific and Asian Studies and the Centre for Archaeological Research, all of the Australian National University. We also received financial and in-kind support the ARC Network for Earth Systems Science, the Rafter Radiocarbon Laboratory (Wellington, New Zealand) and the Waikato Radiocarbon Laboratory (Hamilton, New Zealand). Administrative support was provided by Sharon Donahue and web-support by Amanda Kennedy & Jill Jones. We also thank the session convenors and student volunteers who worked tirelessly to make the conference a success. Authors and reviewers worked hard to prepare and improve the papers we have published here and we also thank Jean Kennedy for editorial assistance. We hope this volume of papers contributes positively to the fostering and further strengthening of archaeological science in the Australasian region and demonstrating how Australasia continues to contribute to the advance of global archaeological science. In this respect it is encouraging that Australia's first postgraduate degree in archaeological science, the Masters in Archaeological Science at ANU, was launched in the year this volume sees publication.

Andrew Fairbairn, Sue O'Connor and Ben Marwick
2008

1

Assessing the frequency distribution of radiocarbon determinations from the archaeological record of the Late Holocene in western NSW, Australia

Simon J. Holdaway[1]
Patricia C. Fanning[2]
Judith Littleton[1]

[1] Department of Anthropology
University of Auckland
Private Bag 92109
Auckland, New Zealand

[2] Graduate School of the Environment
Macquarie University
NSW 2109, Australia

Abstract

When grouped together, radiocarbon determinations from heat-retainer hearths from western New South Wales decrease in frequency with age. One interpretation for this pattern is that it reflects an increase in the frequency of occupation, and perhaps ultimately an increase in Aboriginal populations in this region during the late Holocene. An alternative explanation is that the increase in frequency reflects the differential preservation of the land surfaces on which the hearths are found. According to this explanation, the pre-servation of ancient surfaces itself decreases with time, with the destruction of ancient records of occupation accounting for the relatively recent skew in the age distribution of heat-retainer hearths. We test these hypotheses by comparing the distribution of hearth radiocarbon ages against the distribution of ages obtained from samples associated with buried human remains from western New South Wales. Samples obtained by excavating buried deposits should not be subject to the same range of erosion processes that have affected surface deposits. Therefore the samples from buried deposits are able to act as a control against which the distribution of hearth ages can be evaluated. Results indicate that age estimates obtained from human burials have a distribution different from those obtained from hearths, supporting the conclusion that the decrease in hearth frequency with age is a product of geomorphic preservation rather than cultural change.

Keywords: radiocarbon; hearths; chronology; western New South Wales; human burial

Introduction

The archaeology of the late Holocene in Australia is frequently characterised by, amongst other things, an increase in the number of identified 'sites' (Ross et al. 1992:109), an increase in the number and variety of artefacts found at those sites (Hiscock 1986) and, in western New South Wales (NSW), an increase in the number of heat-retainer hearths (or earth ovens) exposed on the surface (Holdaway et al. 2002). There are two major hypotheses posed to explain these data. On the one hand, the apparent increase in the abundance of artefacts at late Holocene sites may indicate increasing populations (e.g. Beaton 1983), greater occupation spans reflecting in part the existence of more complex social interactions (e.g. Lourandos 1983, 1984, 1985, 1988, 1993), or both — evidence used as part of the argument for a mid to late Holocene intensification in Australia. On the other hand, this pattern may not reflect human behaviour and instead be the result of differential preservation, i.e. more evidence surviving from the recent times compared to the more distant past — the preservation hypothesis (e.g. Bird and Frankel 1993; Fanning et al. 2008; Holdaway et al. 2008)

This issue has proven to be very difficult to resolve, especially in western NSW where the bulk of the archaeological record consists of surface material. Scatters of stone artefacts and associated heat-retainer hearths are visible on the surface because of erosion, the product of overgrazing of domestic stock since the mid to late nineteenth century (Fanning 1994, 1999). It can be very difficult to estimate the age of many of these surface deposits since the same erosion that has exposed them has removed the possibility of determining age through conventional approaches to interpreting archaeological stratigraphy. The challenge for archaeologists is to reliably estimate the age of the surface archaeological record. This will help to determine if the apparent late Holocene increase in the density of the archaeological record results from increases in Aboriginal occupation or whether it is a consequence of the preservation of this record. In this paper, we summarise the results of our investigations of the chronology of a number of locations we have studied in northwestern NSW (Figure 1). We then compare these data with age determinations from burials in southwestern NSW (Figure 2) to help to determine whether intensification or preservation best explains the nature of the late Holocene archaeological record.

Developing a chronology of Aboriginal occupation in wertern NSW

The two components we have used to develop a chronology for the surface archaeological record in western NSW are radio-carbon determinations from heat-retainer hearths and optically stimulated luminescence (OSL)

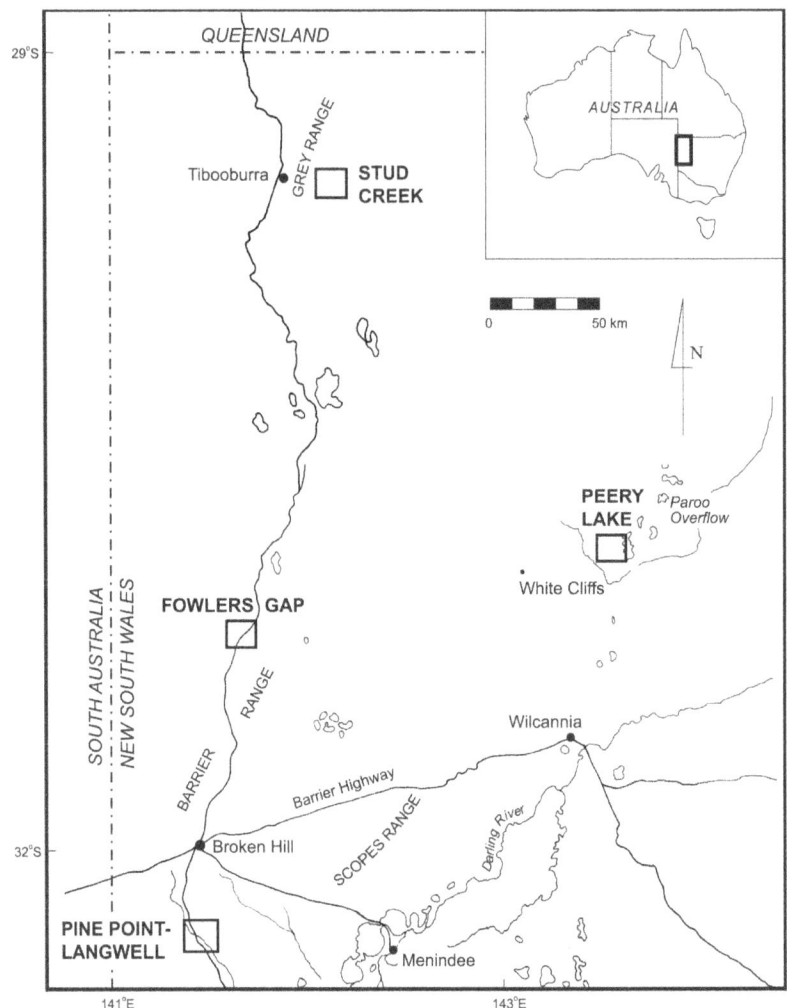

Figure 1. Map of western NSW showing locations of the study areas mentioned in the text.

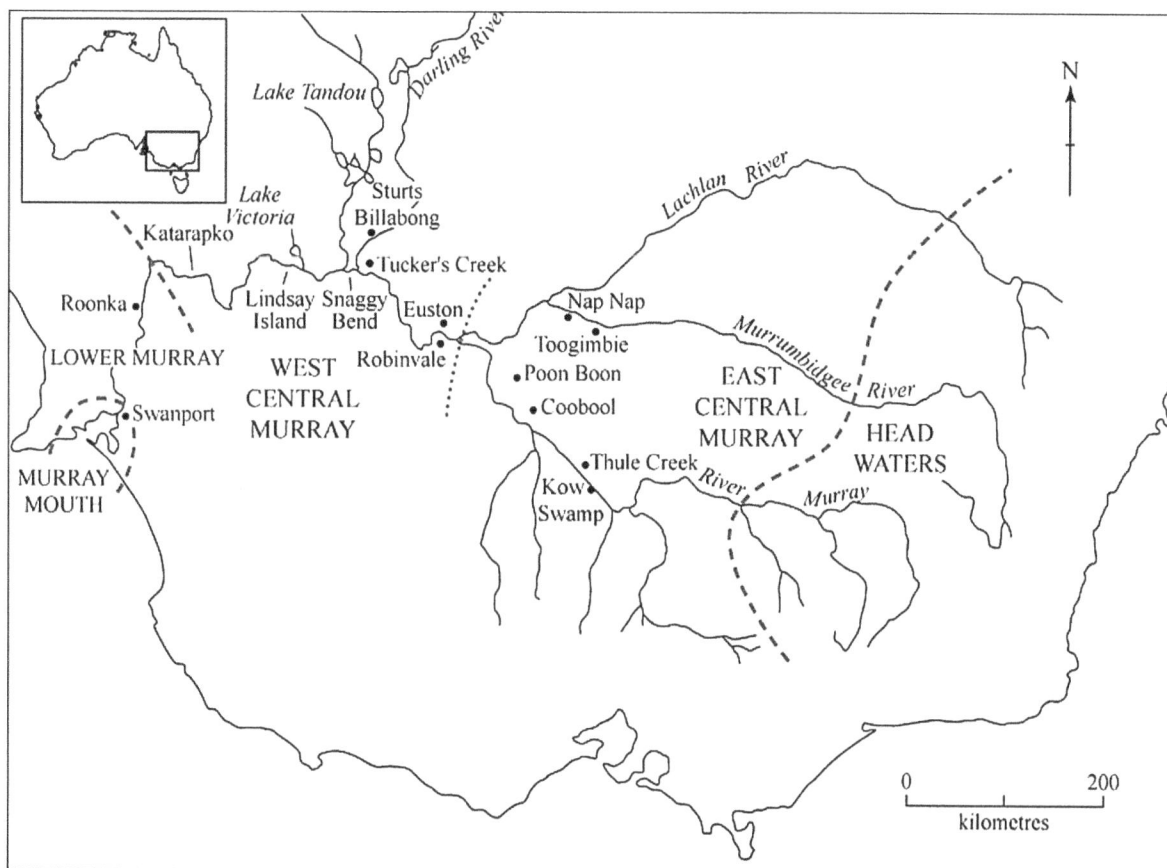

Figure 2. Map of southeastern Australia showing the Lower Darling and Murray Rivers and their major tributaries (from Littleton 2007). Locations where burials forming part of the data illustrated in Figure 4 have been found are indicated.

age estimates from under-lying sediments (Rhodes et al., this volume). The statement that surface deposits lack stratigraphy is correct in an archaeological sense but not true geomorphologically. Over large tracts of western NSW, especially along upland valley floors, erosion has removed much of the surface sediments onto which artefacts were deposited and later buried, leaving the artefacts behind as a lag (Holdaway et al. 2000). The artefacts are today lying unconformably on an older sedimentary unit, the age of which provides the terminus ante quem, or maximum age, for their deposition. Simple application of the laws of superposition says that the artefact deposits cannot be older than the sedimentary unit they are resting on. Individual artefacts in the deposit may be older as they mayhave been manufactured considerably earlier and used many times prior to discard. Here we are concerned only with the maximum age of the artefact deposit.

In some areas of the world, artefacts themselves provide a method for age estimation since they change their form through time. In Australia, however, attempts at seriation (e.g. Hiscock 1986) have not been successful, and there is currently no means of determining the age of late Holocene assemblages with any precision based on artefact typology alone. In western NSW, the remains of heat-retainer hearths provide a chronology since the hearths frequently contain charcoal that is suitable for radiocarbon determinations. Obtaining age determinations from these hearths does not, however, 'date' the stone artefacts associated with these hearths since there is not necessarily a close relationship between the times at which hearths were constructed and the times when the stone artefacts were deposited. It is not possible to use the age estimates for the hearths to 'bracket' stone artefacts as is sometimes the case with stratified deposits. It is possible, however, to use multiple age estimates from hearths to define patterns in the construction of hearths through time and then to compare the nature of artefact assemblage composition against these patterns (e.g. Holdaway et al. 2005). For example, artefact assemblage composition should be different where occupation was continuous versus situations where the radiocarbon hearth chronology indicates a discontinuous pattern of occupation.

We first applied this strategy for determining the chronology of surface artefact scatters at Stud Creek in Sturt National Park (Figure 1). Here, OSL age estimates on sediments infilling the valley floor suggested that stone artefacts and heat-retainer hearths rested unconformably on a surface with an age no older than about 2000 BP (Fanning and Holdaway 2001). Radiocarbon age determinations on the hearths themselves suggested occupation from 250–1700 cal BP but with a gap of about 300 sidereal years between 910 AD and 1260 AD when no hearths were constructed (Holdaway et al. 2002:357). Significantly, this gap in hearth construction coincides with the Medieval Climatic Anomaly (MCA — Stine 1994), a period of warmer average temperatures and hydroclimatic variability, recognised globally in climate proxy records as extending from as early as 800 AD to as late as 1450 AD (Broeker 2001; De Menocal 2001; Eital et al. 2005; Favier Dubois 2003; Herweijer et al. 2007; Huffman 1996; Jones et al. 2001). This suggested the possibility of an environmental correlate for the period with no hearth construction. However, any cause and effect relationship would need to be based on demonstrating the palaeoenvironmental impact of the MCA in Australia and, as yet, there are no climate proxy records from the Australian mainland that demonstrate an MCA signal.

To test the reliability of the dating strategy established at Stud Creek, we investigated multiple locations on Fowlers Gap Arid Zone Research Station, some 200 km south (Figure 1). Here, we collected samples for OSL dating from exposed sections at each of six sampling locations to provide maximum age estimates for the archaeological surfaces. The results supported those from Stud Creek in one sense, in that radiocarbon determinations from the heat-retainer hearths were all more recent than the OSL age estimates for sediment samples, confirming the applicability of the technique as a means for obtaining the terminus ante quem for surface artefact scatters (Fanning et al. 2008; Rhodes et al. this volume). Additionally, the Fowlers Gap results indicated that surface ages could vary considerably depending on the local geomorphology. For example, at the Fowlers Creek (FC) sampling location, a gravelly terrace surface, OSL age estimates indicated that the sediments within about 32 cm of the surface were deposited at least 4340 years ago and perhaps up to 10,920 years ago. In contrast, OSL determinations from an active proximal floodout environment at the Sandy Creek (SC) sampling location were much younger, ranging from modern at the surface to around 4000 years old at 62 cm depth. At a third location (Nundooka — ND), marginal to a bedrock channel, sediments at 52 cm below the modern surface were deposited around 3730 years ago.

When these age estimates were compared with the results of radiocarbon determinations from the heat-retainer hearths, a similar result to Stud Creek was demonstrated, i.e. there is a close relationship between the age of the landform and the age distribution of the heat-retainer hearths. At Fowlers Gap the oldest hearths (around 6000 cal BP) were found at the FC sampling location that also returned the oldest OSL determination. On surfaces associated with more active geomorphic environments, like the SC and ND sampling locations, hearths were much younger (Holdaway et al. 2005).

These results provide a set of answers to the question posed above. At least in the locations we have studied at Fowlers Gap, there is good evidence that the age of the surface and the age of the record that is visible upon it are correlated, i.e. longer duration archaeological records are found on older land-forms. Moreover, relatively ancient records are rare, surviving only where landforms are stable and where the erosion that has exposed the record has not also destroyed it, for example at FC. The geomorphically more active floodplain environments like those at SC and ND are more common at Fowlers Gap than terraces like that at FC, thus, younger archaeological deposits dominate the record. Therefore, at least at Fowlers Gap, the evidence suggests that the prevalence of the archaeological record in contexts dating to the last 2000 years is a function not of people but rather of the geomorphic history of land surfaces that preserve this record [1].

This inference is reinforced when the heat-retainer hearth age estimates from elsewhere in north western NSW are examined (Holdaway et al. 2005). This record does not illustrate continuous occupation during the last 2000 years, but instead shows episodes of hearth construction and use, interspersed with gaps in the record. Our ninety-six radiocarbon determinations from four widely spaced locations

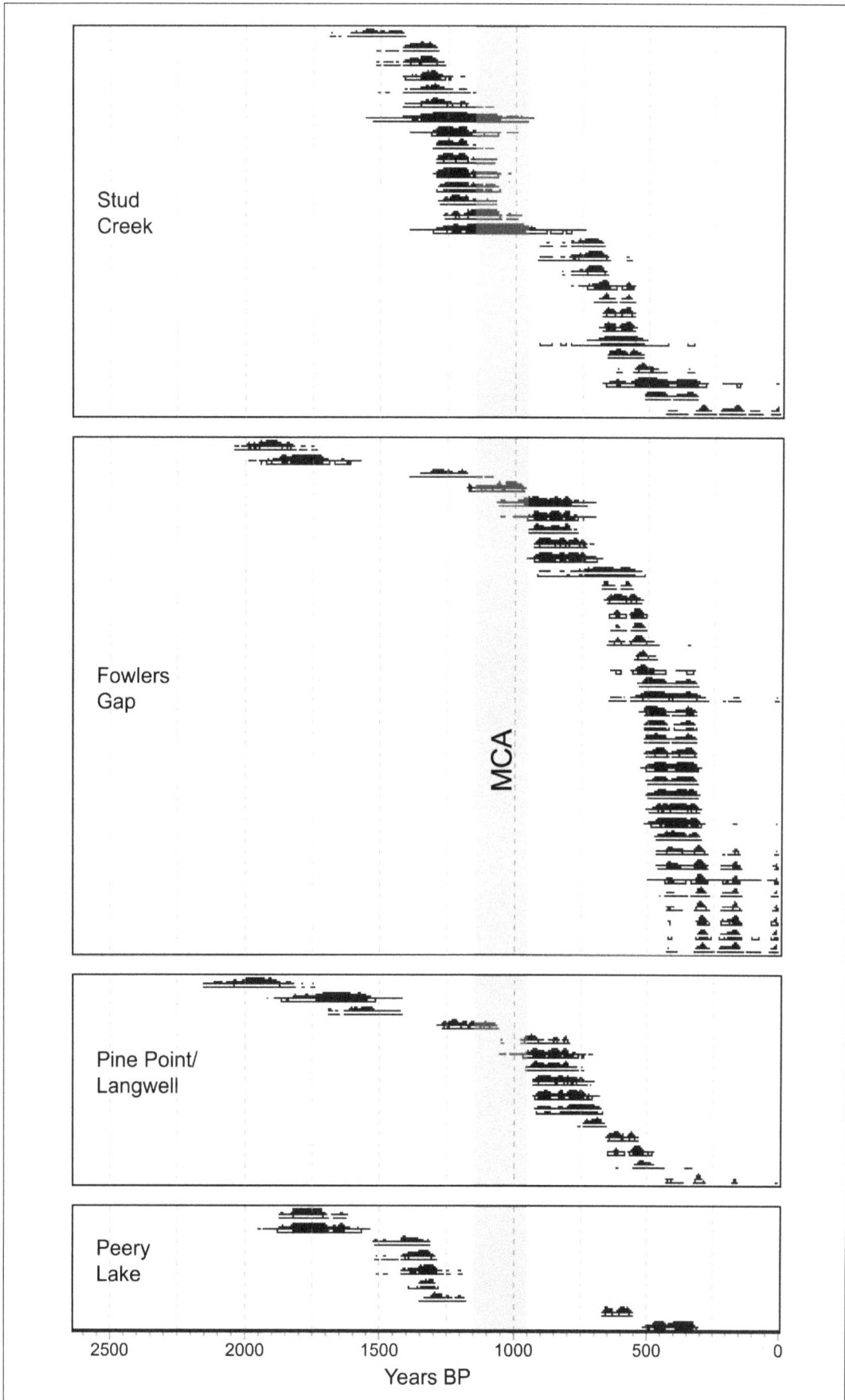

Figure 3. Calibrated radiocarbon determinations from excavated hearths in western NSW (Holdaway et al. 2005: Figure 9). A gap is indicated in hearth construction coincident with the MCA. Calibrated ages were determined using Datelab 2.0 software (Jones and Nicholls 2002) accessing INTCAL98 calibration curve data (Stuiver et al. 1998).

(Stud Creek, Fowlers Gap, Pine Point/Langwell, and Peery Lake; Figure 3) demonstrate a consistent pattern, with a gap in the record that may be coincident with the MCA (although none of the additional sampling locations provided as clear a pattern as that developed for Stud Creek). As indicated in the discussion of the Stud Creek results above, environmental correlates may help to explain the distribution (Holdaway et al. 2002). However, other causes are possible, not the least of which is variation in the production of ^{14}C in the past, as evidenced by variations in the radiocarbon calibration curve. In the rest of this paper, we consider two sets of evidence that may enable us to better understand the distribution of hearth age estimates from western NSW and also help to resolve the intensification versus preservation debate.

Testing the model using dates from burials

The argument derived from the Fowlers Gap results is that hearth preservation and, by extension, the preservation of other parts of the archaeological record, are a function of the geomorphic environment. Hearths that were once buried structures are visible today because of erosion. However, erosion also occurred in the past, to both a greater and a lesser degree, leading to the differential destruction of hearths in different parts of the landscape (Fanning 2002). If this argument is correct then the patterned distribution of hearth ages through time should reflect the preservation conditions largely controlled by geomorphic processes. Equally, archaeological records where the preservation conditions are different should show a different distribution of ages.

As previously noted the bulk of the archaeological record in western NSW consists of surface artefact scatters and associated heat-retainer hearths. However, other evidence for Aboriginal occupation is preserved in quite different contexts. Human burials have survived from the past because of interment underground, thus reducing the impact of surface erosion. This should, in theory, lead to a different record of preservation, one that paints a picture closer to the true regional chronology of human occupation rather than one more closely related to local geomorphic history.

Littleton (2007) has compiled the results of radiocarbon determinations from human burials from a variety of locations in the lower Darling-Murray region to the south of our study area (Figure 2). With one exception, all the ages are directly from bone and the majority of the determinations come from excavation samples. These are generally locations with multiple burials, although included in the sample are a number of single radiocarbon determinations from burials exposed through erosion.

In Figure 4, we compare the probability distribution of the fifty burial age determinations compiled by Littleton (2007) with that for the ninety-six hearth age determinations (Holdaway et al. 2005). While the distribution of human burial age estimates spans the length of the Holocene, the age distribution for hearths is skewed to younger ages, with a large number of determinations clustering within the last two thousand years. Age estimates older than about 6000 BP form smoother peaks in both graphs, a pattern that reflects the relatively larger error terms associated with older age estimates.

What might be accounting for the differences in the two age distributions? Before discussing the most likely reasons, there are two relatively minor differences in the two data sets that need to be considered. First, the hearth determinations are from charcoal samples while the human burials, in the main, are determined from bone apatite samples. The possible effects of this difference in the composition of the two types of samples on the age distributions are unknown. Second, there are half the number of determinations from burials than from hearths (fifty cf. ninety-six). However, samples of this size are large enough to successfully search for long-term patterns.

Notwithstanding these small differences in the two data sets, the resulting probability plots are fundamentally different in shape. Whereas the plot based on burial data shows a variable, but continuous distribution through time, the plot for the hearths diminishes to zero after 2000 radiocarbon years BP, with only small peaks present at earlier times. As argued above, this most likely is a reflection of the low probability of ancient surface preservation. More recently formed surfaces, and therefore a larger number of younger hearths, are more likely to be preserved. In contrast, the act of burial has preserved human

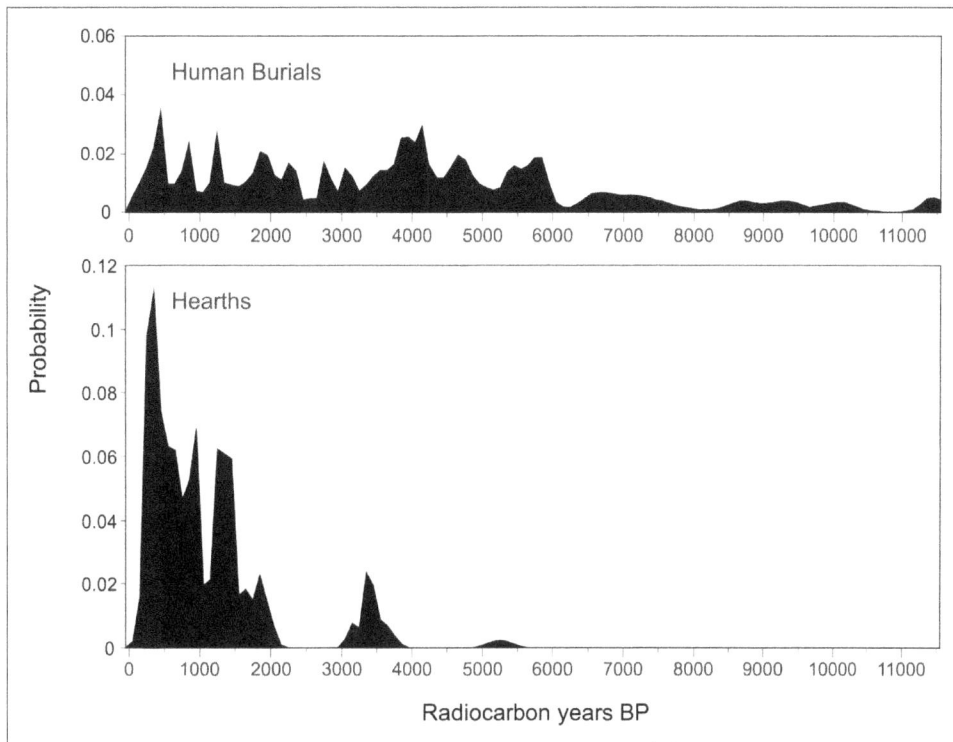

Figure 4. Probability plot for radiocarbon determinations obtained from human burials (upper) and hearths (lower) from western NSW. Probability is calculated per 50 radiocarbon years.

remains that date from a greater range of past time periods. Thus, there is a more even distribution of the human burial dates throughout the Holocene.

Based on this comparison, we suggest that the accumulative hearth record in northwestern NSW, and, by association, the accumulated record of stone artefacts, increases during the last 2000 years because of geomorphic surface preservation rather than greater numbers of Aboriginal people and/or increases in occupation span.

Gaps in the occupation record

Comparison of the ninety-six hearth age estimates from locations across northwestern NSW (Figure 3) suggests that the times when no hearths were constructed may be regional in extent (Holdaway et al. 2005). It is possible, however, that the gaps we detected in the hearth age distributions illustrated in Figure 3 are in some cases an 'artefact' of plateaus in the calibration curve. The burial data discussed above provide a means to assess this, by providing a comparative data set from which to determine whether a similar gap in the age distribution of human burials exists. Box plots of radiocarbon determinations, generated using the Oxcal software, from hearths and from human burials that span the last 2000 years (Figure 5) allow an assessment of the effect of the radiocarbon calibration curve itself on the distribution of calibrated radiocarbon age estimates.

The influence of plateaus on the distribution of calibrated ages is most apparent for the last 600 years where the calibration curve has steep sections bounded by plateaus. To some degree, the distributions of the hearths and burial age determinations permit a test of whether the plateaus in the calibration curve are responsible for the apparent gaps in the radiocarbon record before 600 BP. For the hearth data, for instance, there is a gap in the distribution of ages beginning around 900 years ago (Figure 5a). Evidence for the existence of such a gap is less apparent, however, in the plot for human burials (Figure 5b). At face value, this suggests that while construction of heat-retainer hearths appears to have temporarily ceased in northwestern NSW, people continued to bury their dead during this period further to the south along the lower Darling and Murray Rivers.

The overall lack of precision of the radiocarbon determinations together with the spread of ages from burials over a long time period restricts the types of explanations that can account for the difference between the two plots. Nevertheless, the results obtained from both data sets suggest a method for assessing the reality or otherwise of gaps in the record. If the gaps are an artefact of plateaus in the calibration curve they should be present in both data sets. If they are absent, then behavioural and/or environmental explanations may be called for to explain why no age determinations are forthcoming for particular periods.

An important next step would be to conduct a dating program of hearths located close to some of the human burial

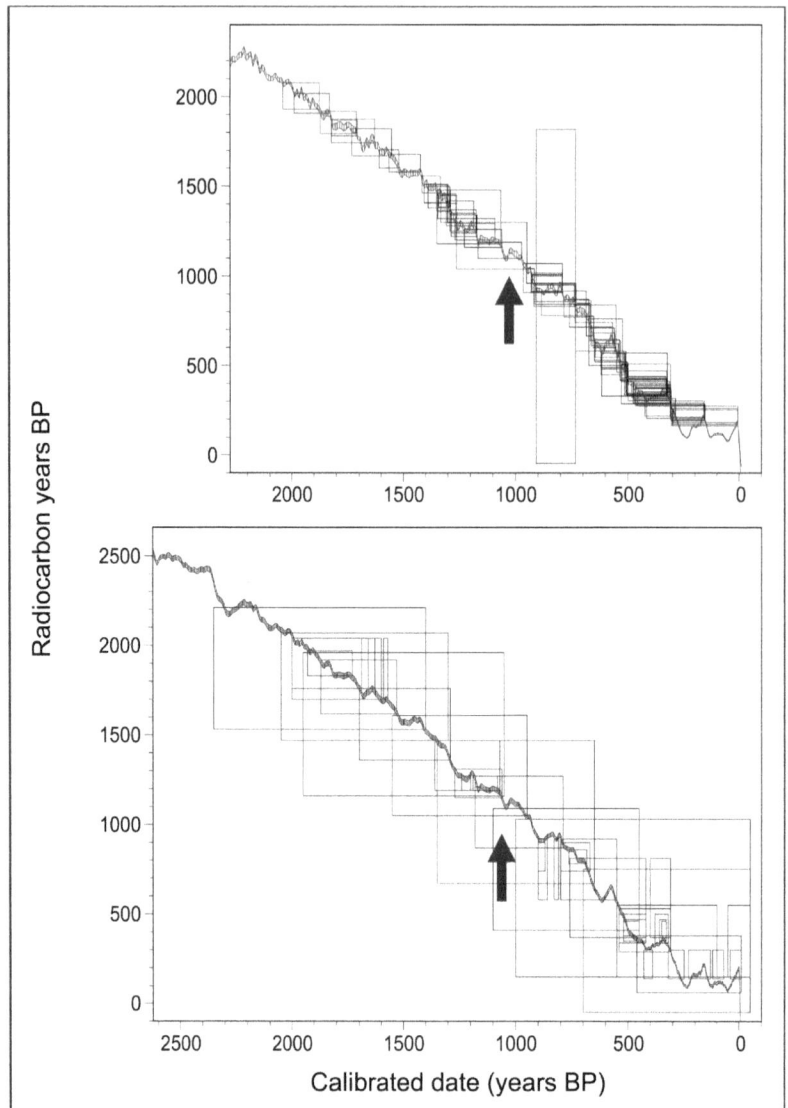

Figure 5. Box plot of radiocarbon determinations from hearths (upper) and human burials (lower) from western NSW. The arrow indicates the time period of the Medieval Climatic Anomaly (MCA). (Atmospheric data from Stuiver et al. 1998; OxCal 3.5 [Bronk Ramsey 1995]).

sites to see whether there is any difference in the age distributions of hearths and human burials when differences in geogrphy are not a factor. We are also about to date a large number of heat-retainer hearths from the vicinity of Peery Lake, near White Cliffs (Figure 1), where the presence of local floods, overflows from the Paroo River, and artesian mound springs are likely to have guaranteed water supplies in the past. Differences in the distribution of hearth ages through time between environmentally distinct areas may help isolate regional differences in hearth construction and lead to more informed interpretations of the presence of gaps in the distribution of hearth and burial ages.

Conclusions

Based on the comparison of heat-retainer hearths from the locations we have studied in northwestern NSW (Holdaway et al. 2005) and burials from the Lower Murray-Darling region in southwestern NSW (Littleton 2007), buried contexts preserve a more complete record of human occupation than material preserved in surface contexts. Therefore, the apparent relative increase in the frequency of archaeological materials during the last two millennia is likely to reflect the preservation of the record rather than an increase in either population size or occupation time. The record that is available for study today reflects the landscape history, particularly the time periods since surfaces were last heavily eroded. Thus, in the majority of surface contexts, there is no correlation between spatial proximity and time: surfaces that differ in age by many thousands of years can exist side by side and accumulate distinct archaeological records.

This has considerable implications for the way archaeologists generate inferences about the past. The results of archaeological survey, for instance, do not translate easily into either settlement patterns or settlement strategies. Although different site types may appear to relate to different environmental or resource zones, our evidence indicates that these differences may simply reflect changes in landscape preservation. In other words, what may appear at first to reflect locations where people undertook different activities related to such things as seasonal variation in resource availability might in fact represent differences in the age of deposits.

The age estimates from hearths (Figure 3) indicate periods when very few hearths were constructed across the northwestern NSW region we have studied. Similar gaps are not apparent in the record of human burials, perhaps reflecting geographical differences in resource availability as well as preservation of the archaeological record. Differences in the radiocarbon ages derived from charcoal from hearths and apatite from burials may be important but are unlikely to be significant. Alternatively, the hearth and burial age estimates may reflect differences in the human reactions to environmental change that affected the creation of one record but not the other. Differentiating among these potential causes will require more research, particularly the investigation of burials and hearths from the same location.

Note

[1] It should also be noted that landscapes at Fowlers Gap, and in much of the Barrier Ranges north of Broken Hill (Figure 1), are dominated by much older surfaces than these (see, for example, Gibson 1997; Hill 2004), but there is little evidence of prolonged Aboriginal occupation of areas that were distant from water sources, save for deposits of flakes and cores at rock outcrops that were utilized as raw material sources (Doelman et al. 2001; Holdaway et al. 2008).

Acknowledgements

This research was funded by an ARC Collaborative Grant to Holdaway and Witter, an ARC Large Grant to Holdaway and Fanning, and a Macquarie University External Collaborative Grant to Fanning. We thank Justin Shiner for permission to include in the analysis radiocarbon determinations from Pine Point/ Langwell, collected as part of his PhD research and funded by an AIATSIS grant. We thank the Broken Hill Local Aboriginal Lands Council, the Wangkamura Cultural Heritage Committee and the Wilcannia Local Aboriginal Lands Council, and the people they represent, for permission to collect the hearth charcoal samples. Radiocarbon determinations on the hearth charcoal were conducted at the Waikato Radiocarbon Laboratory in New Zealand, and we thank Tom Higham, Alan Hogg and Fiona Petchey for their assistance. We would also like to thank Peter Lape and two anonymous referees for useful comments on the submitted paper.

References

Beaton, J. M. 1983. Comment: Does intensification account for changes in the Australian Holocene archaeological record? *Archaeology in Oceania* 18:94–97.

Bird, C. and D. Frankel. 1991. Chronology and explanation in western Victoria and south east South Australia. *Archaeology in Oceania* 26:1–26.

Broeker, W. S. 2001. Was the medieval warm period global? *Science* 291:1497–1499.

de Menocal, P. 2001. Cultural responses to climate change during the late Holocene. *Science* 292:667–673.

Bronk Ramsey, C. 1995. Radiocarbon calibration and analysis of stratigraphy: The OxCal program. *Radiocarbon* 37(2):425–430.

Doelman T., J. Webb and M. Domanski. 2001. Source to discard: Patterns of lithic raw material procurement and use in Sturt National Park, northwestern New South Wales. *Archaeology in Oceania* 36:15–33.

Eital, B., Hecht, S., Machtle, B., Schukraft, G., Kadereit, A., Wagner, G. A., Kromer, B., Unkel, I. and Reindel, M. 2005. Geoarchaeological evidence from desert loess in the Nazca-Palpa region, southern Peru: Palaeoenvironmental changes and their impact on pre-Columbian cultures. *Archaeometry* 47(1):137–158.

Fanning, P. C. 1994. Long term contemporary erosion rates in an arid rangelands environment in western New South Wales, Australia. *Journal of Arid Environments* 28:173–187.

Fanning, P. C. 1999. Recent landscape history in arid western New South Wales, Australia: A model for regional change. *Geomorphology* 29:191–209.

Fanning P. C. 2002. *Beyond the Divide: A new geoarchaeology of Aboriginal stone artefact scatters in western New South Wales, Australia*. Unpublished PhD thesis., Graduate School of the Environment, Macquarie University.

Fanning, P. C. and S. J. Holdaway. 2001. Temporal limits to the archaeological record in arid western NSW, Australia: Lessons from OSL and radiocarbon dating of hearths and sediments. In M. Jones and P. Sheppard (eds), *Australasian connections and new directions: Proceedings of the 7th Archaeometry Conference*, pp 85–104. Auckland: Department of Anthropology, University of Auckland.

Fanning P. C., S. J. Holdaway and E. J. Rhodes. 2008 A new geoarchaeology of surface artefact scatters in western New South Wales, Australia: Establishing spatial and temporal geomorphic controls on the surface archaeological record. *Geomorphology*, 101:524-532.

Favier Dubois, C. M. 2003. Late Holocene climatic fluctuations and soil genesis in southern Patagonia: Effects on the archaeological record. *Journal of Archaeological Science* 30:1657–1664.

Gibson D. L. 1997. Recent tectonics and landscape evolution in the Broken Hill region. *Australian Geological Survey Organisation Research Newsletter* 26:17–20.

Herweijer, C., R. Seager, E. R. Cook and J. Emile-Geay. 2007. North American droughts of the last millennium from a gridded network of tree-ring data. *Journal of Climate* 20:1353–1376.

Hill S. M. 2004. Regolith and landscape evolution of far western New South Wales. In Cooperative Research Centre for Landscape, Environment and Mineral Exploration, *Far Western NSW*, 17 pp. Canberra: AGPS.

Hiscock, P. 1986. Technological change in the Hunter River valley and the interpretation of late Holocene change in Australia. *Archaeology in Oceania* 21:40–50.

Hiscock, P. 1988. *Prehistoric settlement patterns and artefact manufacture at Lawn Hill, Northwest Queensland*. Unpublished PhD thesis, Department of Anthropology and Sociology, University of Queensland.

Holdaway S. J., P. C. Fanning and D. C. Witter. 2000. Prehistoric Aboriginal occupation of the rangelands: Interpreting the surface archaeological record of far western New South Wales, Australia. *The Rangelands Journal* 22:44–57.

Holdaway S. J., P. C. Fanning, D. C. Witter, M. Jones, G. Nicholls, J. Reeves and J. Shiner. 2002. Variability in the chronology of Late Holocene Aboriginal occupation on the arid margin of southeastern Australia. *Journal of Archaeological Science* 29:351–363.

Holdaway S. J., P. C. Fanning and J. Shiner. 2005. Absence of evidence or evidence of absence? Understanding the chronology of indigenous occupation of western New South Wales, Australia. *Archaeology in Oceania* 40:33–49.

Holdaway, S. J., P. C. Fanning and E.J. Rhodes. 2008. Challenging intensification: human-environment interactions in the Holocene geoarchaeological record from western New South Wales, Australia. *The Holocene* 18(3), 403-412.

Holdaway, S.J.,J. Shiner and P.C. Fanning. 2008. Assemblage formation as a result of raw material acquisition in western New South Wales, Australia. *Lithic Technology* 33(1).

Huffman, T.N. 1996. Archaeological evidence for climatic change during the last 2000 years in southern Africa. *Quaternary International* 33:55–60.

Jones, M. and G. Nicholls. 2002. New radiocarbon calibration software. *Radiocarbon* 44(3):663–674.

Jones, P. D., Osborn, T. J. and Briffa, K. R. 2001. The evolution of climate over the last millennium. *Science* 292:662–667.

Littleton, J. 2007 From the perspective of time: Hunter-gatherer burials in southeastern Australia. *Antiquity*, 81:1013-1028.

Lourandos, H. 1983. Intensification: A Late Pleistocene-Holocene archaeological sequence from Southwestern Victoria. *Archaeology in Oceania* 18:81–94.

Lourandos, H. 1984. Changing perspectives in Australian prehistory: A reply to Beaton. *Archaeology in Oceania* 19:29–33.

Lourandos, H. 1985. Intensification in Australian prehistory. In T. D. Price and J. Brown (eds), *Prehistoric hunters and gatherers: The emergence of social and cultural complexity*, pp 385–423. New York: Academic Press.

Lourandos, H. 1988. Palaeopolitics: Resource intensification in Aboriginal Australia and Papua New Guinea. In T. Ingold, D. Riches and J. Woodburn (eds), *Hunters and gatherers: History, evolution and social change*, vol. 1, pp 148–60. Oxford: Berg.

Lourandos, H. 1993. Hunter-gatherer cultural dynamics: Long and short term trends in Australian prehistory. *Journal of Archaeological Research* 1:67–88.

Rhodes E. J., P. C. Fanning, S. J. Holdaway and C. Bolton. 2008. Archaeological surfaces in western NSW: Stratigraphic contexts and preliminary OSL dating of hearths. In A. Fairbairn. and S. O'Connor (eds), this volume.

Ross A., T. Donnelly and R. Wasson. 1985. The peopling of the arid zone: Human-environment interactions. In J. Dodson (ed), *The Naïve lands: Prehistory and environmental change in Austral ia and the southwest Pacific*, pp 76–114. Melbourne: Longman Cheshire.

Stuiver, M., P. J. Reimer, E. Bard, J. W. Beck, G. S. Burr, K. A. Hughen, B. Kromer, F. G. McCormac, J. van der Plicht and M. Spurk. 1998. INTCAL98 radiocarbon age calibration, 24,000-0 cal BP. *Radiocarbon* 40(3):1041–1083.

2

Heat-retainer hearth identification as a component of archaeological survey in western NSW, Australia

Patricia C. Fanning[1]
Simon J. Holdaway[2]
Rebecca S. Phillipps[2]

[1]Graduate School of the Environment,
Macquarie University,
NSW 2109, Australia

[2]Department of Anthropology,
University of Auckland, Private Bag 92109,
Auckland, New Zealand

Abstract

Concentrations of heat-fractured rock are common in some areas of western New South Wales, Australia, and are frequently identified during archaeological surveys as the eroded remains of heat-retainer hearths, a type of earth oven used in the past by Aboriginal people to cook food. However, it can be difficult to consistently identify these features since heat-fractured rocks, like other clasts lying on the surface, can be dispersed or concentrated by a variety of geomorphic processes. In this paper, we describe a method we have developed for hearth identification and description to aid the consistent recording of these features. We recommend the use of a fluxgate gradiometer as a tool to verify that concentrations of heat-fractured rock are indeed the eroded remains of heat-retainer hearths. Using a gradiometer reduces observer variability and ensures that we are not overestimating the number of hearths present by including naturally occurring groups of stones not related to hearth activity. Adoption by others of a common method of hearth identification and classification would provide some degree of standardisation that will aid understanding of the patterns of Aboriginal occupation in the past.

Keywords: geoarchaeology; Australia; Aboriginal; hearth; fluxgate gradiometer

Introduction

Surface artefact scatters in western New South Wales (NSW) (Figure 1) are almost always associated with the remains of one or more heat-retainer hearths, or earth ovens, used in the past by Aboriginal people to cook food (Allen 1972:280–1; Peake-Jones 1988). Hearths are an important part of the archaeological record in this region because they are an easily recognisable indicator of past Aboriginal occupation. For archaeologists, they provide a means for developing a chronology of Aboriginal occupation using radiocarbon determinations from charcoal preserved beneath the fire-cracked rock (e.g. Holdaway et al. 2002, 2005a), and potentially through optically stimulated luminescence of the stone heat-retainers (Rhodes et al., this volume).

Originally, these hearths were constructed by excavating a depression that formed the body of the oven into which stones and then food items could be placed for cooking. Abandonment would in many cases lead to the infilling of the depression, thereby burying and preserving the remains of the hearth. Once buried, hearths were probably difficult to identify even by the people who made them, and our research has thus far shown little evidence of extensive re-use, at the level of precision currently attained by conventional radiocarbon age determinations. However, like the stone artefacts with which they are commonly associated, their visibility in western NSW today is a consequence of erosion processes that are exposing them at the surface, where we can see them, and at the same time causing their destruction. Hearths are no longer being produced, therefore the population is finite and, as a consequence of erosion, declining year by year. It is therefore important for both cultural and scientific reasons to document the record of Aboriginal occupation that they represent before they disappear. In this paper, we describe techniques developed to identify, classify and record the remains of heat-retainer hearths in far north-western NSW. We focus on the use of the magnetic properties of the fire-cracked rock to provide an objective test of the degree of preservation of the hearths and their charcoal. The techniques of hearth identification and classification we describe are a key component of survey methods we have developed to understand the Aboriginal archaeological record in its contemporary landscape setting.

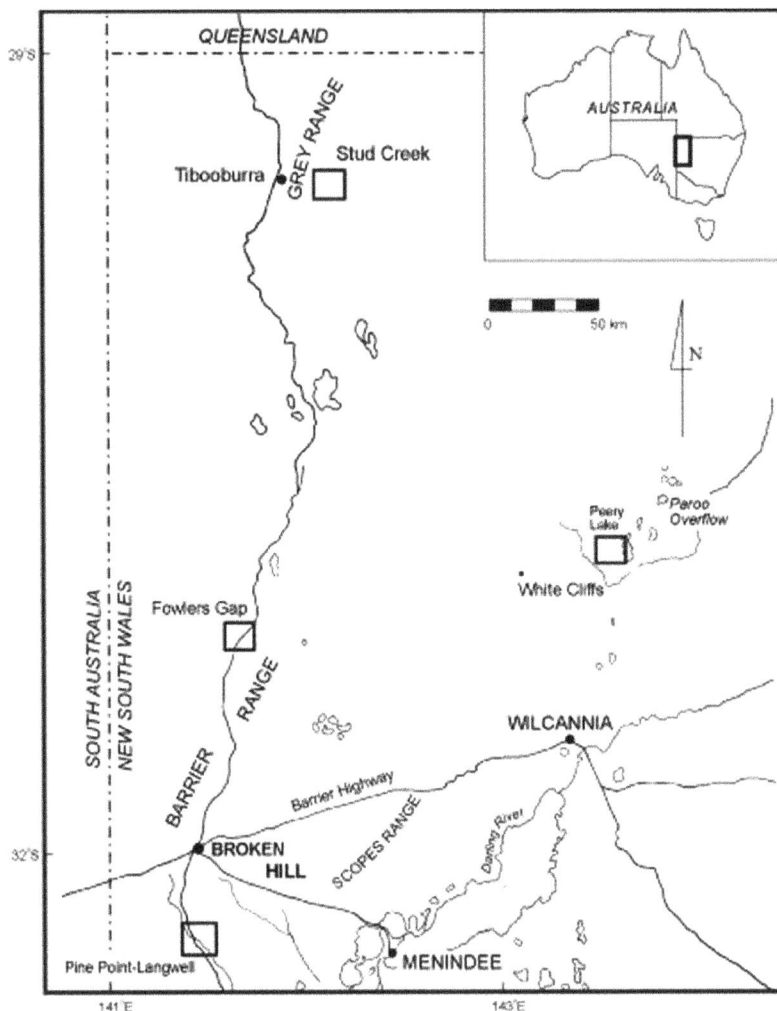

Figure 1. Map of western NSW showing locations mentioned in the text.

Problems with hearth identification

Heat-retainer hearth remains are commonly identified during surface archaeological surveys in western NSW. They are most often found along the banks of creeks, on lake margins, in sand dunes, and exposed on scalds and claypans. At times, hearths may appear in dense concentrations. For example, a recent survey of part of the valley floor of Campbells Creek at Poolamacca, about 80 km north of Broken Hill (Figure 1), revealed at least 220 hearths in an area a little over 500 m in length by 200 m wide (Holdaway et al. 2005b), an average density of one hearth to every 450 m2. This situation is not unusual: a similarly dense distribution was located in 2002 adjacent to Peery Creek in a section of the Paroo Darling National Park (Holdaway et al. 2006). However, there are other parts of the contemporary landscape where heat-retainer hearth remains are either absent or at least not visible on the surface.

Because the density of hearths has something to say about the intensity with which Aboriginal people used different parts of the landscape in the past — the more so since many hearths can provide age estimates, something that is largely impossible when analyzing stone artefacts — it is particularly important that hearth remains are unambiguously identified. Comparing the density of hearths from different locations assumes two kinds of comparability: hearths are equally visible and hearths are equally definable.

The first requires an equal state of erosion such that hearths are exposed to an equal degree. This, however, is obviously not the case, since the contemporary landscape is a mosaic of erosional and depositional surfaces. To date, our approach to this problem has been to map land surfaces in great detail, separating eroded surfaces from those still covered with surface sediment, vegetation, etc. (Fanning and Holdaway 2004). Depositional areas may conceal hearths, but we have not yet attempted a systematic survey of these areas to locate buried hearths, for two reasons. First, test pitting, while potentially applicable, requires very close spacing to identify relatively small features across large areas (Wobst 1983), and is therefore very labour intensive. Second, large scale surface disturbance is increasingly out of favour with both Aboriginal people and heritage managers. Geophysical survey, as discussed below, offers a non-destructive alternative to test pitting but we have not so far explored this option for identifying buried hearths.

The second issue, the ability to determine when a hearth is indeed the remains of a past cooking fire, and not simply a pile of naturally occurring rocks or a low density scatter of fire-cracked rocks, requires consistency in field recording. Up to now, our approach has been to identify concentrations of heat-fractured rock and label these as hearths. Heat-cracked rock has a number of characteristics that permit rapid identification, for example 'pot-lids' may be shed during rapid heating (Purdy 1975 in Hiscock 1985), and crazing patterns (a grid of cracks on the rock surface caused by rapid heating and cooling — Hiscock 1997) may be apparent. Identification is easiest when the heat-fractured stones are concentrated together but becomes progressively harder as erosion disperses the fragments. Identification also depends on the nature of the land surface: hearth remains on surfaces covered with a gravel lag, for instance, can be almost impossible to identify. In addition, erosion may lead to the combining of stones from a number of hearths, the individual identity of which may be very difficult to define.

The Western NSW Archaeology Program hearth survey protocol

In response to the variable state of hearth preservation in western NSW, we developed a hearth survey protocol based on a system initiated by Dan Witter, formerly a NSW National Parks and Wildlife Service archaeologist for western NSW. The survey strategy involves an intensive pedestrian survey with the aim of identifying potential hearths and marking each with a coloured flag (Figure 2). Identification at this stage is based on the presence of any concentration of heat-fractured stones.

The hearth recording team then re-visits each of the marked concentrations and attaches a tag stamped with a number against which all the attributes of each putative hearth are subsequently entered into a database. Attributes recorded for each hearth include:

1. location — AMG co-ordinates determined using differential GPS;
2. dimensions — maximum length and perpendicular width of the cluster of hearth stones, measured with a steel tape;
3. orientation — azimuth of the maximum dimension;
4. number and lithology of hearth stones;
5. presence of any visible charcoal;
6. condition of the hearth (see below).

Figure 2. The remains of heat retainer hearths on an eroded surface in Paroo Darling National Park in western NSW. Hearths identified in the initial intensive pedestrian survey are marked with a coloured surveyor's flag (pink in this image).

A digital image is also taken. Several of these attributes are necessary for entry into the NSW Aboriginal Heritage Information Management System (colloquially, the 'site register'), as hearths are classified as 'sites' under that system (NSW Department of Environment and Conservation 2006).

The condition of each hearth is determined from visual assessment of the degree of disturbance by processes including erosion or bioturbation. Six categories are defined (Figure 3, a to f): buried, partially exposed, intact, disturbed, scattered, remnant. Use of the term 'buried' to describe a category of hearth disturbance may seem inconsistent since this is a classification of the condition of hearths exposed at the surface. However, this term is useful for describing hearths that remain largely buried with only the tips of the fire-cracked rock poking above the surface (Figure 3a).

Partially exposed hearths are relatively common in our study locations. This term is used to describe hearths where a portion of the dense cluster of hearth stones is exposed along an erosion escarpment (Figure 3b), but the bulk of the hearth remains buried. Intact hearths are those where erosion has completely exposed the dense cluster of fire-cracked rock but it has not been dispersed (Figure 3c).

The next three categories — disturbed, scattered and remnant — refer to hearths displaying increasing amounts of disturbance of the heat retainers. Disturbed hearths (Figure 3d) still retain a roughly circular form although stones are more dispersed in comparison to intact hearths. Scattered hearths (Figure 3e) have lost this circular form and the heat retainers are unevenly spread across the surface, with no clear focal point. Remnant hearths are at the last identifiable end point of the disturbance continuum. These hearths retain a circular structure (Figure 3f) but the hearth stones form an open ring or 'halo' with no stones present at the centre. They are often characterised by relief inversion, with the former centre of the hearth pit now forming the apex of a shallow cone with the hearthstones completely dispersed around the edges.

Applying the hearth condition classification

A recent survey of Rutherfords Creek in the Peery Lake section of Paroo Darling National Park (Figure 1) identified 416 potential hearths in an area of approximately 2.5 km². Around half are relatively undisturbed (Table 1), being categorised as buried, partially exposed or intact using the classification described above. The others are damaged, having been classified as either disturbed, scattered or remnant. The three hearths identified as buried are likely to be an underestimate of the true number of buried hearths since a large proportion of the valley floor of Rutherfords Creek is currently buried by modern sediments.

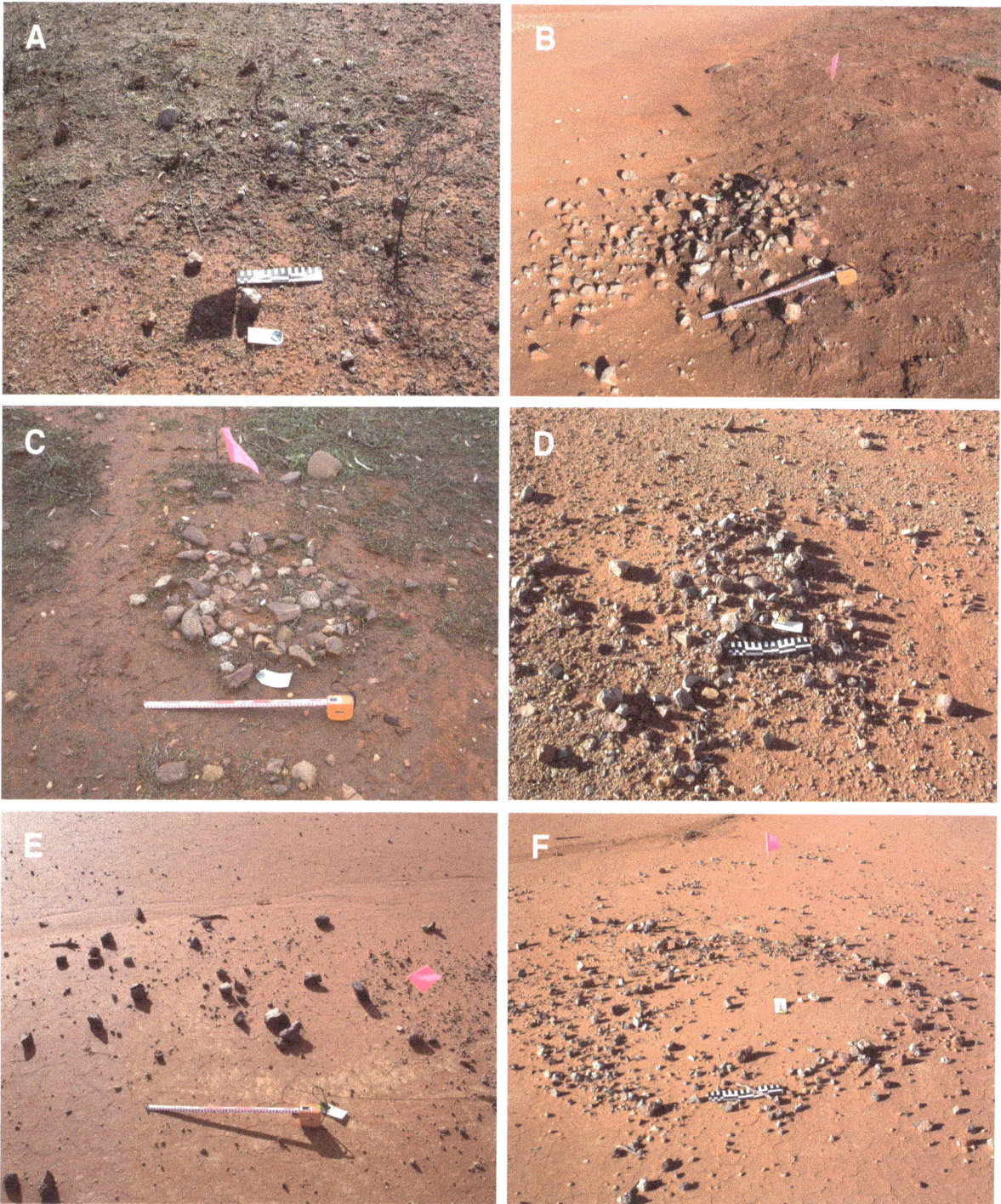

Figure 3. Images of heat retainer hearths from western NSW that illustrate the six categories of hearth condition recognised in the survey protocol:

a. buried
b. partially exposed
c. intact
d. disturbed
e. scattered
f. remnant

Hearth Condition	Number	Proportion (%)	Pearson Correlation Coefficient	P	Spearman Rank Order Correlation Coefficient	P
Buried	3	1	n/a	n/a	n/a	n/a
Partially exposed	179	43	0.4	*<0.01*	0.17	*0.03*
Intact	56	13	0.35	*0.01*	0.42	*<0.01*
Disturbed	96	23	0.23	*0.03*	0.24	0.2
Scattered	62	15	0.38	0.77	0.54	0.68
Remnant	20	5	0.27	0.24	0.48	*0.03*
Total	416					

Table 1. Hearth types recorded along a section of the valley floor of Rutherfords Creek in Paroo Darling National Park in western NSW. Pearson and Spearman correlation coefficients are parametric and non-parametric measures of the strength of the relationships between the gradiometer reading and number of hearth stones for identified hearths of each condition type. Significant results are indicated by *italics*.

The proportion of the Rutherfords Creek valley floor that is eroded can be used to estimate the true density of hearths, including those still buried, using two techniques. First, we use digital aerial photographs and image analysis software to estimate the area of the valley floor covered by scalds. We also measure the length and breadth of scalds as part of our artefact survey strategy, and thus can calculate the area of exposed surfaces.

However, the second part of the calculation for hearth density in Rutherfords Creek requires that the count of hearths is accurate. Here we face the problem of inconsistency in hearth identification. Are the hearths we have categorised as disturbed, scattered and remnant actually hearths, or are they simply concentrations of heat-fractured stones grouped together as the result of natural processes? A more objective means of identifying hearth remains is required to overcome this problem.

Gradiometer survey as an objective method of hearth identification

Our solution was to employ a fluxgate gradiometer to provide a more objective measure of the potential for the stones to have been heated in situ, from which we could be more confident of positive hearth identification. The gradiometer works by recording minute distortions in the earth's magnetic field at particular locations. We used a Geoscan Research Fluxgate Gradiometer FM36 (Figure 4) to search for near-surface magnetic anomalies. The gradiometer returns positive and negative readings in nanoTeslas (nT), with a background value of around zero. In accordance with the discussion in Jones and Munson (2005), it is likely that the readings taken at Rutherfords Creek derive from two sources. First, heating of rocks and soil in hearths can increase the magnetic susceptibility of minerals leading to the preferential flow of a geomagnetic field through these minerals and, as a consequence, locally higher field strength. In addition, if minerals are heated above a critical temperature (the Curie temperature, $120°C$ - $675°C$) they will acquire permanent magnetism (termed 'thermo-remnant magnetism' or TRM). The gradiometer measures both sources.

Unlike conventional approaches, where gradiometer measurements are taken in a systematic grid across a surface, we were not concerned here with identifying buried hearths. Instead, to further investigate hearths identified during our initial pedestrian survey, we employed the gradiometer in 'survey' mode (Clark 2000). At each location, the gradiometer was zeroed against a background surface that lacked heat-fractured stones. Readings were taken along a transect aligned north-south across a potential hearth, the operator noting the maximum reading (either positive or negative) along the transect.

Figure 4. Rebecca Phillipps calibrates the Geoscan Research Fluxgate Gradiometer FM36 used in this survey.

Figures 5a to 5e illustrate the differences between the hearth reading and the background reading for hearths in the different categories described above. Buried hearths are not included in the analysis as there were too few readings to graph the results. For both partially exposed and intact hearths, the differences are quite large, up to +65 nT and −95 nT in some cases. But for disturbed, scattered and remnant hearths the differences between the background and the hearth readings are much smaller, as we would expect if the heat-fractured rocks were in low concentrations.

We tested the relationship between the number of hearth stones and the magnitude of the gradiometer readings using both parametric and non-parametric tests. There is a significant correlation in both cases (Table 1), suggesting that TRM is contributing significantly to the gradiometer readings over the hearths. This relationship is clearest for partially exposed, intact and disturbed hearths. For the scattered and remnant categories, the statistical tests indicate that there is only a weak relationship between the numbers of hearth stones and the gradiometer readings, which would be expected given the higher degree of dispersion of heat-fractured rock for the hearths in these two categories. The significant result for the Spearman rank correlation test on the remnant hearths (Table 1) is probably because the 20 hearths of this type recorded were those with the highest density of hearth stones.

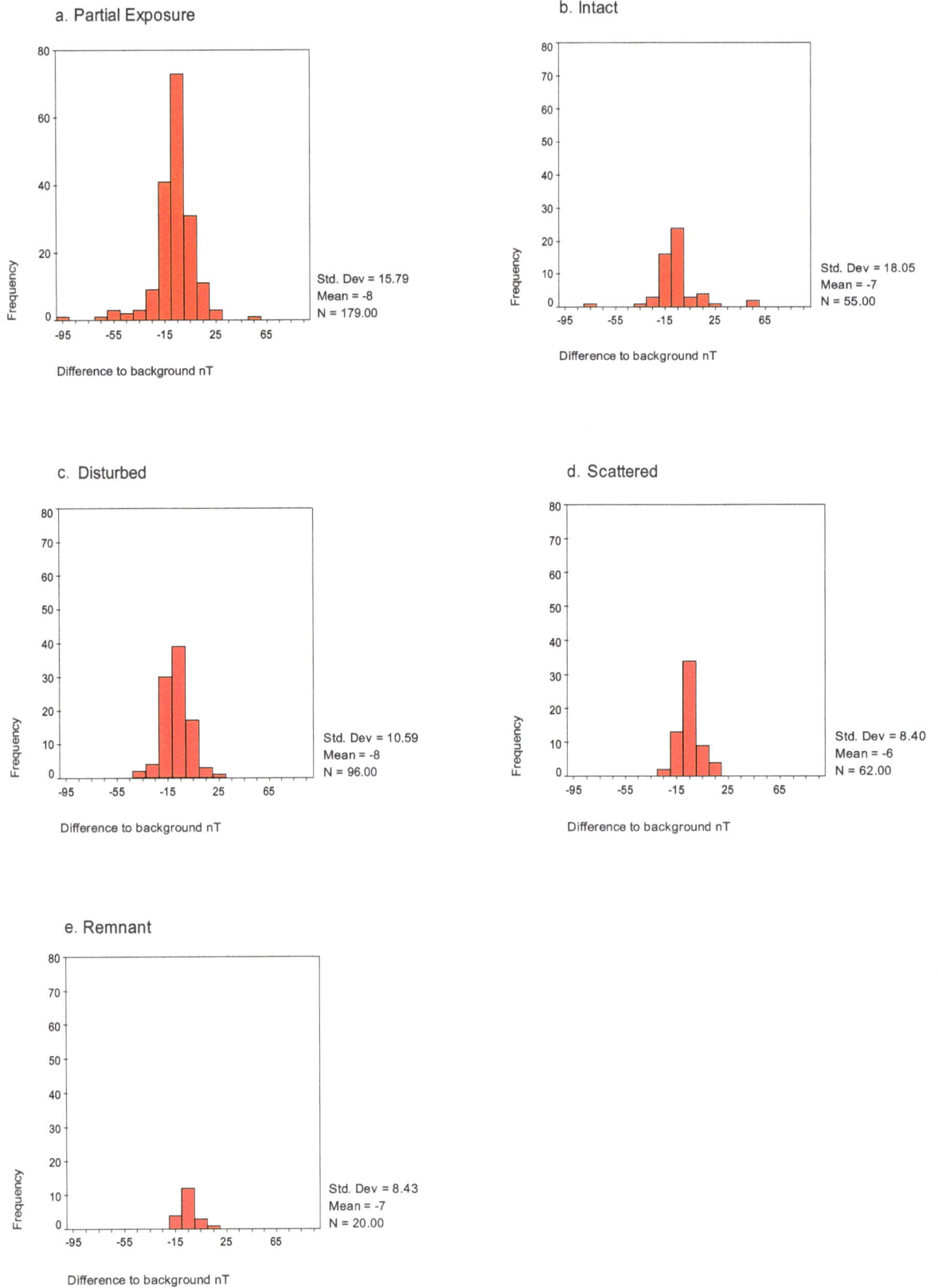

a. Partial Exposure

b. Intact

c. Disturbed

d. Scattered

e. Remnant

Figure 5. Histograms showing the differences in the TRM between the hearths and the background for each of the hearth condition categories identified in the survey:

a. partially exposed;   b. intact;   c. disturbed;   d. scattered;   e. remnant.

Discussion

The trial to test the effectiveness of using a gradiometer survey to identify fire-affected rock, described above, produced promising results. Those hearths with the densest concentration of heat-fractured stones produced magnetic readings exhibiting the largest differences from the background magnetism. The correlation between the number of hearth stones and the strength of the difference to background for the different hearth types suggests that much of the signal read by the gradiometer relates to TRM. If so, then a case can be made that the gradiometer is an effective tool for objectively identifying hearths in a continuum that runs from the 'intact' through to 'remnant' categories (Table 1).

The primary issue here is not just accurately identifying former hearths. It is more about objectively determining a point in the progressive destruction of hearths by erosion at which the location where the hearth once existed can no longer be defined. The hearths that are now categorised as remnant were once buried. As erosion progresses, defining the number and precise location of hearths becomes more difficult, and calculating the original density of hearths on a valley floor is problematic.

Our experiments show that use of a gradiometer in survey mode provides an objective measure of hearth 'intactness' that can be used with the condition classification described above. Based on our experience at Rutherfords Creek, if the reading over a collection of heat-fractured stones varies very little from background, i.e. within ±5 nT, then it is excluded from further survey. In imposing these somewhat arbitrary threshold values we may well be excluding some clusters that once were hearths, but we are certainly not including those that were not. Hence, we can be sure that we are not overestimating the number of hearths present in this location by including such phenomena as naturally occurring groups of stones not related to hearth activity. Use of the gradiometer also provides us with an objective measure that allows us to take account of both inter- and intra-observer variability, a source of measurement variance that previously has proved significant in our analyses of stone artefacts (Gnaden and Holdaway 2000).

Use of the gradiometer at Rutherfords Creek was a pilot study and we plan to expand its use in the future. In particular, we need to take measurements on stone clusters that we know were never part of a hearth, to better understand how values from these features compare to our intact hearth category. We also need to undertake controlled experiments by creating our own hearths to determine the absolute effects on the gradiometer readings of the degree of stone scatter, and the effects of different lithologies. Controlled experiments might also allow us to determine whether or not hearth reuse can be detected.

Conclusions

The eroded remains of heat-retainer hearths in western NSW are an important aspect of the archaeological record because they are highly visible, are easy to interpret and can be dated. Unfortunately their visibility is a result of processes that are also leading to their destruction. Erosion and bioturbation are progressively leading to partial exposure then destruction of the original hearth, as the heat retainers are scattered. Eventually the lack of integrity is such that hearth identification becomes problematic. Hearths are protected by heritage conservation legislation, but this only ensures protection against destruction through development or deliberate vandalism, not 'natural' erosion or damage by domestic stock. Hearths are a diminishing resource and our research suggests that archaeologists have a limited time to study this resource before it is gone forever. Our hearth classification scheme, if universally adopted, provides standardisation of hearth identification that will allow comparison between separate archaeological surveys, and monitoring of changes in hearth condition over time.

The techniques described here provide an objective means for identifying hearths based on combining a classification of condition and use of a fluxgate gradiometer, and therefore provide a way to estimate the number and condition of the hearths that remain in particular locations. Results suggest that differences in the gradiometer readings relative to local background are largely due to remnant thermo-magnetism in the heat retainers. Those hearths with high numbers of heat-fractured stones concentrated in a small area provide the largest deviations from background. Those concentrations of hearth stones that give values at or below ± 5nT should probably not be considered hearths.

When combined with a detailed assessment of stone artefacts (e.g. Holdaway et al. 2004) and a program of obtaining age estimates for the hearths (e.g. Holdaway et al. 2005a; Fanning et al. 2008), study of heat-retainer hearths offers the opportunity to detail both the chronology and intensity of Aboriginal use of place in the past.

Acknowledgements

This project was funded by an ARC Discovery Grant to Fanning, Holdaway and Rhodes. We thank the Wilcannia Local Aboriginal Lands Council and the Wilcannia Community Working Party and the people they represent for permission to work on their traditional lands and to collect samples of charcoal from the hearths. We particularly thank Murray Butcher, Warlpa Thompson and Bilyara Bates and students from the University of Auckland for their help with fieldwork. We also thank Badger Bates and Sarah Martin for their help and support. The NSW Department of Environment and Climate Change gave permission for the research to be conducted on Paroo Darling National Park, and we are grateful to departmental officers for their assistance, particularly Phil Purcell and Christian Hampson. Thanks also to Peter Hiscock for helpful comments on the submitted paper.

References

Allen, H. 1972. Where the crow flies backwards: Man and land in the Darling Basin. Unpublished PhD thesis. Canberra: Department of Prehistory, Research School of Pacific Studies, Australian National University.

Clark, A. 2000. *Seeing beneath the soil: Prospecting methods in archaeology*. New York: Routledge.

Fanning, P. C. and S. J. Holdaway. 2004. Artifact visibility at open sites in western New South Wales. Australia. *Journal of Field Archaeology* 29(3–4):255–271.

Fanning. P. C., S. J. Holdaway and E. J. Rhodes. 2008. A new geoarchaeology of Aboriginal artefact deposits in western NSW, Australia: Establishing spatial and temporal geomorphic controls on the surface archaeological record. *Geomorphology* 101:524-532.

Gnaden, G. and S. J. Holdaway. 2000. Understanding observer variation when recording stone artifacts. *American Antiquity* 65:739–748.

Hiscock P, 1985. The need for a taphonomic perspective in stone artefact analysis. *Queensland Archaeological Research* 2:82–95.

Hiscock, P. 1997. Taphonomy of artefacts. http://arts.anu.edu.au/arcworld/resources/intro/tapho.htm

Holdaway, S. J., P. C. Fanning, D. Witter, M. Jones, G. Nicholls and J. Shiner. 2002. Variability in the chronology of Late Holocene Aboriginal occupation on the arid margin of southeastern Australia. *Journal of Archaeological Science* 29:351–363.

Holdaway, S. J., P. C. Fanning and J. Shiner. 2005a. Absence of evidence or evidence of absence? Understanding the chronology of Indigenous occupation of western New South Wales, Australia. *Archaeology in Oceania* 40:33–49

Holdaway, S. J., P. C. Fanning, E. J. Rhodes and Broken Hill Local Aboriginal Lands Council. 2005b. A geoarchaeological and geochronological assessment of the surface archaeology of the Campbells Creek area, 'Poolamacca' Station, western NSW. Unpublished report for NSW Department of Environment and Conservation.

Holdaway, S. J., P. C. Fanning and J. Shiner. 2006. Geoarchaeological investigation of Aboriginal landscape occupation in Paroo Darling National Park, Western NSW, Australia. *Research in Anthropology and Linguistics — e* 1. University of Auckland: Department of Anthropology.

Holdaway, S. J., J. Shiner and P. C. Fanning. 2004. Hunter-gatherers and the archaeology of the long term: An analysis of surface stone artefact scatters from Sturt National Park, New South Wales, Australia. *Asian Perspectives* 43(1):34–72.

Jones, G. and G. Munson. 2005. Geophysical survey as an approach to the ephemeral campsite problem: Case studies from the Northern Plains. *Plains Anthropologist* 50:193–224.

NSW Department of Environment and Conservation. 2006. Aboriginal Heritage Management System. http://www.nationalparks.nsw.gov.au/npws.nsf/Content/Aboriginal+Heritage+Information+Management+System. Accessed 26th May 2006.

Peake-Jones, K. 1988. *To the desert with Sturt: A diary of the 1844 expedition*. Adelaide: South Australian Government Printer.

Rhodes, E. J., P. C. Fanning, S. J. Holdaway and C. Bolton. 2008. Archaeological surfaces in western NSW: Stratigraphic contexts and preliminary OSL dating of hearths. In A. Fairbairn and S. O'Connor (eds), Proceedings of the 8th Australasian Archaeometry Conference, December 2005, Canberra.

Wobst, M. 1983. We can't see the forest for the trees: Sampling and the shapes of archaeological distributions. In J. Moore and A. Keene (eds), *Archaeological hammers and theories*, pp 37–85. New York: Academic Press.

3

Persistent places: An approach to the interpretation of assemblage variation in deflated surface stone artefact distributions from western New South Wales, Australia

Justin Shiner

Archaeology Program
La Trobe University
Victoria
Australia 3086

Abstract

The construction of models of past settlement systems in the Australian arid zone are based on serious misunderstandings of the formation of deflated archaeological deposits. This in turn leads to the application of inappropriate interpretative frameworks that often ignore chronological contexts and assume that spatially separate deposits are contemporary, and demonstrate consistency in human behaviour through time. This denies archaeologists the opportunity to explore the temporal aspects of deflated records, both in terms of chronology and the management of stone artefact manufacture across space and through time. Radiocarbon determinations from heat-retainer hearths and stone artefact assemblage data from the Pine Point/Langwell area of western New South Wales (NSW) near Broken Hill are used to demonstrate the concept of archaeological deposits as persistent places across the landscape. This combined with an analysis of variability in stone artefact assemblage composition provides an alternative framework to synchronic models.

Keywords: deflated deposits, persistent place, chronology

Introduction

Heat-retainer hearths and stone artefacts dominate the surface archaeological record of semi-arid western New South Wales. These most frequently occur as deflated and spatially extensive distributions of varying density with occasional hearths. They typically lack clear and readily definable boundaries. The deflation of the artefacts has resulted in the loss of vertical integrity and relative chronological relationships between artefacts. Consequently, it is difficult to group artefacts into assemblages for analysis. The definition of assemblages in these contexts rarely, if ever, has anything to do with fine-scale temporal and spatial behavioural patterns. Rather, assemblages are often defined according to similarities in landscape context, broader temporal context or geomorphologic boundaries (Holdaway et al. 2000). In this sense assemblages are collections of artefacts that may represent multiple behavioural episodes rather than discrete events. The formational characteristics of these deposits pose numerous methodological and theoretical challenges for archaeologists interested in the study and interpretation of the surface archaeological record from the Australian arid zone.

Surface deposits contain numerous artefacts often representing multiple stages of core reduction, tool manufacture and discard. Based upon assemblage composition alone it is difficult to classify these deposits into distinct site types that correspond with the perceived function of the location within a settlement system. For instance, based on ethnographic observation Cane (1984) argued that assemblages with high proportions of scrapers represented woodworking locations. But Cane's own investigations demonstrated that high proportions of scrapers were also recorded at sites where ethnographic information indicated that woodworking was not the primary activity.

The raw material types present reflect the exploitation of both local and non-local sources. Where chronological contexts have been established (Holdaway et al. 2002, 2005; Shiner 2004), these indicate multiple episodes of hearth construction. Although it is impossible to establish direct temporal relationships between any single artefact and hearth, the multiple age determinations from the hearths point to numerous episodes of occupation during which artefacts may have been discarded. From this perspective the deflated surface deposits of western NSW are palimpsests of material that may have accumulated from an unknown number of behavioural events. These events may be spread over several hundred or more years. Therefore, these deposits suggest the repeated use of place through time and not a single one-off behavioural event.

These factors mean that synchronic interpretations of site function within a single land-use model is not an appropriate interpretative framework for deflated surface deposits. Interpretations of the record that do not consider the temporal dimension of the formation of surface deposits ignore the possibility that the artefacts found on a common eroded surface may have been discarded at different times during the past. The behavioural context under which these were discarded may also have varied. This leads to questions regarding the nature of these occupations and the factors that may result in the reuse of places.

Schlanger (1992) used the term 'persistent place' to describe those areas of the landscape that are the focus of repeated activity through time. Persistent places fall into two categories. The first is associated with features of the natural environment that may attract human occupation, e.g. swamps, waterholes, rock outcrops etc. The second reflects the type of materials and features that humans create in the occupation of a location. In the case of western NSW these might be stone arrangements, heat-retainer hearths or raw material caches. Once established these features will structure the future use of the location, either through reuse or avoidance (see also Fletcher 1995). The role of features may also change during subsequent occupations. With this the duration and character of occupation may also vary through time. Another factor to consider is that the preserved pattern of site distribution and assemblage composition might not be representative of the prehistoric pattern of human activity. As Waters and Kuehn (1996) note, landscapes are dynamic and continually changing, sites are destroyed over time and this fragments the record of cultural systems. This is particularly true in western NSW where significant landscape change has occurred with the arrival of pastoral and mining land use (Fanning 2002).

Even though persistent places may not attract permanent settlement, they may attract long-term episodic use. The notion of persistent place is useful for investigating the long-term histories of individual locations because it acknowledges the role of multiple behavioural events in the accumulation of the archaeological record. Within this context, assemblage accumulation represents multiple processes that may have very different temporal trajectories, rather than the result of a synchronic functional-environmental relationship between discard and place use. The composition of assemblages and the spatial structure of artefact distributions reflect the long-term repeated use of locations. The notion of persistent places is a useful framework for investigating the formation of deflated surface archaeological distributions across western NSW.

The surface archaeological record of Pine Point and Langwell

Pine Point and Langwell are sheep grazing properties approximately 50 kilometres south of Broken Hill (Figure 1). The properties straddle the transitional zone between the foothills of the Barrier Range in the north to the alluvial sand plain of the Murray-Darling in the south. Following the methods of Holdaway et al. (2000), land system classifications were used to divide the study area into four separate areas. Aerial photographs were then used to identify locations within each of the land systems that were likely to contain surface archaeological deposits. Extensive scalding, proximity to drainage systems and topography were the key criteria used to identify these locations. Each identified location was then ground checked to assess the potential for detailed recording.

In common with other areas of western NSW (e.g. Holdaway et al. 2000; Holdaway et al. 2004), surface stone artefact distributions across Pine Point and Langwell have lost their vertical integrity through the erosion of topsoil resulting in the deflation of the artefacts onto the scalded A2 soil horizon. Hearths appear as either clustered or dispersed distributions of stone and clay heat retainers that represent the base of the hearth pit. In some instances the heat retainers serve to 'cap' charcoal concentrated in sediments on the hard baked bottom of the hearth fire pit. To be selected a location had to have an extensive distribution of stone artefacts and heat-retainer hearths that were likely to retain charcoal for radiometric analysis.

Ten locations meeting these criteria were selected for investigation. Preliminary artefact recording and hearth survey were conducted at each location. Following this initial phase of investigation further recording was restricted to four main locations where charcoal-bearing hearths and extensive stone artefact scatters were abundant (Table 1). Two of these, CN1 and CN3, are in the Conservation/Fowlers land system and are situated on an alluvial terrace adjacent to Pine Creek, the

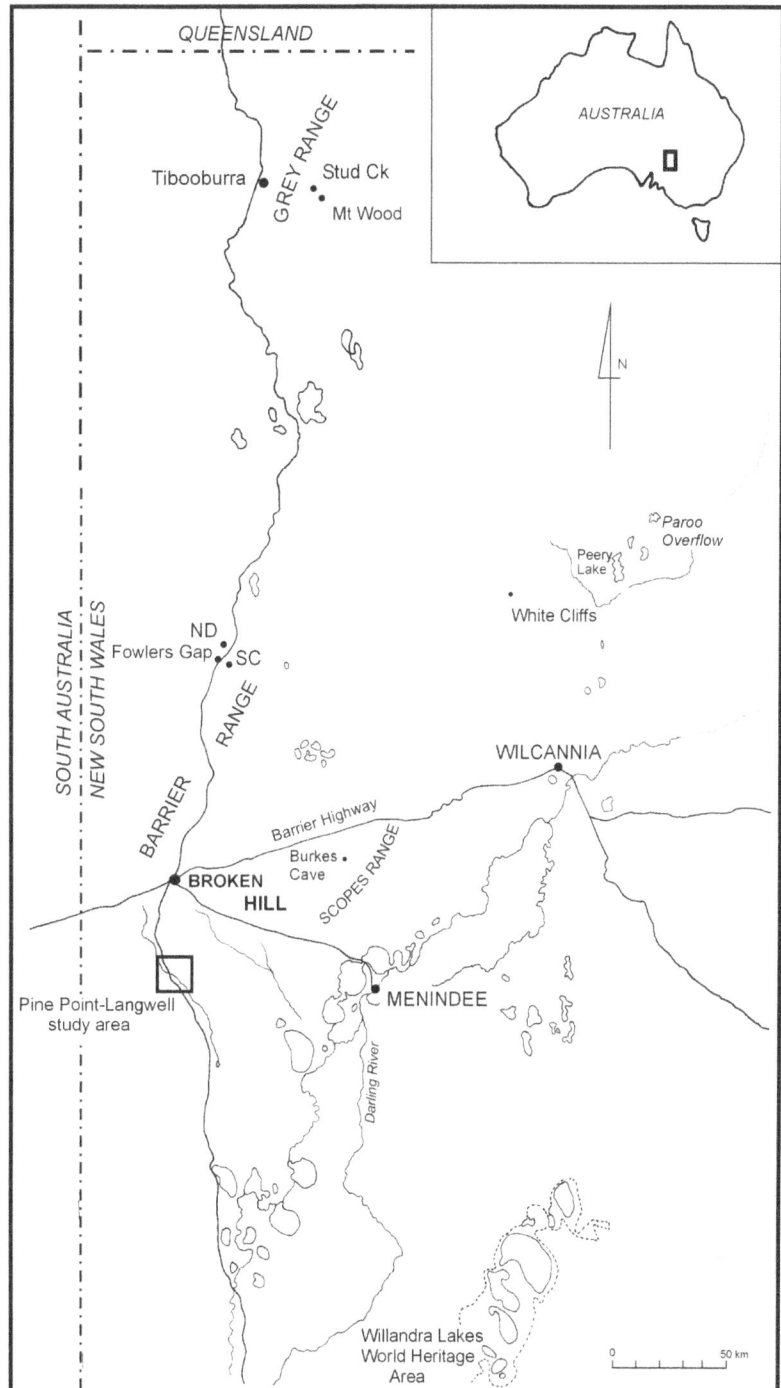

Figure 1. Location of the Pine Point/Langwell study area and other locations mentioned in the text.

largest drainage channel in the study area. The other two, KZ1 and KZ2, are in the Kars land system and are adjacent to Rantyga Creek, the second largest drainage channel in the study area. KZ1 covers an alluvial terrace and distal flood plain on both sides of Rantyga Creek. KZ2 is situated on a scalded sandy rise on the northern side of Rantyga Creek. The locations are all situated within the immediate vicinity of the confluence of Rantyga Creek with the larger Pine Creek. The distance from the furthest two locations KZ1 and CN3 is four kilometres.

Land System	Sampling Location	Description	Sampling Area m^2	Artefacts	Hearths
Conservation/ Fowlers	CN1	Discontinuous hardpan scald situated on top of a major terrace of Pine Creek, approximately 600 metres downstream of the Pine Creek–Rantyga Creek confluence.	19355	8788	CNH7, CNH23 to CNH25, CNH42, CNH55
Conservation/ Fowlers	CN3	Extensive hardpan scald situated on top of terrace of Pine Creek, starting 1.2 kilometres metres downstream of the Pine Creek–Rantyga Creek confluence.	12813	4904	CNH32 to CNH36 and CNH56
Kars	KZ1	Series of hardpan scalds on distal flood plain of Rantyga Creek, approximately three kilometres upstream of confluence with Pine Creek.	11795	3419	KZ1H2, KZ1H3, KZ1H7
Kars	KZ2	Series of hardpan exposures on top of a sandy rise and distal flood plain adjacent to Rantyga Creek, approximately 1.5 kilometres upstream of confluence with Pine Creek.	7203	11192	KZ2H25

Table 1. Characteristics of the archaeological sampling areas.

The four locations chosen for study are not discrete and bounded sites. Rather they represent areas of high ground exposure (resulting from erosion and a lack of vegetation) where extensive distributions of stone artefacts and hearths are visible. The issue of site definition has been widely discussed in such contexts (e.g. Ebert 1992; Thomas 1975). In western NSW, Holdaway et al. (1998) suggested that site boundaries are difficult if not impossible to define and that archaeologists should consider factors relating to the geomorphic context of artefact distributions when attempting to define assemblages. The same approach is adopted here. The Pine Point/Langwell assemblages were defined according to their geomorphic context. In the case of CN1 and CN3 this was the major terrace running parallel to the northern bank of Pine Creek. For KZ1 and KZ2, artefact recording was confined to the hard pan surfaces on the fringe of the distal flood plain. The assemblages represent samples of larger artefact distributions.

The proximity of the artefact assemblages to the Pine Creek–Rantyga Creek confluence suggested the possibility that the chronological record would indicate multiple phases of occupation. Although significant environmental change has occurred with pastoral land use, the creek confluence is a major landscape feature that appeared to be a focus of past human activity (numerous stone artefacts and hearths). It is possible that the four locations may represent persistent places with characteristics similar to those identified by Schlanger. To investigate this required the establishment of an occupational chronology and the analysis of stone artefact assemblage composition.

Chronological context

The remains of 122 heat-retainer hearths were recorded during intensive pedestrian survey of the Pine Creek–Rantyga Creek confluence. Information on the excavation and recording of the hearths has been published elsewhere (Holdaway et al. 2005), and is briefly summarised here. Hearths were classified into three groups according to their relative degree of preservation. Of those with partially exposed and clustered heat retainers, 30 were selected for excavation, based on their spatial proximity to the stone artefact recording areas. The excavated hearths consisted of a sandy matrix with a cluster of heat retainers and varied amounts of mostly fragmented charcoal. No structural evidence of hearth reuse or multiple lenses of charcoal were found.

Sufficient charcoal for radiocarbon determinations was recovered from 16 hearths. The hearths are distributed along the alluvial gullies of Rantyga and Pine Creeks. Hearths KZ1 H2, H3, and H7 are associated with the KZ1 location. KZ2 H25 is associated with KZ2 location. CN H7, H23, H24, H25, H42 and H55 are associated with the CN1 location. Hearths CN H32, H33, H34, H35, H36 and H56 are associated with the CN3 location. In Table 2 the determinations are listed in chronological order from youngest to oldest.

Lab Number	Result (year BP)	δ C13	Calibrated 2 σ ages (y BP (probability)	Hearth ID
Wk-9994	261±49	−23.8±0.2	160–166 (0.053) 287–315 (0.514) 408–420 (0.107)	CN H 42
Wk-12322	458±58	−24.1±0.2	463–545 (0.682)	CN H 34
Wk-12323	481±55	−23.5±0.2	477–534 (0.686)	CN H 35
Wk-12324	516±52	−23.0 ± 0.2	510–554 (0.572) 609–620 (0.102)	CN H 36
Wk-12326	584±46	−22.2±0.2	544–564 (0.212) 587–590 (0.025) 599–645 (0.47)	KZ2 H 25
Wk-12325	771±46	−24.6±0.2	737–795 (0.344) 812–829 (0.098) 862–906 (0.235)	CN H 56
Wk-9995	848±69	−23.4±0.2	687–794 (0.542) 813–827 (0.057) 866–886 (0.078)	KZ1 H 2
Wk-10282	886±47	−23.9±0.2	737–795 (0.344) 812–829 (0.098) 862-906 (0.235)	CN H 55

Lab Number	Result (year BP)	δ C13	Calibrated 2 σ ages (y BP (probability)	Hearth ID
Wk-10280	910±52	−23.0±0.2	765–779 (0.071) 787–798 (0.059) 809–839 (0.162) 842–912 (0.384)	CN H 23
Wk-10832	959±51	−22.1±0.2	793–814 (0.137) 826–867 (0.26) 884–929 (0.274)	KZ1 H 7
Wk-12320	967±62	−21.9±0.2	794–814 (0.133) 826–867 (0.257) 884–931 (0.287)	CN H 25
Wk-12319	1002±48	−24.5±0.2	797–809 (0.094) 835–848 (0.075) 911–963 (0.517)	CN H 24
Wk-12321	1247±56	−23.9±0.2	1092–1108 (0.071) 1126–1160 (0.158) 1169–1193 (0.132) 1197–1241 (0.259) 1245–1261 (0.082)	CN H 33
Wk-10281	1653±52	−23.2±0.2	1421–1430 (0.028) 1489–1497 (0.024) 1515–1613 (0.553) 1623–1625 (0.006) 1674–1689 (0.061)	CN H 32
Wk-9993	1747±76	−23.5±0.2	1550–1734 (0.678)	CN H 7
Wk-10831	2004±73	−23.2±0.2	1871–2044 (0.678)	KZ1 H 3

Table 2. Radiocarbon determinations from Pine Point/Langwell (after Holdaway et al. 2005).

A full analysis of the Pine Point/Langwell radiocarbon determinations is presented in Holdaway et al. (2005), the feature of which is a Bayesian analysis of temporal patterning within the total pool of 16 determinations. This indicated that, rather than forming a continuous sequence, the determinations cluster into five main phases of hearth construction. The oldest is a single determination from KZ1 H3 at approximately 2000 cal BP. The second group consists of two hearths with determinations between 1500 and 1700 cal BP. The third group is one hearth with a determination between 1050 and 1300 cal BP. Seven determinations make up the fourth group, with determinations spanning 750 to 950 cal BP. The fifth and final group consists of five determinations spanning 550 to 350 cal BP. Together the radiocarbon determinations from the 16 hearths indicate a chronology of late Holocene occupation across the four study localities and a temporal framework for the investigation of stone artefact assemblage formation.

Measuring the intensity of raw material utilisation

Artefact assemblages usually consist of both short and long use-life artefacts. Short use-life artefacts are those that have performed little or no work (Shott 1989, 1995). Characteristically, these are unretouched flakes and other debitage produced during core reduction and tool manufacture (Holdaway et al. 2004; Shiner et al. 2005). These have a high probability of discard and the majority of assemblages are composed of large numbers of these items. Long use-life artefacts mainly consist of retouched tools that can be ranked according to the degree to which they have been resharpened. The least resharpened tools are utilised flakes, which exhibit non-invasive retouch. These are usually flakes that have been used to perform work without prior edge modification. More heavily resharpened tools consist of denticulates, notches, scrapers and burren and tula adzes that have been deliberately modified before use. The most heavily resharpened item in arid zone assemblages is the tula adze, which following repeated resharpening is often, but not always, discarded as a worked-out 'slug' (Gould 1980; Hayden 1979, Holdaway et al. 2004). In some cases, the use-life of these items may exceed the occupation duration of a single location.

The discard of artefacts is a time-dependent process and the composition of assemblages is an indirect result of occupation duration (Bamforth and Becker 2000). As occupation becomes longer there is a greater probability that long use-life artefacts will be created and discarded (Holdaway et al. 2000).

Differences in the intensity of raw material reduction offer one promising approach to the investigation of assemblage composition because increased occupation duration may lead to the more intensive utilisation of material available within the immediate context of an occupied location (Dibble 1988; Dibble and Rolland 1992; Elston 1990). This can be investigated through the study of assemblage composition. Assemblages that exhibit high flake-to-core ratios, a low proportion of cortical artefacts, a decrease in flake and core size, and the presence of heavily worked tools and cores reflect more intensive reduction of raw materials (Dibble et al. 1995). These characteristics, combined with the concept of time-dependent artefact discard, provide a method for investigating the intensity of place use from stone artefact assemblages, without recourse to functional interpretations.

Raw material utilisation

In common with surface stone artefact assemblages from other areas of western NSW (Holdaway et al. 2000; Shiner et al. 2005; Witter 1992), the Pine Point/Langwell assemblages are dominated by quartz and silcrete. There are clear differences in the proportion of raw materials between the two CN and the two KZ assemblages (Table 3). All four assemblages are dominated by quartz, but the proportion of quartz in the CN assemblages is considerably greater than in the KZ assemblages. Approximately 80% of artefacts in the CN1 and CN3 assemblages are made from quartz. There are two types of silcrete — clast (quartz grains present in the matrix) and non-clast (quartz grains rare or absent in the matrix). The two types of silcrete combined account for 16.3% of raw materials at CN1 and 14.1% at CN3.

Core Type	CN1	CN3	KZ1	KZ2
Clast	1091 (12.4)	544 (11.1)	936 (27.4)	3526 (31.5)
Non-clast	347 (3.9)	145 (3)	248 (7.3)	575 (5.1)
Quartz	7228 (82.2)	4107 (83.7)	2135 (62.4)	6895 (61.6)
Other	122 (1.4)	108 (2.2)	100 (2.9)	196 (1.7)
Total	8788 (100)	4904 (100)	3419	11192 (100)

Table 3. Raw material number and percentage (in parentheses) per assemblage.

The KZ1 and KZ2 assemblages demonstrate a pattern of raw material abundance different from the CN1 and CN3 assemblages. Quartz accounts for 62.4% of raw materials at KZ1 and 61.6% at KZ2. The two types of silcrete combined account for 34.7% at KZ1 and 36.6% at KZ2. Therefore, the proportion of silcrete in the KZ assemblages is considerably greater than in the CN assemblages. As with the CN assemblages, silcrete is dominated by clast material. The proportion of non-clast silcrete is slightly greater at KZ1 (7.3%) compared to KZ2 (5.1%). The category 'other materials' includes crystal quartz, chert, hornfels, ironstone, quartzite, sandstone and schist, and these make up only a small percentage of raw material.

Differences in raw material access are likely to account for some of the variability in the relative proportions of raw materials between the assemblages. Quartz is available as fist-sized cobbles in creek beds and as gibber pavements within immediate vicinity of all the assemblages, and is classified as a local raw material. Clast silcrete outcrops with evidence of human use occur in the low hills approximately six kilometres north of Pine Creek. The characteristics of one of these assemblages are analysed in Shiner (2006). No sources of non-clast silcrete were identified in the study area. Both types of silcrete are regarded as a non-local raw material because sources of each are not available within the immediate vicinity of any of the assemblages. The two KZ assemblages contain the highest proportions of silcrete, and are located two to three kilometres in a straight line from the silcrete outcrops. The CN assemblages are located five to six kilometres in a straight line from the silcrete outcrops.

Access to raw materials is clearly an important factor in assemblage variability. Exposure of raw material sources is unlikely to have been a limiting factor because the landscapes of Pine Point and Langwell are primarily erosion dominated, and outcrops and gibber pavements are widely distributed. Alternatively, access to raw material sources is likely to have varied with factors associated with the duration of occupation, and mobility reduction as occupation duration increases. The role of these factors can be assessed through the comparison of key technological indices related to the intensity of raw material reduction. These methods have been widely used in both Australia (Holdaway et al. 2000, 2004; Shiner 2004; Shiner et al. 2005; Veth 1993) and elsewhere (Bamforth and Becker 2000; Dibble 1995) and are discussed in the following section.

Core form

Core form is indicative of the techniques used to reduce nodules. For example, non-specialised core forms with platforms flaked from only one direction (unifacial) and with only one or two negative flake scars (test), suggest non-intensive core reduction strategies. Those with platforms flaked from two or more directions (bifacial and multi-platform) indicate core rotation and a concern with extending the reduction life of nodules. There is some possibility that sample size is influencing the proportion of different core forms represented in each assemblage. For example, the largest assemblage, KZ2, also generally has the greatest number of different core forms. Despite this, the number of artefacts in each assemblage is sufficiently large to allow confidence that the patterns are also representative of behavioural characteristics e.g. occupation duration, rather than sample size alone.

The proportion of different clast silcrete core forms in each assemblage is presented in Table 4. Eight different core forms are represented, with eight at KZ2, five at KZ1, four at CN1 and three at CN3. Microblade, nuclear-tool and radial core forms were only recorded at KZ2, while flake-blank forms were confined to KZ1 and KZ2. Proportions of test core forms vary from 9.4% at CN1 followed by 3% at KZ1 and 2.6% at KZ2. They are not represented at CN3.

Core Type	CN1	CN3	KZ1	KZ2
Bifacial	4 (12.5)	2 (9.5)	6 (18.2)	31 (27)
Flake blank			1 (3)	8 (7)
Microblade				1 (0.9)
Multiple	1 (3.1)	4 (19)	9 (27.3)	5 (4.3)
Nuclear tool				1 (0.9)
Radial				3 (2.6)
Test	3 (9.4)		1 (3)	3 (2.6)
Unifacial	24 (75)	15 (71.5)	16 (48.5)	63 (54.8)
Total	32 (100)	21 (100)	33 (100)	115 (100)

Table 4. Frequency and proportion of clast silcrete core types per assemblage.

Unifacial cores are the single most common clast silcrete core form in all of the assemblages, varying from 75% at CN1, 71.5% at CN3, 54.8% at KZ2 and 48.5% at KZ1. At KZ1 the combined proportion of rotated forms (45.5%) is smaller than that of unifacial forms (48.5%), although this is still a relatively high proportion of rotated forms and is much greater than that for the other three assemblages. The proportion of rotated forms decreases from 31.3% at KZ2 to 28.5% at CN3 and 15.6% at CN1.

The proportion of non-clast silcrete core forms is presented in Table 5. Eight different forms are represented across all assemblages, with six forms in the KZ1 and KZ2 assemblages, four in CN1 and two in CN3. One bipolar core was recorded at CN1, while one flake-blank core was recorded at KZ1 and two at KZ2. One radial core was recorded at both KZ1 and KZ2. Test core forms were also rare, with only one example recorded at KZ1. Two nuclear-tool cores were recorded at KZ2. No microblade cores were recorded in this material.

Core Type	CN1	CN3	KZ1	KZ2
Bifacial	2 (20)		4 (36.4)	5 (21.7)
Bipolar	1 (10)			
Flake blank			1 (9.1)	2 (8.7)
Multiple	1 (10)	1 (20)	2 (18.2)	1 (4.3)
Nuclear tool				2 (8.7)
Radial			1 (9.1)	1 (4.3)
Test			1 (9.1)	
Unifacial	6 (60)	4 (80)	2 (18.2)	12 (52.2)
Total	10 (100)	5 (100)	11 (100)	23 (100)

Table 5. Frequency and proportion of non-clast silcrete core types per assemblage.

Unifacial cores are the single most common non-clast core form in all assemblages except KZ1, varying from 80% at CN3, followed by 60% at CN1, 52.2% at KZ2 and 18.2% at KZ1. At KZ1 the proportion of bifacial cores is twice that of unifacial cores (36.4% compared to 18.2%) and the combined proportion of rotated core forms is considerably higher than that of unifacial forms at KZ1 (54.6% compared to 18.2%). This pattern does not follow at CN1, CN3 and KZ2. These assemblages show similar patterns to those identified for clast silcrete.

Table 6 presents the proportion of quartz core forms. As in the clast silcrete component, microblade forms are confined to KZ2. Radial cores occur only at CN1, CN3 and KZ1. Bipolar cores are uncommon in all assemblages. Nuclear-tool forms account for less than 5% of quartz core forms in all of the assemblages. Test core forms are relatively uncommon across all assemblages. The highest proportion of test cores occurs at CN1 (8.5%) and CN3 (6.1%). Unifacial cores are the most common quartz core form in all assemblages, varying from 51.1% at KZ1 to over 66% at CN3. Again the proportion of rotated core forms (bifacial and multiple combined) is highest at KZ1 (35.2%). The low proportion of rotated core forms indicates that extending the life of quartz cores was not a priority.

Core Type	CN1	CN3	KZ1	KZ2
Bifacial	92 (18.3)	49 (16.7)	37 (26.6)	118 (28.2)
Bipolar	6 (1.2)	2 (0.7)	6 (4.3)	19 (4.5)
Flake blank	5 (1)	8 (2.7)	1 (0.7)	11 (2.6)
Microblade				2 (0.5)

Core Type	CN1	CN3	KZ1	KZ2
Bifacial	92 (18.3)	49 (16.7)	37 (26.6)	118 (28.2)
Bipolar	6 (1.2)	2 (0.7)	6 (4.3)	19 (4.5)
Flake blank	5 (1)	8 (2.7)	1 (0.7)	11 (2.6)
Microblade				2 (0.5)
Multiple	46 (9.1)	19 (6.5)	12 (8.6)	11 (2.6)
Nuclear tool	19 (3.8)	3 (1)	6 (4.3)	6 (1.4)
Radial	2 (0.4)	1 (0.3)	1 (0.7)	
Test	43 (8.5)	18 (6.1)	5 (3.6)	19 (4.5)
Unifacial	290 (57.7)	194 (66)	71 (51.1)	233 (55.6)
Total	503 (100)	294 (100)	139 (100)	419 (100)

Table 6. Frequency and proportion of quartz core types per assemblage.

Minimum number of flakes (MNF) to core ratio

The MNF to core ratio is the most basic measure of core reduction intensity and is calculated by summing the total number of flakes with a platform (complete and proximal flakes), together with half the longitudinal splits (Holdaway and Stern 2004). As core reduction proceeds, the number of flakes produced increases relative to the number of cores (Dibble 1995). Longer occupation by less mobile groups will limit opportunities to replenish raw material stocks and result in more complete reduction of cores and the increased production of flakes.

Plotting the ratio for each raw material type (Table 7) indicates both similarities and differences among the assemblages. In all of the assemblages the ratio for quartz is lowest, indicating that this local material was the least intensively reduced. The quartz ratio also shows the least variability between assemblages. KZ2 has the highest ratio with eight and CN1 the lowest with 7.1, while the ratio is 7.3 for both CN3 and KZ1.

Material	CN1	CN3	KZ1	KZ2
Clast	20.8	17.5	16.5	18.3
Non-clast	23.7	21	14.5	25.4
Quartz	7.1	7.3	7.3	8

Table 7. MNF (minimum number of flakes) to core ratio per raw material and assemblage.

The ratio for clast silcrete is lower than that for non-clast silcrete in each of the assemblages except KZ1. The CN1 assemblage has the highest ratio for clast silcrete, followed by KZ2, CN3 and KZ1. The higher ratio at CN1 and the lower ratio at KZ1 are consistent with a distance-decay relationship, but this is not the case for either the KZ2 or CN3 assemblages. The ratios for non-clast silcrete demonstrate a greater amount of variability between the assemblages than those for clast silcrete and quartz. KZ2 has the highest ratio for non-clast silcrete, followed by CN1 and CN3.

None of the assemblages demonstrate a clear pattern of intensive core reduction in all of the raw material categories. KZ1 has the lowest value for clast and non-clast silcrete and the second lowest for quartz, suggesting that reduction was less intensive relative to the other assemblages. This may reflect the closer proximity of KZ1 to major clast silcrete sources than is true for the other assemblages. Both CN1 and KZ2 have the highest ratios for clast and non-clast silcrete, indicating that silcrete core reduction was most intensive in these assemblages. In addition, KZ2 has the highest ratio for quartz. CN3 exhibits more intensive core reduction than KZ1, but this is not as intensive as either CN1 or KZ2. The implication of the overall patterning in MNF to core ratios is that KZ2 saw the longest occupation of the four assemblages.

Non-cortical flake to cortical flake ratio

Increased core reduction also leads to a decrease in the proportion of cortical surfaces on flakes and cores (Dibble 1995). The MNF to core ratios suggested that non-clast silcrete was generally the most intensively worked material. Thus it is expected that the non-cortical to cortical complete flake ratio will be highest for this material. Results for this ratio (Table 8) indicate that this is not the case, but rather the ratio is highest for clast silcrete across all the assemblages.

Material	CN1	CN3	KZ1	KZ2
Clast	12.3	11.7	9	8.6
Non-clast	9.9	11.2	8.1	7.6
Quartz	0.8	0.2	1.4	1

Table 8. Non-cortical to cortical complete flake ratio per raw material and assemblage.

Non-clast silcrete has the second highest ratio, followed by quartz. The ratio is consistent for quartz in the four assemblages and points to the reduction of local cortical nodules. The same cannot be said for clast silcrete. Clearly clast silcrete, although available within the wider area, was not utilised in the same way as local quartz and was less likely to be available as fist-sized gibber nodules. Clast silcrete nodules were transported to the locations as partially decortified cores. This is further supported by the relative proportion of cortical complete flakes to non-cortical complete flakes (Tables 9–11). Quartz exhibits a pattern different from both the silcretes. Cortical complete flakes are common and a large proportion of the flakes have greater than 50 percent cortex. This indicates on-site reduction of cortical nodules.

Cortex Category	CN1	CN3	KZ1	KZ2
None	431 (92.5)	258 (92.1)	358 (89.9)	1336 (89.5)
1–50%	19 (4.1)	15 (5.4)	31 (7.8)	104 (7)
50–99%	13 (2.8)	6 (2.1)	9 (2.3)	50 (3.4)
Complete	3 (0.6)	1 (0.4)	0 (0)	2 (0.1)
Total	466 (100)	280 (100)	398 (100)	1492 (100)

Table 9. Frequency and percentage (in parentheses) of clast silcrete complete flakes with different amounts of cortex.

Cortex Category	CN1	CN3	KZ1	KZ2
None	139 (90.8)	67 (91.8)	89 (89)	189 (88.3)
1–50%	12 (7.8)	3 (4.1)	6 (6)	18 (8.4)
50–99%	2 (1.3)	3 (4.1)	5 (5)	6 (2.8)
Complete	0 (0)	0 (0)	0 (0)	1 (0.5)
Total	153 (100)	73 (100)	100 (100)	214 (100)

Table 10. Frequency and percentage (in parentheses) of non-clast silcrete complete flakes with different amounts of cortex.

Cortex Category	CN1	CN3	KZ1	KZ2
None	1212 (44.6)	698 (41.3)	413 (58)	1222 (50.4)
1–50%	823 (30.3)	572 (33.9)	189 (26.5)	815 (33.6)
50–99%	537 (19.7)	329 (19.5)	97 (13.6)	321 (13.2)
Complete	147 (5.4)	90 (5.3)	13 (1.8)	66 (2.7)
Total	2719 (100)	1689 (100)	712 (100)	2424 (100)

Table 11. Frequency and percentage (in parentheses) of quartz complete flakes with different amounts of cortex.

The KZ1 and KZ2 ratios for quartz are higher than the ratios for CN1 and CN3, suggesting more intensive reduction of quartz cores in the KZ assemblages, a result supported by the MNF to core ratio for KZ2, but not KZ1. The patterns emerging at KZ2 and to a lesser extent KZ1 cannot be attributed to differential raw material access because both locations are the same distance from the nearest silcrete sources. CN1 and CN3 show similar values for both types of silcrete. This result supports the relatively high values of the MNF to core ratio and suggests that silcrete reduction was intensive in these two assemblages. KZ1 and KZ2 have similar results for both types of silcrete. This result is surprising because the MNF to core ratio suggested that core reduction was more intensive at KZ2 than KZ1.

Non-cortical core to cortical core ratio

The non-cortical to cortical core ratio provides another measure of core reduction intensity. Values for this ratio by assemblage and raw material are presented in Table 12. Clast silcrete has the highest ratio for each of the assemblages, followed by non-clast silcrete, except for CN3, where no cores with cortex were recorded. Quartz has the lowest ratio in all the assemblages, with the ratio not exceeding one, indicating that there are more cortical than non-cortical cores.

Material	CN1	CN3	KZ1	KZ2
Clast	2.6	3.2	3.7	2.1
Non-clast	1.5	0	2.7	1.3
Quartz	0.2	0.1	0.6	0.3

Table 12. Non-cortical to cortical core ratio per raw material and assemblage.

The highest ratios for the three raw material types all occur in the KZ1 assemblage. This is unexpected because the MNF to core and the non-cortical flake to cortical flake ratios suggested that KZ1 cores are the least intensively worked, but it does fit with the high proportion of rotated core forms at KZ1, which suggest a greater likelihood of cortex removal resulting from the flaking of multiple surfaces. The results for the CN1 and KZ2 assemblages are also ambiguous. The MNF to core and the non-cortical to cortical complete flake ratios suggested that core reduction was intensive at CN1. This is not supported by the non-cortical to cortical core ratio, which indicates a low proportion of decortified cores at CN1. There is a similar result for KZ2, but the pattern is less clear. The MNF to core ratio was high at KZ2, but the non-cortical to cortical complete flake ratio low. The low proportion of decortified cores at KZ2 supports the non-cortical to cortical complete flake ratio, but is in disagreement with the MNF to core ratio. CN3 follows a pattern consistent with the non-cortical to cortical complete flake ratio.

Unmodified flake to tool ratio

The unmodified flake to tool ratio is the simplest measure of tool production. Low values for this ratio indicate that proportionally more flakes in an assemblage are modified into tools. Proportionally fewer quartz flakes are modified into tools in the CN1, CN3 and KZ2 assemblages compared to both types of silcrete (Table 13). The KZ1 assemblage is an exception. Here the ratio for quartz is less than that for most clast silcrete components. The ratio in all assemblages is lowest for non-clast silcrete indicating that there are fewer unretouched flakes relative to retouched flakes on this material than in any other material. Non-clast silcrete appears to have been favoured for tool production.

Material	CN1	CN3	KZ1	KZ2
Clast	8.3	10.5	11.7	13
Non-clast	3.8	2.8	4.3	3.9
Quartz	13.9	18	9.5	19.9

Table 13. Flake to MNT (minimum number of tools) ratio per raw material and assemblage.

In general, tool production is least intensive at KZ2. The ratio for clast silcrete and quartz at KZ2 is the highest of the assemblages; the value for non-clast silcrete is the second highest behind KZ1. The CN1 and CN3 assemblages have the lowest values for both types of silcrete, but there is only a marginal difference between the non-clast value at CN1 and KZ2. Additionally, there is no clear pattern between the silcretes in the CN1 and CN3 assemblages, with the value for clast lowest at CN1 and the value for non-clast lowest at CN3. For quartz, the value at CN1 and CN3 is much higher than that at KZ1.

Discussion

As with much of the archaeological record of western NSW, the surface archaeological distributions across Pine Point and Langwell Stations cannot be interpreted as ethnographic slices of time. Instead, they represent a time-averaged (Stern 1994) record of archaeological deposition and geomorphic process. Interpretations should take account of the time-accumulative nature of the record. The radiocarbon chronology from the heat-retainer hearths indicates multiple episodes of occupation during the last 2000 years. Further, inconsistencies in the technological indices point to the variable nature of assemblage formation. Some of these inconsistencies may reflect the variable forms in which the raw materials are available as well as the variable nature of the behavioural processes responsible for assemblage accumulation. These behaviours may include artefact recycling, and artefact removal as well as variable occupation spans. This will be further discussed below.

Consistent patterns of reduction intensity for both types of silcrete are difficult to identify in the assemblages. In all four assemblages the non-cortical to cortical complete flake and the non-cortical to cortical core ratios are highest for clast silcrete rather than non-clast silcrete. These results do not follow the MNF to core ratio that generally suggested non-clast cores were the most intensively worked. The MNF to core and the non-cortical to cortical complete flake ratios suggest that core reduction is most intensive in the CN1 and KZ2 assemblages, but this is not supported by the non-cortical to cortical core ratio. From this it is difficult to draw straightforward conclusions about the intensity of clast and non-clast silcrete core reduction. Instead, the variability hints at the complex nature of assemblage formation, and suggests that the Pine Point/Langwell assemblages do not represent a single process of silcrete acquisition and reduction through time. Rather, the assemblages represent multiple raw material management processes. The re-occupation of the four locations through time also raises the possibility that at least some of the artefacts were reused on numerous occasions. The implication of this is that not only were many of the artefacts, and especially those made of either type of silcrete, manufactured elsewhere, but also they may have experienced multiple episodes of reduction and use through time at the locations.

The reduction of quartz across the assemblages shows far greater consistency compared to either of the silcretes. Quartz cores are the least rotated, indicating reduction from predominantly one platform surface. Further, the MNF to core ratio suggest that quartz is the least intensively worked material, and the non-cortical to cortical complete flake and the non-cortical to cortical core ratio support this. Quartz is also generally the least intensively utilised material for the production of tools. This consistency suggests a more limited set of behaviours are represented in the acquisition and reduction of quartz compared to the silcretes. Quartz is a local raw material to all of the assemblages and occurs as fist-sized rounded cobbles on valley slopes and creek beds. The size and form of the nodules may also constrain the reduction of quartz thus producing a more uniform pattern than the two types of silcrete.

At least some aspects of variability among the assemblages reflect differential access to raw materials. For example, the non-intensive reduction of quartz reflects to some degree the abundant sources of this material within the immediate vicinity of the assemblages. Measures of silcrete reduction are full of inconsistencies: while there are some aspects of a distance-decay relationship, assemblages located furthest from possible silcrete sources do not demonstrate a clear pattern of more intensive reduction. This may be interpreted in a number of ways: silcrete was not always transported to the locations from the closest sources, each location has variable occupation histories, and a single place was rarely used the same way through time.

The surface archaeological record of Pine Point/Langwell was investigated as a series of places with individual use histories. These were revealed through the investigation of assemblage formation over a 2000 year period of occupation which was indicated by the radiocarbon ages from 16 heat-retainer hearths. While this is not a definitive chronological record, it provides a good chronological framework consistent with other hearth dating programmes across western NSW (e.g. Holdaway et al. 2002, 2005). The results of this analysis reveal both variability and consistency in assemblage composition across a relatively small area of the landscape. While some aspects represent responses to the distribution and form of lithic raw material sources, others are indicative of variability in the intensity of occupation over the long-term. Different locations exhibited varied occupational signatures. For instance, measures of raw material utilisation suggest that occupation intensity at KZ1 and CN3 was less intensive than at KZ2 and CN1.

The Pine Point/Langwell locations can be regarded as persistent places in the sense that they document multiple episodes of occupation over the last 2000 years. Although it is not possible to link any single artefact with a specific period of occupation as indicated by the radiocarbon chronology, the presence of hearths with multiple age determinations within each of the artefact assemblages, and around the broader Pine Creek–Rantyga Creek confluence indicate multiple visits. Within this context artefact discard would also have occurred on multiple occasions, and some of this may have coincided with the occupational episodes represented by the hearths. The boundaries of these occupations and the exact reasons behind the reuse of the four locations can never be known. In any case, these are likely to be many and complex. What we do know is that the Pine Creek–Rantyga Creek confluence was the focus of repeated occupation. As applied by Shiner et al. (2005) in relation to assemblages from Burkes Cave, Fowlers Gap (SC and ND locations) and Stud Creek, the analysis of stone artefact assemblage composition within temporal and economic frameworks is one method of assessing the persistence of occupation within palimpsest deposits.

Conclusion

Analysis of the composition of the Pine Point/Langwell assemblages indicates both consistencies and inconsistencies in the reduction and utilisation of lithic raw materials. Some of these consistencies are argued to reflect the character and distribution of the wider lithic landscape. In general, there is a distance-decay relationship in the reduction of silcrete. This relationship is not evident in all measures of reduction intensity. Variation in measures of core reduction is interpreted to reflect the variable nature of occupation through time at each of the locations in both duration and frequency. Over the time span represented in the Pine Point/Langwell occupational chronology, multiple behavioural patterns resulted in internal assemblage variability. These patterns are consistent with persistent but varied use of the locations during the last 2000 years.

The variable patterns identified in the Pine Point/Langwell assemblages raise some interesting questions about interpretative scales. Depending upon the perspective adopted, the Pine Point/Langwell assemblages may represent either accumulation over too long a timescale or over too short a timescale. Those seeking ethnographic-scale interpretations would argue that the Pine Point/Langwell assemblages are coarse grained with few clear behaviours represented. Others, who are interested in clear average patterns of assemblage composition representative of the dominant types of behaviour, may contend that the assemblages do not represent enough accumulation through time, hence variable and inconsistent intra-assemblage patterns. Viewed as persistent places that were utilised on multiple occasions during the last 2000 years, the Pine Point/Langwell assemblages point to the complex and varied processes responsible for the formation of the archaeological assemblages we define in the present.

Acknowledgements

Thank you to Andrew Fairbairn and Sue O'Connor for the opportunity to publish this paper in the Archaeometry volume. I wish to acknowledge the assistance of Simon Holdaway, Patricia Fanning and Peter Sheppard in supporting my research. The Broken Hill Local Aboriginal Land Council supported the project and consented to the excavation of the heat-retainer hearths. In particular I thank Raymond O'Donnell. Lisa Shiner and Bridget Mosley assisted during the 2002 field season. This work was funded by a University of Auckland Doctoral Scholarship, an AIATSIS grant, and a University of Waikato Radiocarbon Dating Laboratory grant. The Harvy family of Pine Point and the Harrison family of Langwell allowed me to undertake research on their properties. I also thank Val Attenbrow and two anonymous referees for their helpful comments.

References

Bamforth, D. B. and M. S. Becker. 2000. Core/biface ratios, mobility, refitting, and artifact use-lives: A Paleoindian example. *Plains Anthropologist* 45(173):273–90.

Cane, S. 1984. *Desert camps: A case study of stone artefacts and behaviour in the Western Desert.* Unpublished PhD thesis, Department of Prehistory, Australian National University.

Dibble, H. L. 1988. Typological aspects of reduction and intensity of utilization of lithic resources in the French Mousterian. In H. L. Dibble and A. Montet-White (eds), *Upper Pleistocene prehistory of Western Eurasia*, pp 181–97. Philadelphia: University of Pennsylvania Museum.

Dibble, H. L. 1995. Raw material availability, intensity of utilization, and Middle Paleolithic assemblage variability. In H. Dibble and M. Lenoir (eds), *The Middle Paleolithic site of Combe-Capelle Bas (France)*, pp 289–315. Philadelphia: University of Pennsylvania Museum.

Dibble, H. L. and N. Rolland. 1992. On assemblage variability in the Middle Paleolithic of Western Europe: History, perspectives, and a new synthesis. In H. L. Dibble and P. Mellars (eds), *The Middle Paleolithic: Adaptation, behavior and variability*, pp 1–25. Philadelphia: University of Pennsylvania Museum.

Dibble, H. L., B. Roth and M. Lenoir. 1995. The use of raw materials at Combe-Capelle Bas. In H. Dibble and M. Lenoir (eds), *The Middle Paleolithic Site of Combe-Capelle Bas (France)*, pp.259–287. Philadelphia: University of Pennsylvania Museum.

Ebert, J, I. 1992. *Distributional archaeology*. Albuquerque: University of New Mexico Press.

Elston, R. E. 1990. A cost-benefit model of lithic assemblage variability. In R. G. Elston and E. E. Bundy (eds), *The archaeology of James Creek Shelter*, pp 153–164. Salt Lake City: University of Utah Anthropological Papers 115.

Fanning, P. 2002. *Beyond the Divide: A new geoarchaeology of Aboriginal stone artefact scatters in Western NSW, Australia*. Unpublished PhD thesis, Macquarie University.

Fletcher, R. 1995. *The Limits of settlement growth, a theoretical outline*. Cambridge: Cambridge University Press.

Gould, R. A. 1980. *Living archaeology*. Cambridge: Cambridge University Press.

Hayden, B. 1979. *Palaeolithic reflections: Lithic technology and ethnographic excavations among Australian Aborigines*. Canberra: Australian Institute of Aboriginal Studies.

Holdaway, S., D. Witter, P. Fanning, R. Musgrave, G. Cochrane, T. Doelman, S. Greenwood, D. Pigdon and J. Reeves. 1998. New approaches to open site spatial archaeology in Sturt National Park, New South Wales, Australia. *Archaeology in Oceania* 33:1–19.

Holdaway, S., P. C. Fanning and D. C. Witter. 2000. Prehistoric Aboriginal occupation of the rangelands: Interpreting the surface archaeological record of Far Western New South Wales. *Rangelands Journal* 22(1):44–57.

Holdaway, S. J., P. C. Fanning, M. Jones, J. Shiner, D. C. Witter and G. Nicholls. 2002. Variability in the chronology of Late Holocene Aboriginal occupation on the arid margin of Southeastern Australia. *Journal of Archaeological Science* 29:351–363.

Holdaway, S. J., J. Shiner and P. Fanning. 2004. Hunter-gathers and the archaeology of the long-term: An analysis of surface stone artefacts from Sturt National Park, Western New South Wales, Australia. *Asian Perspectives* 43:34–72.

Holdaway, S and N. Stern. 2004. A record in stone: *The study of Australia's flaked stone artefacts*. Canberra: Museum Victoria and Aboriginal Studies Press, Australia Institute of Aboriginal and Torres Strait Islander Studies.

Holdaway, S. J., P. Fanning and J. Shiner. 2005. Absence of evidence or evidence of absence? Understanding the chronology of indigenous occupation of western New South Wales, Australia. *Archaeology in Oceania* 40(2):33–49.

Schlanger, S. H. 1992. Recognising persistent places in Anasazi settlement systems. In J. Rossignol and L. Wandsnider (eds), *Space, time and archaeological landscapes*, pp 91–112. New York: Plenum Press.

Shiner, J. I. 2004. *Place as occupational histories: Towards an understanding of deflated surface artefact distributions in the West Darling, New South Wales, Australia*. Unpublished PhD thesis, The University of Auckland.

Shiner, J. I. 2006. Artefact discard and accumulated patterns in stone artefact assemblage composition in surface archaeological deposits from Pine Point and Langwell Stations, western New South Wales. *The Rangeland Journal* 28(2):183–195.

Shiner, J., S. Holdaway, H. Allen and P. Fanning. 2005. Stone artefact assemblage variability in Late Holocene contexts in Western New South Wales: Burkes Cave, Stud Creek and Fowlers Gap. In C. Clarkson and L. Lamb (eds), *Lithics 'Down Under': Australian Perspectives on Lithic Reduction, Use and Classification*, pp 67–80. Oxford: Archaeopress.

Shott, M. J. 1989. On tool-class use lives and the formation of archaeological assemblages. *American Antiquity* 54(1):9–30.

Shott, M. J. 1995. How much is a scraper? Curation, use rates, and the formation of scraper assemblages. *Lithic Technology* 20(1):53–72.

Stern, N. 1994. The implications of time-averaging for reconstructing the land-use patterns of early tool-using Hominids. *Journal of Human Evolution* 27:89–105.

Thomas, D. H. 1975. Nonsite sampling in archaeology: Up the creek without a site? In J. W. Mueller (ed.), *Sampling in archaeology*, pp 61–81. Tucson: University of Arizona Press.

Veth, P. M. 1993. *Islands in the interior: The dynamics of prehistoric adaptations within the Arid Zone of Australia*. Ann Arbor: International Monographs in Prehistory.

Waters, M. R. and D. D. Kuehn. 1996. The geoarchaeology of place: The effect of geological processes on the preservation and interpretation of the archaeological record. *American Antiquity* 61(3): 483–497

Witter, D. 1992. *Regions and resources*. Unpublished PhD thesis, Department of Prehistory, Australian National University.

4

Developing methods for recording surface artefacts *in situ* on nineteenth and twentieth century sites in Australia

Samantha Bolton

Archaeology M405
University of Western Australia
35 Stirling Highway
Crawley WA 6009

Abstract

Surface artefacts are a common type of archaeological deposit in Australia. During field recording of ten sites along the Mundaring-Kalgoorlie historical settlement corridor, Western Australia, in 2004 and 2005, methods were developed to record *in situ* surface artefacts at sites from the late nineteenth and early twentieth centuries. The types of features and artefacts recorded included structural remains; large areas where individual features could not be distinguished; artefact scatters containing glass, ceramic and metal among other materials; bottle glass dumps; and individual artefacts. Detailed spatial data about the location of the archaeological material and surrounding topography were recorded using differential GPS. Using the archaeological and historical data, an analysis of when the sites were occupied, how they were used and by whom, was conducted. The research considers whether these methods are able to address the difficulties inherent in recording large, complicated surface historical sites.

Keywords: surface sites, recording methods, Mundaring, Kalgoorlie, Goldfields Water Supply Scheme, Bullabulling

Introduction

The majority of archaeological artefacts found in Australia are from surface scatters (Burke and Smith 2004: 202; Holdaway et al. 1998: 1). Despite this, most archaeological research continues to concentrate on excavated, and therefore collected, material, particularly in historical archaeology (Crook et al. 2002; Murray 2002: 11). Studies in historical archaeology that have recorded surface material in situ include the Central Australia Archaeology Project (Birmingham 1997) and Paterson's work in the south-western Lake Eyre Basin (Paterson 1999, 2003, 2005). Many studies of prehistoric sites are also of surface scatters and methods have been developed to record these sites.

There has been much discussion about how much information can be recorded at surface sites, and some of the associated problems (e.g. see Lewarch and O'Brien 1981; papers in Sullivan 1998; Redman 1987; Wandsnider and Camilli 1992). As part of a study on prehistoric sites in Sturt National Park, western New South Wales, Holdaway et al. (1998: 3) identified four main difficulties applicable to recording surface sites. First, the lack of chronological control because of the absence of stratigraphy; second, the difficulty in determining site boundaries and features demarcated by a group of artefacts; third, identifying and interpreting artefacts in the field; and fourth, the problem of obtaining a representative sample from sites where there is uneven exposure or visibility. Because these problems are also relevant to historic sites they need to be taken into account when recording them.

Different research questions require different approaches, and this paper presents one way to record a large number of surface sites from the historic period. The aim was to record the archaeological information in situ, and to determine whether it was possible to obtain the minimum required information (as defined by the research questions) without collecting the archaeological material. A case study from late nineteenth and early twentieth century Western Australia is used to discuss the methods, and whether they can help overcome the problems identified by Holdaway et al. (1998) at sites representing the historic period.

The archaeological project

Archaeology in the School of Social and Cultural Studies, University of Western Australia and the National Trust (WA) were awarded an ARC Linkage Grant to conduct archaeological investigations of nineteeth and twentieth century society along the Mundaring-Kalgoorlie settlement corridor following the Goldfields Water Supply Scheme, and this research is part of that project. The National Trust (WA) manages many of the sites associated with the Goldfields Water Supply Scheme and is currently undertaking a major heritage project of the region, the Golden Pipeline Project, conducting interpretation and conservation of sites associated with the scheme. The region between Perth and Kalgoorlie is important in the development of Western Australia's European history, particularly in the search for pastoral land and gold. While there have been numerous historical and sociological studies of the area, usually focusing on Kalgoorlie, there has been very little archaeological work done in this region post-European settlement.

The aim of the Mundaring-Kalgoorlie Settlement Corridor archaeological project is to conduct archaeological investigations of nineteenth and twentieth century society between Mundaring, east of Perth, and Kalgoorlie, Western Australia, following the Goldfields Water Supply Scheme (Figure 1) to determine what kinds of settlements evolved along the migration route. The study period is from 1830–1914, as this area of Western Australia was settled by Europeans in 1830. 1914 serves as a convenient cut-off point, mainly due to technological changes in manufacturing methods around the time of the First World War.

The overall research questions relate to how the settlement sites that were occupied for various periods of time differed, and how those living and working along the corridor utilised the available resources at both temporary and permanent sites. This includes who was living there; the spatial organisation of the sites; how long they were occupied for; what material culture was used; and how this information relates to the historical record. Chronological and spatial information about the site from the archaeological data, and functional information about the artefactual material, help to answer these questions by identifying occupation and activity areas, and the period and nature of their occupation.

The study region is approximately 560km long, and is within an arbitrary boundary of 15km either side of the pipeline. Archaeological sites within this region were chosen for study based on several criteria. Sites chosen had archaeological material present from the study period of 1830–1914; were not heavily disturbed by later use; and were selected to cover a range of different site uses, including railway stations, town sites and water condensers.

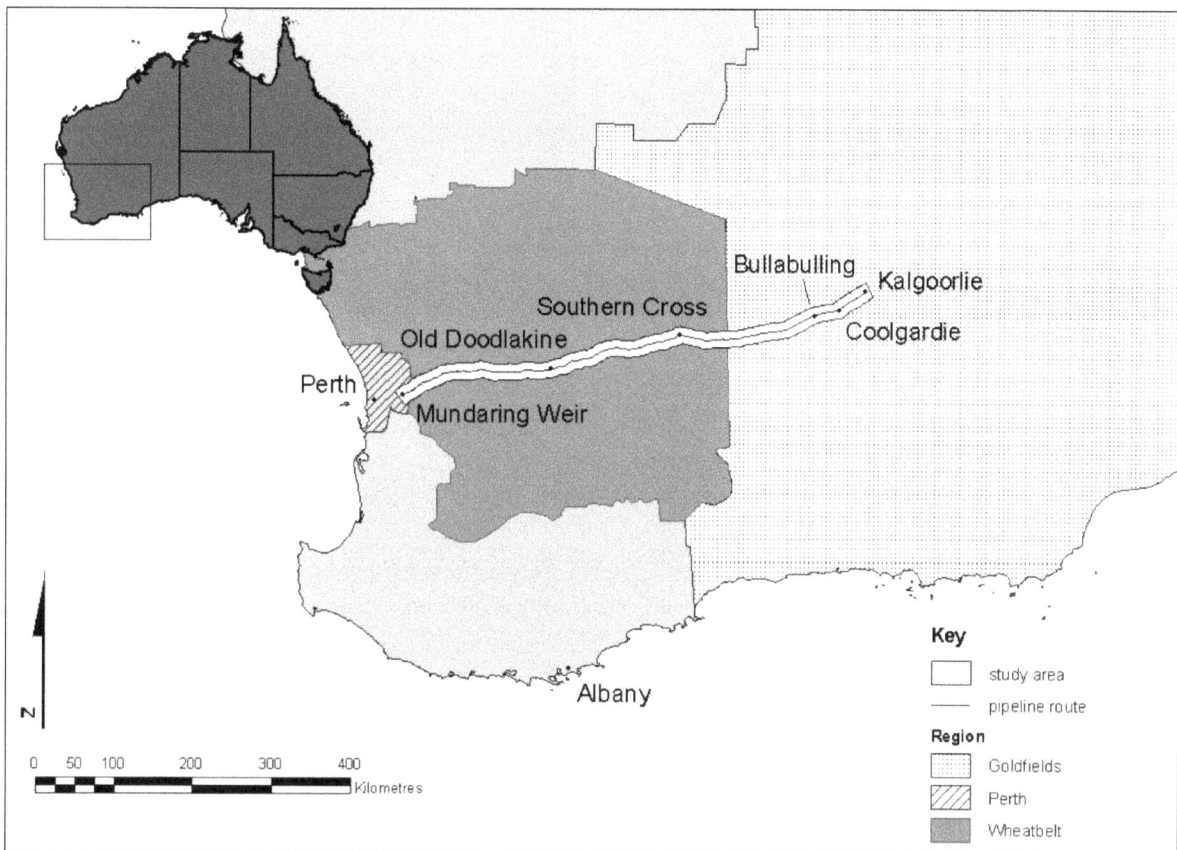

Figure 1. Location of study area including administrative regions and sites mentioned in text

The majority of the study area is within two modern administrative regions known as the Wheatbelt and the Goldfields (Figure 1). The Wheatbelt is an agricultural area, with low-moderate rainfall, and the sites in this region are usually either on agricultural land or have been classed as public reserves. East of Southern Cross, the Goldfields are in a semi-arid environment. Most of the sites are within national parks, although some are on leased or unallocated Crown Land. The two major land uses have significantly affected the preservation of sites and artefacts within each region.

Most sites in the Wheatbelt were destroyed by later occupation, and only one (Old Doodlakine) had enough undisturbed archaeological material to be included in the study. Sites in the Goldfields contained surface material and shovel test pits were excavated to confirm that there was little sub-surface archaeological material. The artefactual material across all of the sites included surface scatters of cans, glass, ceramics and metals, as well as isolated artefacts and artefacts common on late nineteenth and twentieth century sites, for example clay pipe fragments and charcoal.

Methods

The methods were designed to record the material in situ in a way that was both simple and accurate. Because the research questions included a comparison of different sites along the route, as many as possible needed to be recorded in the time available. Systematic sampling of each site was used to identify trends in spatial distribution of the archaeological material and characterise the assemblages, with minimal disturbance.

The methods for this project were adapted from those developed by Birmingham and Wilson for the Central Australia Archaeology Project (CAAP), during which material from a similar time period and in a similar environment was recorded (Birmingham 1997: 4–6). Surveys in 1995 and 1996 for the CAAP identified 100 sites comprising over 5000 features, which is comparable in scale to this project. The approach used by the CAAP were developed to record information from as many features 'as possible, with consistent data recording on each for systematic entry into a GIS [Geographic Information Systems] database' (Birmingham 1997: 4). To achieve this Birmingham and Wilson used survey and feature recording, which also formed the basis of data recording for this project.

For the sites recorded along the Mundaring-Kalgoorlie settlement corridor, site survey was conducted to determine the extent and the location of archaeological material. The survey methods varied, both between and within sites, according to topography and ground coverage. In order to determine the site boundary and the extent of the artefactual material within that boundary, transects, ranging from 5m to 20m according to visibility, were walked. Also, discrete areas, normally delineated by features such as the pipeline and railway line, were defined, and individual team members recorded all features within them. As well as the systematic survey, all sites were walked several times during the process of feature and GPS recording, which provided an extra check that all features were noted and recorded.

A site chosen for intensive study, as per the criteria outlined above, was divided into features which ranged from structures, scatters of artefacts, both discrete and disperse, to isolated artefacts. All features were designated a type (Table 1) and recorded, resulting in a qualitative and quantitative record of each based on previously established terminology conventions.

Feature type	Definition
general cover	a spread of artefacts over a large area such that individual depositional events cannot be discerned
scatter	a discrete area of artefacts that consists of several different material types, for example glass, ceramics and metal
cluster	a discrete area of artefacts that contains one material type and is the result of a single event, e.g. a bottle breaking
single artefact	single artefact
arbitrary point	arbitrary point
structure – regular	usually a four sided structure, such as a building
structure – non-regular	any other structure
fill	a depression containing artefacts, opposite to a heap
heap	a raised area of artefacts, opposite to a fill
other	a feature type not covered by other definitions

Table 1. Feature types and definitions

The feature types provide a means of characterising the distribution of artefacts, interpreting how the site was used, and how the archaeological record was formed. For example, a scatter, which is the most common feature type, is often formed as a result of several events over a period of time, whereas a cluster is the result of a single event. Assigning a type to a given feature does not mean that different types cannot be compared. The comparison is of artefact distribution, date of manufacture and function, which are independent of the feature type.

The edge of a feature was defined as the point where the concentration of artefacts effectively drops to zero. This was not always easy to determine as it was sometimes unclear whether the spread of artefacts related to the feature being recorded or an adjacent one. It was necessary to make assumptions, and sometimes draw arbitrary boundaries between two adjacent features. By defining the extent of the feature and classifying it as a particular type the recorder was forced to think about the feature and what it actually constituted.

As most of the sites were in a semi-arid environment, surface visibility was generally high (>90% for individual features). However, it is inevitable that artefacts were missed, particularly very small items such as pins, sew-on hooks and eyelets, and items that are hard to see. For example, throughout the study region, there was a very low number of clay pipe fragments (less than ten over all of the sites recorded) and one of the reasons for this may be due to visibility. Clay pipes are commonly found on nineteenth and early twentieth century sites in Australia (Lawrence 2006: 368; Courtney and McNiven 1998: 44), and it was expected that there would be a higher number on these sites. Although there was high surface visibility, there was a bias towards larger artefacts and those that are more brightly coloured, such as glass and ceramics.

In order to determine artefact distribution and activity areas across the site, the artefacts were counted usually according to raw material (e.g. glass colour, ceramic type), although some artefacts were counted according to type (e.g. can, insulator, copper grommet). Zones of different artefact concentration were identified within a feature and the artefacts were counted within each zone. If a feature had greater than 100 artefacts then a representative area, usually a 1m x 1m square, was chosen from each concentration zone as a sample for the artefact count. From a sketch plan of the feature, including the different zones, the surface area was determined and the total artefact count was calculated for that feature based on the sample squares.

The artefact classifications of raw material or artefact type were based on common objects found on similar sites (e.g. Birmingham 1997; Paterson 1999) and from observations made during earlier visits. Raw material is one of the primary levels of information about an artefact that can be recorded, and does not require any level of interpretation (see Crook et al. 2002; Brooks 2005 for a discussion of the issues related to artefact catalogues and interpretation of artefacts). Broad inferences can be made about the original intended function of an artefact based on its raw material. For example, curved coloured glass is associated with vessels, as opposed to flat window glass. Dark olive (black) and olive glass are associated with alcohol bottles, amber is associated with beer, medicines or chemicals, and tinted glass is associated with food storage bottles. Although these classifications are not absolute, it is possible to get an overview of the distribution of artefacts across a site based on these counts. Further refinement of artefact use and activity areas was done by more detailed recording of selected artefacts as described below.

Unlike glass or ceramic which could be easily differentiated based on other aspects of their material type such as colour or ware, metals, in particular iron, needed to be separated by artefact type in order obtain a count that could be used for further analysis. These types included cans, nails, gun cartridges, scrap iron and copper grommets. Again, these were broad categories chosen in order to gain an overview of the type of material present. Artefacts such as telegraph insulators and electrodes were also counted according to type as it was an easy way to distinguish them from other materials. By using these categories for the counts, it was possible to obtain a detailed map of the distribution of artefacts across the site, in order to help answer the research questions.

Artefacts with identifying features that allowed them to be dated were recorded in further detail for a later functional analysis. The artefacts included any items with inscriptions, cans, buttons, nails, unusual artefacts or artefacts that could not be easily identified on site. The attributes recorded about these artefacts were material (e.g. glass colour, metal type); form (bottle, jar, can etc.); technology (hand or machine made); shape; modifications; inscriptions; dimensions; and minimum item count. A set of established attribute codes was used to record these attributes. When there were no codes for a given artefact, the recorder gave a detailed description with either a diagram and/or a photograph.

To identify recycled and reused artefacts (e.g. see Stuart 1993; Busch 1987 for discussions of bottle reuse) modifications to the artefact were noted. However, unless there are any obvious modifications, it is often almost impossible to determine whether or not an artefact has been reused, and if so, what for. Therefore, the assumption was made that each artefact was used according to its primary intended function, and it was that function that was recorded.

The attributes recorded provided information to answer questions about what artefacts were used, where and when they were made, and how, if at all, they were adapted for other uses. The codes enabled this information to be recorded in the field in a relatively fast and systematic way, so the information recorded by different people was directly comparable, therefore addressing the problem of identifying and interpreting artefacts. Additionally, diagnostic information, such as inscriptions, technology, form and colour, provided a *terminus ante quem* for occupation of the site and a given area, thereby giving a level of chronological control.

The survey component consisted of recording the location of each feature using a differential global positioning system (dGPS). The GPS location of each feature was combined with the sketch plan of the feature using GIS software (ArcGIS) to create a site plan. The artefact count was then linked to the individual feature points on the site plan in order to give a visual representation of artefact distribution (e.g. Figure 2). This information is used to provide overall maps of the site and its boundary, different activity areas, and how occupation of the site varied over time.

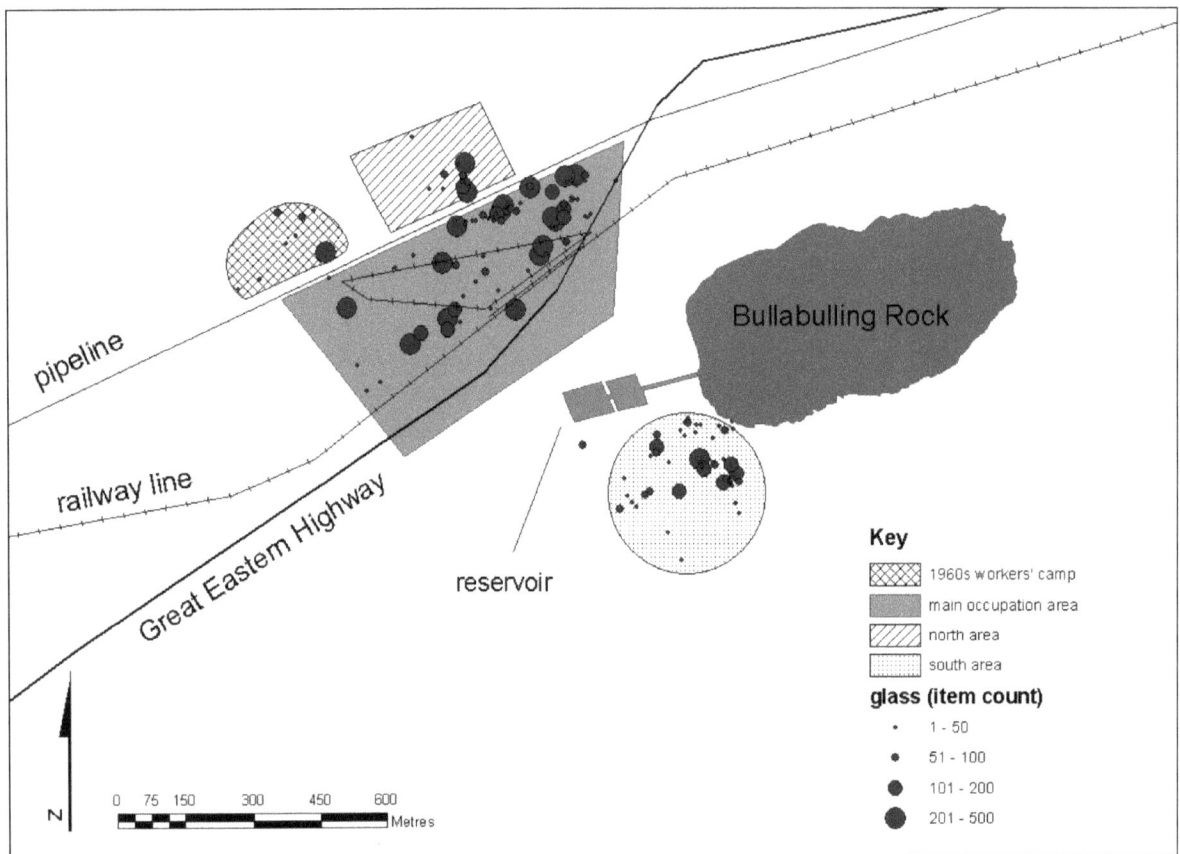

Figure 2. Site plan of Bullabulling showing distribution of glass

Case study

A case study of one of the sites in the study, Bullabulling, illustrates how these methods can overcome the problems associated with surface site recording to produce information with chronological control and answer the research questions. Bullabulling was a town site and railway station established in 1893 as a watering point, for trains and people (RICH 2001: 7) (Figure 1). The site is located at the base of a 15m high granite outcrop and was built as a rock catchment, which is a wall of upright granite slabs, approximately 0.5m high, built around the base of a granite outcrop designed to catch the water flowing off the rock and channel it into a reservoir. Its population reached a maximum of 54 in 1903 (RICH 2001: 9) and, following the closure of the railway in the 1960s, the population dramatically declined, to the current level of two.

Four major zones of occupation were identified based on the archaeological material (Figure 2, Figure 3). To the north and the south were two areas containing material almost exclusively produced before 1910, suggesting that these areas were abandoned by this date. The artefacts in the northern area included hole-in-cap cans, common before 1910 (Rock 1984), which were recorded as individual objects. In the southern area, among other artefacts, there was glass predominantly from the 1890s (see discussion below, Figure 4) and a large number of torpedo, or egg-shaped, bottles, common in the nineteenth century (Jones and Sullivan 1989: 72). To the northwest of the site is a 1960s workers' camp (Rosa Minozzi pers. comm. 2004; RICH 2001:9) which was not included in the study as it was outside the study period.

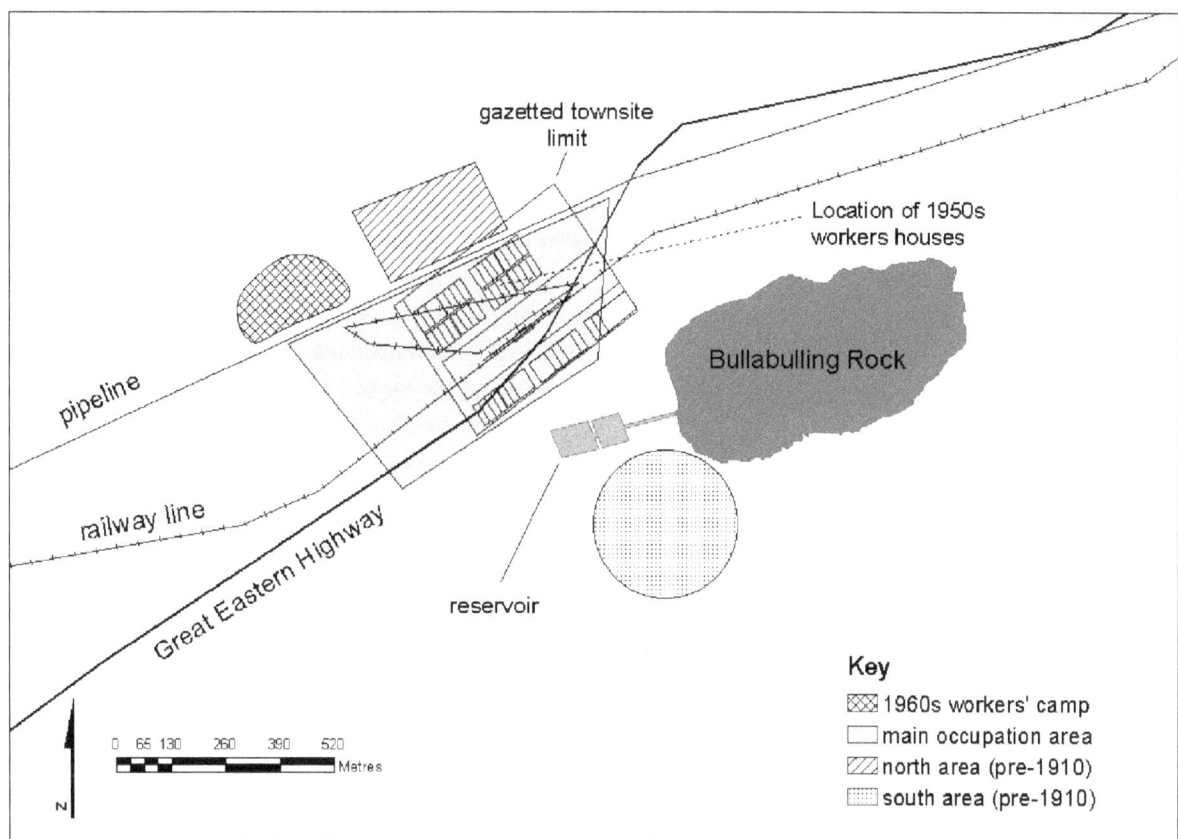

Figure 3. Site plan of Bullabulling showing areas of archaeological deposits

From the artefact counts and the location of the features, the distribution of artefact types can be seen (Figure 2). Figure 4 is a graph of the distribution of glass according to colour and indicates that the colours commonly associated with older bottles (e.g. dark olive and amethyst) are more common in the northern and southern areas, and those associated with more modern bottles (e.g. amber, which at this site came only from items that were produced after 1920) are most common in the central area. Using glass colour to date the manufacture of bottles is not definitive, however it can serve as a useful guide to overall trends (Jones and Sullivan, 1989:12; also see Boow, 1991; Lockhart, 2006; Bolton, 2005; Burke and Smith, 2004 for some guides to the dating of glass based on colour).

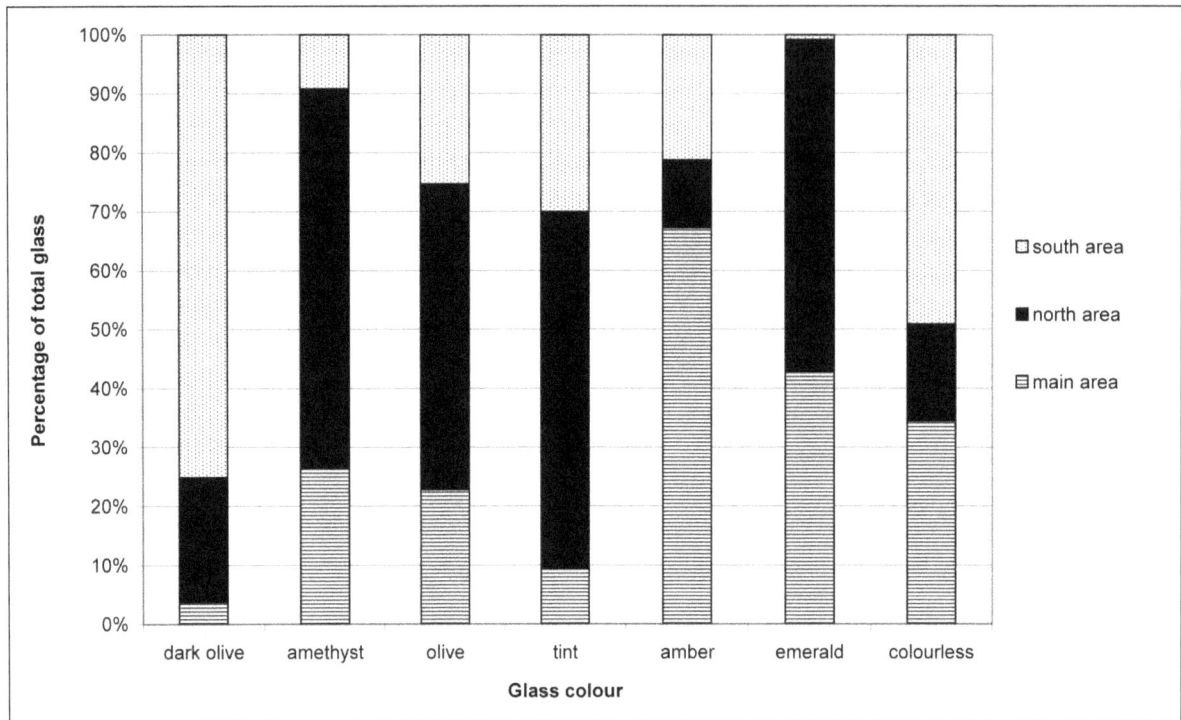

Figure 4. Distribution of glass at Bullabulling according to colour

From the amount and date range of all of the artefactual material, recorded using the artefact counts and individual object recording, and the location of the structures, it can demonstrated that the centre, where the gazetted townsite and railway station were (RICH 2001: 14), was the main focus of occupation. The artefacts here included material ranging in production from the late nineteenth century until the 1960s, and, as an example, Figure 2 shows that there is more glass in this central area.

Based on the distribution of archaeological material, the following interpretation can be made. Initial occupation occurred in the southern area, when the rock catchment was built. The focus then moved to the main area in the centre, around the railway station. At first this occupation included the area to north, however once the pipeline was built the area was cut-off and abandoned. Unlike other sites, there was no evidence for any ladders over the pipeline, so it was probably too difficult to constantly climb over.

At Bullabulling the archaeological material was used to determine different zones of occupation. Without stratified material it can be difficult to determine change over time, however, the combination of knowing when certain elements were built and the distribution of all artefacts from a given time period can be used to answer questions about changes in the site's use over time.

Discussion

The first problem raised by Holdaway et al. (1998) is the lack of stratigraphy on surface sites, and therefore chronological control. Historical records and technological changes in the manufacture of artefacts allow sites to be dated, and this inevitably makes it possible to date historical sites with a finer resolution than prehistoric sites. However, historical records for a given site do not always exist and there may be a time lag between the manufacturing date and the use of an artefact (Adams 2003). Also, artefacts were often recycled and re-used. Despite these problems, based on the distribution of archaeological material at Bullabulling, it was possible to identify different zones of occupation. Allowing for time lag, the zones were based on the clear demarcation of artefacts from different time periods.

The second problem identified by Holdaway et al. that activity areas are difficult to identify because feature boundaries are not clear is dealt with by using arbitrary boundaries. Features often do not have clear boundaries, and at some point a decision has to be made about their extent. By recording different zones of artefact concentration and looking at the site as a whole, activity areas can be identified and interpretations made about the use of the site.

The problem of identifying and interpreting an artefact's use for applies to both historic and prehistoric artefacts. When recording artefacts in the field, less information is often recorded about them, for example it is difficult to weigh artefacts in the field, and the artefacts cannot be checked. Whether it is necessary to record that information depends on what the research questions are. In this project in situ recording was sufficient as the questions related to the spatial organisation of the site, what was being used, when and how, and by whom. By identifying the artefacts that have diagnostic information and recording only those, and by standardizing the recording as much as possible, time is spent in the field recording the information required to answer the research questions.

The third problem about exposure and visibility of features and artefacts varies between and within sites. This was dealt with by conducting a systematic survey of the site, followed by feature recording and the GPS survey, ensuring that the location and composition of all exposed features were recorded.

In addition to the problems identified by Holdaway et al. other factors affect surface sites. These include a greater potential for disturbance by environmental, animal and human factors, and when recording surface sites in situ, there is a bias towards larger artefacts, such that small personal items, including pins, eyelets as well as clay pipe fragments, may not be seen and recorded.

Regarding the greater potential for disturbance by environmental, animal and human factors, surface artefacts from the historical archaeological period are highly visible. They are prone to be blown by the wind, washed away during heavy rain, kicked about by passing animals and picked up, moved, and taken away by artefact collectors and the curious. Numerous bottle dumps have been identified at the sites recorded, and because they are very visible, without fail, all have been raided by collectors. However, features recorded at several sites have clear spatial patterns that indicate they have not been significantly disturbed.

There will be a bias towards larger artefacts such that small items may not be recorded. Nevertheless, small artefacts, including buttons, eyelets and sew-on hooks, have been found in this project and recorded at a number of features, demonstrating they are not being missed altogether. By studying the distribution of artefacts using the GIS, and incorporating other factors such as topography and ground coverage, it may be possible to predict where the small artefacts are likely to be located.

Conclusion

Chronological controls provided by the presence of material of a known period of production allow the interpretation of site functions over time. Unlike prehistoric sites, historical archaeological sites usually feature some datable material. Changes in manufacturing technology, the use of makers' marks, and the introduction of new items and new materials, enable the production of many individual artefacts to be dated. Using the archaeological and historical records it is possible not only to provide a *terminus ante quem* for dating the occupation of a site as a whole, but different components of the site.

The artefacts are also used to identify different activity areas such as a workshop of domestic area. Not only can an interpretation be made about what areas of a site were used at different times, it is also possible to determine how, if at all, these functions changed over time. By studying the distribution of artefacts from earlier periods of occupation as compared to those from later periods, it may be possible to determine how the site was used over time and the rate of change.

The methods described here are one way of recording archaeological data from historic period sites in situ. Further testing and development will ultimately determine whether they are successful.

Surface artefacts are a major component of the archaeological record in Australia for sites occupied in all time periods. There are numerous difficulties inherent in recording surface artefacts, mostly relating to dating and preservation of sites. Yet they are important sources of archaeological information, and it is necessary to continue to develop and refine methods to record them, such that questions about a site can be answered.

Acknowledgements

This project is funded by an ARC Linkage Grant awarded to Alistair Paterson, Archaeology, UWA and the National Trust (WA). Thanks to Alistair Paterson and Jane Balme, Archaeology, UWA; Anne Brake, Kelly Rippingale, Diana Frylinck and the National Trust (WA); Department of Archaeology and Natural History, ANU; Mike Stewart, Department of Spatial Sciences, Curtin University; Cameron Waters, GlobalPOS; Geoff Glazier, OmniSTAR; Barry Hooper, DEC (previously CALM); Andrew Wilson, University of Sydney; Laurie and Rosa Minozzi, Bullabulling Tavern; Lara Bosi, Kelly Fleming, Shaun Mackey, Daniel Monks, Stafford Smith, Norbert Svatos and Vicky Winton. Jane Balme, Shaun Mackey, Peter Davies and two anonymous reviewers provided invaluable comments on earlier versions of this paper. All errors and omissions are my own.

References

Adams, W. H. 2003. Dating historical sites: the importance of understanding time lag in the acquisition, curation, use, and disposal of artifacts. *Historical Archaeology* 37(2):38–64.

Birmingham, J. 1997. Fieldwork in contact archaeology, Central Australia. In C. Petrie and S. Bolton (eds), *In the Field*, pp 1–12. Sydney: Sydney University Archaeological Methods Series.

Bolton, S. 2005. Purple Haze: evidence for a later date for solarized amethyst glass. *Australian Archaeology* 60:51–52.

Boow, J. 1991. *Early Australian Commercial Glass*: Manufacturing Processes. Sydney: Heritage Council of NSW.

Brooks, A. 2005. *An Archaeological Guide to British Ceramics in Australia, 1788–1901*. The Australasian Society for Historical Archaeology and The La Trobe University Archaeological Program.

Burke, H. and C. Smith. 2004. *The Archaeologist's Field Handbook*. Australia: Allen and Unwin.

Busch, J. 1987. Second time round: A look at bottle reuse. *Historical Archaeology* 21(1):67–80.

Courtney, K. and I. J. McNiven. 1998. Clay tobacco pipes from Aboriginal middens on Fraser Island, Queensland. *Australian Archaeology* 47:44–53.

Crook, P., S. Lawrence and M. Gibbs. 2002. The role of artefact catalogues in Australian historical archaeology. *Australasian Historical Archaeology* 20:26–38.

Holdaway, S., D. C. Witter, P. Fanning, R. Musgrave, G. Cochrane, T. Doelman, S. Greenwood, D. Pigdon and J. Reeves. 1998. New approaches to open site spatial archaeology in Sturt National Park, New South Wales, Australia. *Archaeology in Oceania* 33:1–19.

Jones, O. and C. Sullivan. 1989. *The Parks Canada Glass Glossary for the Description of Containers*, Tableware, Closures, and Flat Glass. Quebec, Canada: National Historic Parks and Sites, Canadian Parks Service, Environment Canada.

Lawrence, S. 2006. Artifacts of the modern world. In J. Balme and A. Paterson (eds), *Archaeology in Practice: A student guide to archaeological analyses*, pp 362–388. Malden, USA: Blackwell Publishing.

Lewarch, D. E. and M. J. O'Brien. 1981. The expanding role of surface assemblages in archaeological research. In M. B. Schiffer (ed), *Advances in Archaeological Method and Theory*, pp 297–342. New York: Academic Press.

Lockhart, B. 2006. The color purple: Dating solarized amethyst glass containers. *Historical Archaeology*. 40(2):25–56

Murray, T. 2002. But that was long ago: Theory in Australian historical archaeology 2002. *Australasian Historical Archaeology* 20:8–14.

Paterson, A. G. 1999. Confronting the sources: *The archaeology of culture-contact in the South-Western Lake Eyre Basin, Central Australia*. Unpublished PhD thesis, Dept. of Prehistoric and Historical Archaeology, Faculty of Arts, University of Sydney, Sydney, NSW.

Paterson, A. G. 2003. The texture of agency: An example of culture-contact in central Australia. *Archaeology in Oceania* 38:52–65.

Paterson, A. G. 2005. Early pastoral landscapes and culture contact in Central Australia. *Historical Archaeology* 39(3):28–48.

Redman, C. L. 1987 Surface collection, sampling, and research design: A retrospective. *American Antiquity* 52(2):249–265.

Research Institute for Cultural Heritage (RICH). 2001. *Conservation Plan for Bullabulling Township and Railway Catchment* — DRAFT. Unpublished Draft Report prepared for National Trust of Australia (WA), Perth.

Rock, J. T. 1984. Cans in the countryside. *Historical Archaeology* 18(2):97–111.

Stuart, I. 1993. Bottles for jam? An example of recycling from a post-contact archaeological site. *Australian Archaeology* 36:17–21.

Sullivan, A. P., III (ed). 1998. *Surface archaeology*. Albuquerque, New Mexico: University of New Mexico Press.

Wandsnider, L. and E. L. Camilli. 1992. The character of surface archaeological deposits and its influence on survey accuracy. *Journal of Field Archaeology* 19(2):169–188.

5

Late Quaternary environments and human occupation in the Murray River Valley of northwestern Victoria

A. L. Prendergast[1]
J. M. Bowler[2]
M. L. Cupper[2]

[1]Geoscience Australia
Canberra 2601, Australia

[2]School of Earth Sciences
University of Melbourne
Victoria 3010, Australia

Abstract

A multidisciplinary approach has been employed to study the environmental and cultural evolution of the Victorian Mallee. Regional geomorphic mapping of relict landforms and stratigraphic analyses reveal evidence of ongoing climatic oscillations in the central Murray Valley, where the Murray River system interacts with aeolian Mallee landscapes. Analysis of landforms using a land systems approach provides new insights into patterns of fluvial-aeolian interaction over the last glacial cycle. Five land systems are identified within the study area; three of these preserve evidence of palaeochannel activity markedly different from the present Murray River system. Fluvial morphology evolved over the late Quaternary from wide, laterally-migrating channels associated with source-bordering dunes to narrower, more sinuous regimes. An extensive archaeological record overprints the region, with radiocarbon and optically stimulated luminescence (OSL) dating revealing human presence in this landscape from at least 15,000 cal BP.

Keywords: Australia; Mallee; Quaternary; environmental change; geomorphology; archaeology; oxygen isotopes; luminescence dating; radiocarbon dating; Alathyria jacksoni.

Introduction

The semi-arid central Murray Valley (Figure 1) preserves a wealth of evidence with which to reconstruct the environmental and cultural evolution in southeastern Australia. The lack of ice cover during glacial periods has ensured that a relatively complete record of climate change is preserved in sediments and landforms (Bowler and Magee 1978). In addition, the semi-arid fringes of southeastern Australia preserve one of the longest records of human occupation on the continent (Mulvaney and Kamminga 1999; O'Connell and Allen 2004).

This paper aims to investigate the impact of climate change over the last glacial cycle on the geological and archaeological records of the Mallee region. A detailed geomorphic and sedimentological study of fluvial, aeolian and lacustrine landforms provides the basis for interpreting climatic trends through the proxy of an evolving land system. In addition, archaeological survey and excavation reveals human interactions with this changing landscape. This evidence is integrated to construct a model of environmental evolution in the Mallee over the last glacial cycle.

Materials and methods

Land systems mapping

Land systems (Christian and Stewart 1952; Gerard 1981) and sediment mapping were carried out through satellite image (Landsat TM) and aerial photograph interpretation (River Murray Mapping Digital Colour Infrared Orthophotomap series, Murray-Darling Basin Commission 1996), supplemented with detailed field traverses. Stratigraphic logs were obtained by cleaning vertical sections along river banks and gullies, and by digging trenches. Soil development and sediment composition were assessed in the field and later confirmed through thin section analysis.

Archaeological investigation and dating

Archaeological survey was conducted in tandem with land systems mapping. North to south transects were surveyed at approximately one kilometre intervals from the base of the Woorinen Land System to the southern edge of the Murray River. The Woorinen Land System and the northern sideof the Murray River were not surveyed. Several previously unrecognised sites were identified (Figure 2, Table 3).

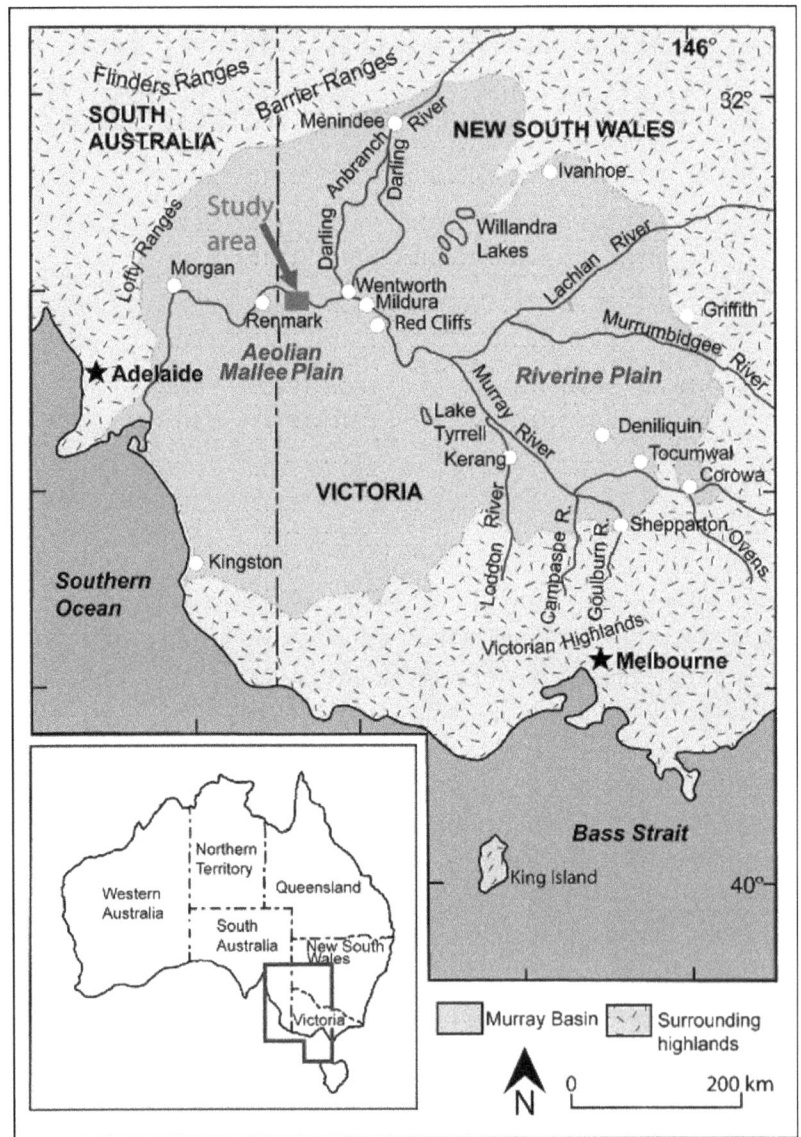

Figure 1. Map of the Murray Basin showing the location of the study area, the major drainage networks and the division between the Aeolian Mallee Plain and the Riverine Plain (adapted from Brown and Stevenson 1989).

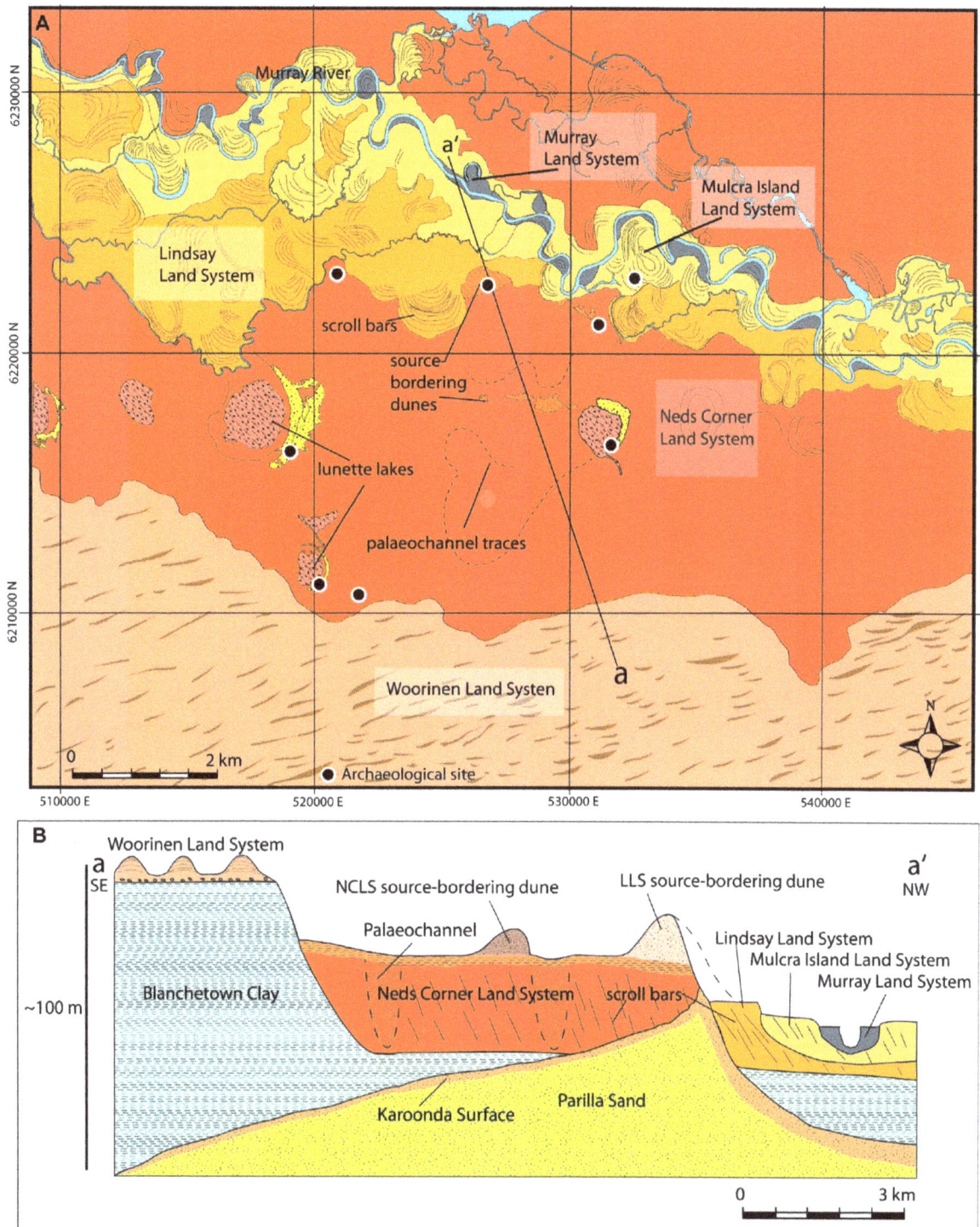

Figure 2.
A: Land systems map of the study area showing the locations of archaeological sites. From south to north the five land systems defined within the study area are: Woorinen Land System, Neds Corner Land System, Lindsay Land System, Mulcra Island Land System and Murray Land System. Within the land systems are land units of lunette lakes, palaeochannels and source bordering dunes.
B: Schematic southeast to northwest cross section (a-a') of the study area. Vertical scale is exaggerated.

Two 1 m x 1 m trenches were excavated in a previously unrecorded shell midden (the Homestead Midden: 34°08'08.2''S,141°21'40.5''E). All sediments were sieved and freshwater mussel shells (*Alathyria jacksoni*) were collected for radiocarbon dating at the Waikato Radiocarbon Dating Laboratory, New Zealand.

Four OSL samples were obtained from above and below archaeological horizons by hammering 40 mm diameter stainless steel tubes into freshly cleaned vertical sediment faces. The 90–125 μm sediment fractions were retained for OSL dating at the University of Melbourne using standard procedures (Galbraith et al. 1999). Palaeodoses were calculated using a single aliquot regenerative dose (SAR) protocol using procedures described in Stone and Cupper (2003). Cosmic-ray dose rates were determined from established equations (Prescott and Hutton 1994).

Land systems mapping

Five land systems were defined via affinities in topography, pedology, sedimentology and vegetation. Each of these land systems was subdivided into land units, such as palaeochannels, lunettes or linear dunes. They are described sequentially from south to north in progression from highest to lowest elevation (Figure 2, Table 1). This broadly corresponds to a chronological description from the oldest to the youngest units.

Woorinen Land System
The Woorinen Land System comprises land units of east-west linear dunes and sand plains, which flank the southern margins of the study area. This system extends southwards to become part of the Central Mallee Land System described by Rowan and Downes (1963). The Woorinen Land System forms the uppermost topography of the study area, occurring around ten metres above the Neds Corner Land System (Figure 2, Table 1).

Several alternating phases of soil formation and aeolian activation have previously been identified within linear dune sequences throughout the Mallee (Churchward 1961; Bowler and Magee 1978). The last phase of dune activation occurred between ~25–15 ka, during and just after the last glacial maximum (Bowler and Magee 1978; Bowler 1980, 1986a; Wasson 1989; Readhead 1990). The chronology of linear dune palaeosol formation serves as a benchmark that is used in this study to correlate to the fluvio-aeolian-lacustrine units defined within the study area.

Neds Corner Land System
The Neds Corner Land System (Rowan and Downes 1963), situated around 10 m below the Woorinen Land System, extends throughout the study area as a large, flat terrace (Figure 2, Table 1). Although it is the dominant feature of the Murray Valley in this area, its soil-sedimentary components have not previously been analysed. The palaeochannel-floodplain and lunette-lake land units of the Neds Corner Land System preserve the best evidence of combined fluvial, aeolian and lacustrine interactions in the region.

Based on its relative position in the landscape, as well as its high pedogenic carbonate content (Figures 2, 3, Table 1), the Neds Corner Land System is interpreted to be the oldest of the four fluvial terraces identified in this study. Meander wavelengths and channel widths of the palaeochannels are substantially larger than in the presently active Murray Land System (Table 2). Two lunette lakes, Lake Wallawalla and Little Lake Wallawalla, occur as broad, low relief, oval depressions incised into the surface of the Neds Corner Land System (Figure 2). As the lakes overlie fluvial meander loops, their formation postdates the final phase of fluvial activity on the Neds Corner Land System.

Land System	Land Units	Dominant Process	Surface Expression	Sediments	Soils / Degree Of Pedogenesis	Dominant Vegetation	Archaeology	Age
Murray Land System	floodplain, billabongs and river channels	fluvial	mildly undulating surface with some flooding channels	compacted light grey medium clay	very little soil development	River Red Gum	surface scatters of artefacts	**Holocene** (~10 ka to present)
Mulcra Island Land System	floodplain and palaeomeander scrolls	fluvial	mildly undulating channel floodplain sequences	medium grey gilgai clay	mild pedal development	Black Box with some grasses and Lignum	midden deposits on floodplain surface	**Late Pleistocene,** (post 15 ka)
	billabongs	fluvial	arcuate depression, sometimes filled with water	fine medium grey overbank clays	mild pedal development	Black Box and Red Gum	not observed	**Late Pleistocene** (post 15 ka)
Lindsay Land System	floodplain and palaeomeander scrolls	fluvial	mildly undulating channel floodplain sequences	dark grey gilgai clay and sandy bedload	mild pedal development	Lignum with some Black Box	not observed	LGM (Pre 15 ka)
	LLS source-bordering dunes	aeolian	red sandy rises	fine to medium, well-sorted red sand	carbonate pedogenesis below 30 cm	various Chenopod species	middens, stone artefacts, burials and hearths	LGM (Pre 15 ka)
Neds Corner Land System	floodplain and palaeochannels	fluvial	predominantly flat surface	fine red silty sand	secondary carbonate, duplex soils	Chenopod shrubland	hearth stones and stone artefacts on surface	> 40 ka
	NCLS Source-bordering dunes	aeolian	irregularly shaped sandy rises	fine, well-sorted red brown sand	clay cutan development. carbonate at depth	Blue Bush, Old Man Saltbush	middens, burials hearth stones and stone artefacts on surface	> 40 ka
	lunette lakes	aeolian/lacustrine	ovoid depressions, 4-5 m below surrounding plains flanked by arcuate dunes on eastern margins	pelletal clay, gypseous grey-brown sandy clay	some carbonate and gypsum. weak pedal development	Blue Bush, Dillon Bush, native grasses	grinding stones, hearth stones, burials and stone artefacts on surface	LGM? (~20 ka?)
Woorinen Land System	linear dunes and swales	aeolian	regular rises with rounded crests	fine red brown sand	paleosols and carbonate development	Mallee eucalypts	not observed	active throughout last and previous glacial cycles

Table 1. Summary of the major land systems and land units mapped in the Central Murray Valley, northwestern Victoria.

	Murray Land System	Mulcra Island Land System	Lindsay Land System	Neds Corner Land System
Meander geometry	highly sinuous	moderately sinuous	laterally migrating	laterally migrating
Meander wavelength	600 m	800 m	1000 m	1200 m
Channel width	60 m	75 m	100 m	150 m
Fluvial load	suspended load	mixed load to suspended load	bedload	bedload
Pedology	negligible	soil forms in peds (1-5 cm)	more pedal development than MILS, some carbonate accumulation in source-bordering dunes	significant dispersed and blocky carbonate accumulation
Mineralogy	quartz, clay, feldspar, mica	quartz, gilgai clay, feldspar, mica	quartz, gilgai clay feldspar, mica significant carbonate	quartz, clay, feldspar, mica, and carbonate below 30 cm / quartz with clay cutans above 30 cm
Elevation above present Murray River system	present river level	1-2 m	2-4 m	6-8 m
Dominant vegetation	River Red Gum	Black Box trees	Lignum	Chenopod shrubland

Table 2. Summary of the fluvial units observed in the Central Murray Valley, northwestern Victoria: MLS is the Murray Land System, MILS is the Mulcra Island Land System, LLS is the Lindsay Land System, NCLS is the Neds Corner Land System.

Lindsay Land System

The Lindsay Land System is a fluvio-aeolian system of floodplains, palaeochannels and source-bordering dunes. It sporadically outcrops between the Neds Corner Land System and the Mulcra Island Land System, with an average terrace width of 500 m. Alternating vegetation bands of Black box trees and Lignum preserve evidence of large meander scrolls on Lindsay and Mulcra Islands (Figure 4a, 4d, Tables 1, 2).

Mulcra Island Land System

The Mulcra Island Land System typically sits 1–2 m above the Murray Land System, 1 m below the Lindsay Land System (where present) and 2–4 m below the Neds Corner Land System (Figure 2, Table 1). Its width ranges from 50 to 900 m. The system is essentially composed of floodplain, palaeomeander scrolls and billabongs (Table 2).

Murray Land System

The Murray Land System includes the River Murray and its presently active floodplain. The terrace occupies the lowest point in the landscape, typically 1–2 m below the Mulcra Island Land System (Figure 2, Table 1). Floodplain width varies from 10 to 200 m (Table 2) and is vegetated with River Red Gum woodland.

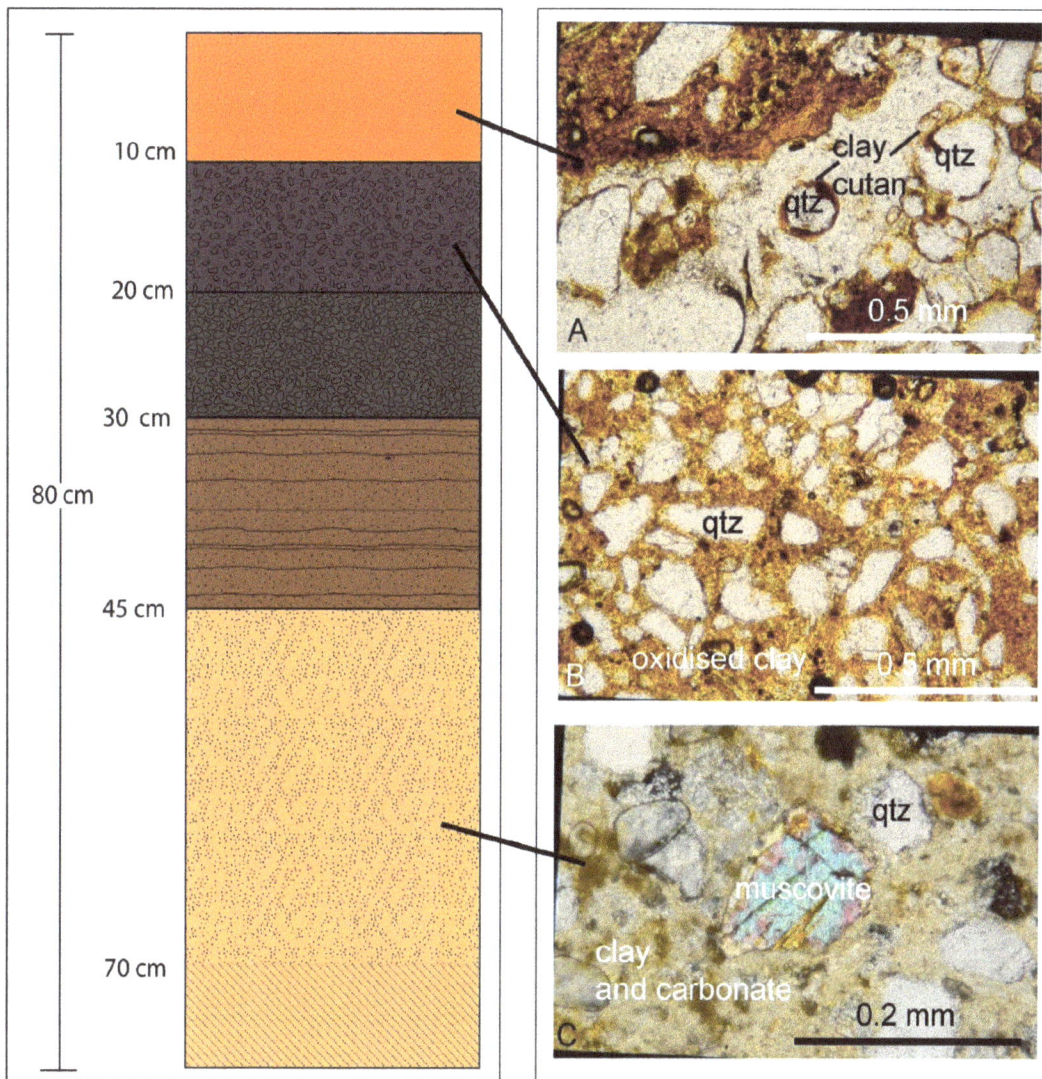

Figure 3. Interpretative stratigraphic log from the Neds Corner Land System floodplain (34°10'30''S, 141°19'30.6''E) with thin section photomicrgraphs.

A: Polycyclic clay cutans surrounding quartz grains in the upper 30 cm aeolian-dominated horizon;
B: Typical suspension-load quartz grains surrounded by clay cutans in the upper 30 cm aeolian-dominated horizon;
C: A muscovite flake amidst clay-rich pedogenic carbonate from the lower pedogenised fluvial horizon. For further thin section description and interpretations see Table 3.

Figure 4. Photomicrographs and field photographs from the lower three land systems.

A: Thin section of typical gilgai floodplain sediments from the Lindsay Land System. Note the abundance of quartz and clay with significant muscovite (crossed polars);
B: A thin section viewed under plain light showing pedal development in the upper clay-rich sediments of the Mulcra Island Land System. The peds are a reflection of cyclic wetting and drying regimes;
C: Plain light thin section of grey clay alluvium from the Murray Land System terrace. Note the dominance of clay over quartz and the occurrence of muscovite which is characteristic of these fluvial sediments;
D: A large meander scroll bounded by a source-bordering dune (34°09′26.2′′S, 141°16′24.3′′E) of the Lindsay Land System;
E: Typical gilgai cracking clays from a meander scroll in Lindsay Land System (the ruler is 30 cm long).

Archaeology and dating

Climatic conditions influence where humans live by affecting the distribution of water, food, shelter and the availability of raw materials. Geomorphic and stratigraphic evidence sheds light on the environments inhabited by people in the past. Correlating this evidence with analysis of the distribution of cultural material can reveal the human response to changing environments. In the Mallee, this response is reflected in the archaeological record of burials, middens and artefact scatters.

Archaeological survey

Archaeological material was recorded at eight locations in the study area. Most of these were on topographic highs such as source bordering dunes and lunettes associated with the Neds Corner and Lindsay Land Systems (Figure 2, Table 3).

Site	Location	Landform	Archaeological material					
			Freshwater Mussels	Freshwater Snails	Stone Artefacts	Hearth Stones	Grinding Stones	Burials
Homestead Midden	34°08'08.2"S, 141°21'40.5"E	source-bordering dune	*	*	*	*		*
Potterwalkagee Gully	34°09'21.4"S, 141°20'42.5"E	source-bordering dune	*					
Tip Site	3°09'21.2"S, 141°20'41.9"E	source-bordering dune	*		*	*	*	*
Lindsay Island Dune	34°08'32.8"S, 141°16'51.2"E	source-bordering dune				*	*	
Lake Wallawalla	34°11'51.3"S, 141°12018.6"E	lunette						*
Little Lake Wallawalla	34°12'22.8"S, 141°12'17.2"E	lunette			*	*		*
Subdued Lunette	34°10'49"S, 141°20'32.6"E	lunette			*	*		
Central Rise 1	34°12'20.1"S 141°12'8.2"	source-bordering dune			*	*		

Table 3. Archaeological material recorded in the study area.

Several different types of archaeological material were found at each location as follows:

- *Heat-retainer hearths* were recorded at six of the eight locations, all elevated above the level of the plain and always associated with stone artefacts such as flakes and grinding stones. Due to the lack of rock outcrop in the study area, the heat retainers are predominantly baked clay balls (Figure 5a), either *in situ* or as scattered fragments.
- *Burials:* three locations containing burials were identified in the study area (Figure 2, Table 3), each including multiple internments. The Potterwalkagee Gully and Homestead Midden locations also contained shell middens and artefact scatters.
- *Grinding implements:* sandstone grinding implements (mullers and millstones) were found at three sites (Figure 5b, Table 3), all of which had associated hearths. The lack of suitable outcrops of sandstone nearby implies that these implements were imported.
- *Stone artefact scatters:* predominantly of ferruginised silcrete, were found at four sites in the study area (Figure 2, Table 3), generally associated with hearths. Assemblages comprised worked flakes (average length 5 cm from 20 measured flakes, Figure 5c) and some cores. The most likely source for the stone is the Berribee silcrete quarry to the west of the study area, which is an outcrop of the Karoonda weathering profile above the Parilla Sands (Figure 2b) (Grist 1995).
- *Shell middens:* three middens were identified in the study area (Figures 2, 5d, Table 3). All were associated with stone artefact scatters and burials. The Homestead Midden (discussed below) was the largest and densest midden recorded.

Figure 5. Archaeological deposits found during archaeological survey of the study area.

A: An *in situ* hearth from the Subdued Lunette site;
B: Sandstone grinding implement (muller) from the Lindsay Island Dune;
C: A silcrete flake from the Homestead Midden;
D: Freshwater mussels (*Alathyria jacksoni*) from the Tip Site midden. For locations of archaeological sites see Figure 2 and Table 3.

Archaeological excavation

The Homestead Midden occurs on a source-bordering dune associated with the Lindsay Land System (Figure 2). The dimensions of the midden are approximately 400 m east–west and 20 m north–south. Four human burials were observed eroding out of the dune.

Trench 1 was excavated to a depth of 110 cm (Figure 6). The archaeological deposits (~70 % *Alathyria jacksoni* freshwater mussel shells, ~10% *Notopala hanlei* freshwater snail shells, ~20 % sediment, scattered silcrete flakes and clay heat retainers) sit unconformably over a reworked carbonate horizon of the dune surface. This indicates that the cultural material was deposited after stabilisation of the Lindsay Land System, when the Mulcra Island Land System was active (Figure 6). Ferruginised sand stratigraphically overlies the midden, although erosion has exposed some sections at the surface.

Radiometric radiocarbon ages (Table 4) were determined from two excavated Alathyria jacksoni shells. The shells yielded similar ages of 15,000 ± 750 cal BP (Wk-15263) and 15,100 ± 750 cal BP (Wk-15262).

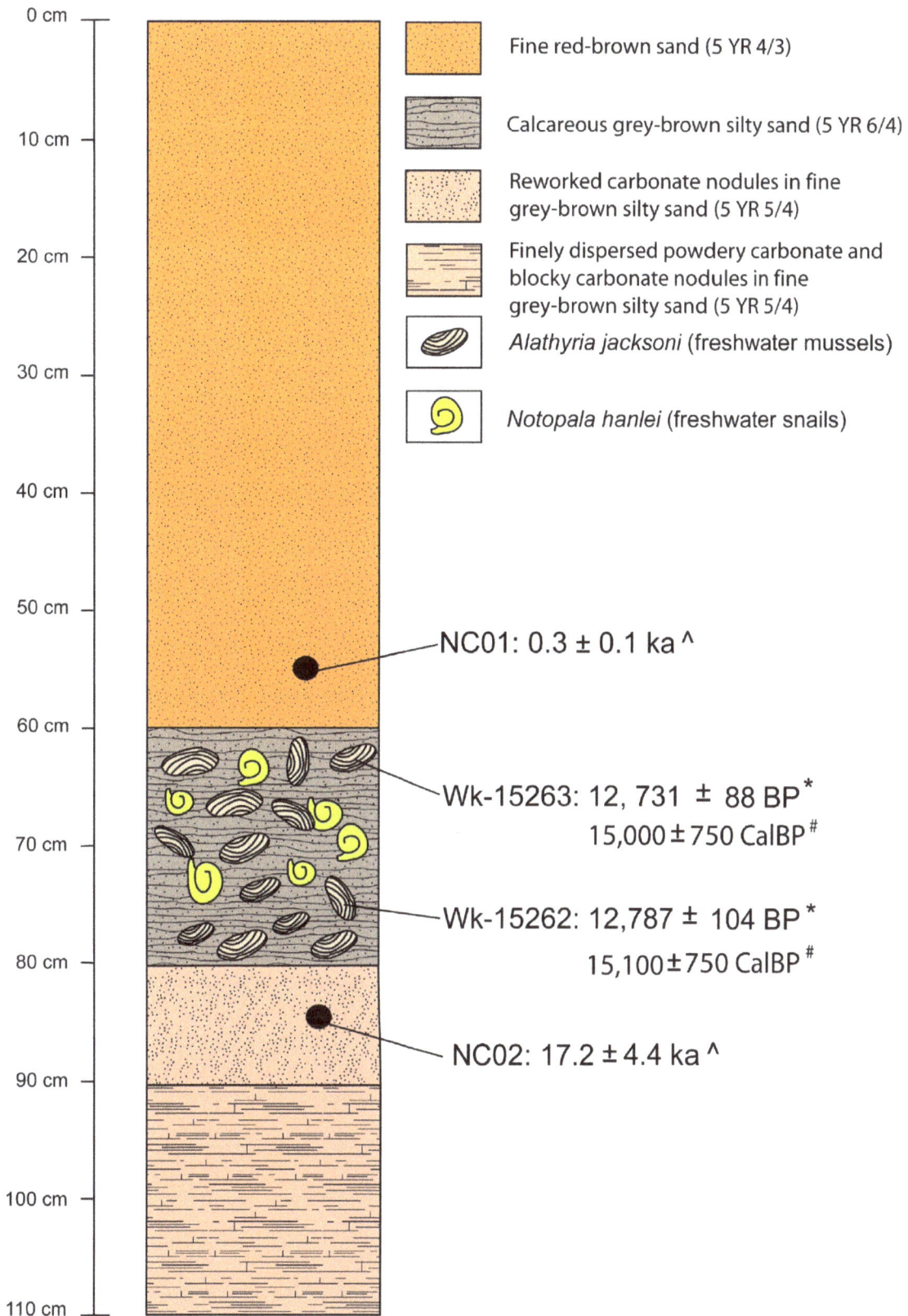

Figure 6. Stratigraphic log of the trench 1 excavation at the Homestead Midden, into source-bordering dune sediments.

^ OSL age (Table 6)
* Uncalibrated radiocarbon years before present (Table 5)
# Calibrated radiocarbon years before present (Table 5).

Site	Material	Lab Reference Number	Depth (cm)	Uncalibrated Radiocarbon Age[a]	Calibrated Radiocarbon Age[b]	Geomorphic Context	$\delta^{13}C$ (‰)
Homestead Midden trench 1	*Alathyria jacksoni* shell (aragonite)	Wk-15262	70 cm	12,787 ± 104 BP	15,100 ± 750 CalBP	source-bordering dune	- 9.7 ± 0.2
Homestead Midden trench 1	*Alathyria jacksoni* shell (aragonite)	Wk-15263	60 cm	12,731 ± 88 BP	15,000 ± 750 CalBP	source-bordering dune	- 9.1 ± 0.2

a: mean ± 1 σ uncertainty
b: mean ± 2 σ uncertainty

Table 4. Conventional and calibrated radiocarbon ages from the Homestead Midden, Ned's Corner Station

Two sediment samples overlying archaeological deposits from the Homestead Midden (Figure 6) and nearby Potterwalkagee Gully (Table 3) yielded identical OSL ages of 0.3 ± 0.1 ka (Figure 6, Table 5). These young ages may indicate reworking of the dune surface immediately after European contact. The ages from the upper and lower horizons of the Homestead Midden (0.3 ± 0.1 ka and 17.2 ± 4.4 ka respecti vely) are in stratigraphic agreement with the radiocarbon ages (15,000 ± 750 cal BP and 15,100 ± 750 cal BP), whilst an age of 11.1 ± 1.9 ka from Potterwalkagee Gully midden sediments is slightly younger than the excavated midden deposits.

Discussion: Environmental evolution and archaeology of the central Murray Valley

The soils, sediments and landforms of the central Murray Valley preserve evidence of environmental evolution throughout the last glacial cycle. Incorporation of archaeological evidence allows an evaluation of the human response to environmental changes as these human populations would have needed to adapt their settlement, hunting and foraging activities to the changing environment as the landscape evolved through successive arid and humid cycles.

The fluvial activation of the Neds Corner Land System: >40 – ~24 ka
The Neds Corner Land System preserves evidence of multiple phases of environmental evolution in its fluvial, aeolian and lacustrine landforms. The oldest units are floodplain and palaeochannel sequences with morphologies markedly different from the modern Murray fluvial system (Table 2). These differences imply a contrasting hydrological regime, likely to reflect divergent climatic conditions (Drury 1967). A change in channel morphology also connotes a change in fluvial load. While the present regime is characterised by suspended-load, low flow rates and sinuous channels, the larger, laterally-migrating channels of the Neds Corner Land System may have resulted from bedload-dominated flow regimes (Page and Nanson 1996). This interpretation is consistent with the coarser sandy channels and source-bordering dunes observed on the Neds Corner land surface (Table 1). In Australia, source-bordering dunes are associated only with relict channel systems (Bowler and Magee 1978; Mabbutt 1980; Williams et al. 1991; Page et al. 2001). Their formation is a result of reduced vegetation, high bedload and a strong component of seasonal flow (Bowler 1986a; Page et al. 1991; Page and Nanson 1996; Bullard and McTainsh 2003).

Laboratory Number	Locality	Coordinates	Depth (cm)	Water[a] (%)	K[b] (%)	Th[b] (ppm)	U[b] (ppm)	a radiation[c] (Gy ka^{-1})	b radiation[d] (Gy ka^{-1})	g radiation[e] (Gy ka^{-1})	Cosmic-ray radiation[f] (Gy ka^{-1})	Total dose rate (Gy ka^{-1})	Equivalent dose[g] (Gy)	Optical age (ka)
NC01	Homestead Midden	34~°08'42.0''S, 141°19'20.0''E	45	5 ± 1	1.19 ± 0.02	8.55 ± 0.07	1.30 ± 0.03	0.03 ± 0.01	1.18 ± 0.04	0.54 ± 0.04	0.19 ± 0.02	1.94 ± 0.07	0.6 ± 0.2	**0.3 ± 0.1**
NC04	Potterwalkagee Gully	34°09'21.4''S, 141°20'42.5''E	25	5 ± 1	0.74 ± 0.02	4.07 ± 0.03	0.55 ± 0.06	0.03 ± 0.01	0.68 ± 0.03	0.44 ± 0.03	0.20 ± 0.02	1.35 ± 0.05	0.3 ± 0.1	**0.3 ± 0.1**
NC03	Potterwalkagee Gully	34°09'21.4''S, 141°20'42.5''E	50	5 ± 1	1.04 ± 0.02	7.05 ± 0.06	0.72 ± 0.05	0.03 ± 0.01	0.97 ± 0.04	0.55 ± 0.04	0.19 ± 0.02	1.74 ± 0.06	19 ± 3	**11.0 ± 1.5**
NC02	Homestead Midden	34~°08'42.0''S, 141°19'20.0''E	85	5 ± 1	1.08 ± 0.02	6.66 ± 0.05	0.84 ± 0.04	0.03 ± 0.01	1.01 ± 0.04	0.63 ± 0.05	0.18 ± 0.02	1.85 ± 0.06	32 ± 7	**17.1 ± 4.0**

a: estimated time-averaged moisture contents, based on measured field water values (% dry weight).

b: obtained by INAA (Becquerel Laboratories, Menai).

c: assumed internal alpha dose rate.

d: derived from INAA radionuclide concentration measurements using the conversion factors of Adamiec and Aitken (1998), corrected for attenuation by water and beta attenuation.

e: derived from field gamma spectrometry measurements using the conversion factors of Adamiec and Aitken (1998), corrected for attenuation by water.

f: calculated using the equation of Prescott and Hutton (1994), based on sediment density, time-averaged depth and site latitude, longitude and altitude.

g: including a ± 2 % systematic uncertainty associated with calibration of the laboratory beta-source.

Table 5. OSL ages from the Homestead Midden and Potterwalkagee Gully, Ned's Corner Station

The degree of pedogenesis in the Neds Corner Land System fluvial horizons, together with the accumulation of ~30 cm of aeolian dust above the floodplain surface, suggest that this fluvial activity is substantially older than the other land systems identified by this study (Figure 3). Although this sequence has not been dated, its relative age can be inferred by comparison with equivalent fluvial sequences in the adjacent Riverine Plain. Channels of the Green-Gully Tallygaroopna system (Bowler 1978), like the Neds Corner channels, are wide, laterally migrating, bedload-dominated and associated with source-bordering dunes. Both sequences are highly pedogenised. Channel widths and meander wavelengths are comparable with those of the Neds Corner fluvial system, suggesting that both channel sequences represent the same regime of greater seasonal discharge. Bowler (1978) estimated the Green-Gully Tallygaroopna channels to be at least 40,000 years old. This assumption has been substantiated by recent OSL dating of these sequences (Dr Tim Stone, pers. comm.).

Climatic conditions at this time were more arid than today. The apparent contradiction between drier conditions and increased surface flow is a function of enhanced effective precipitation: as vegetation decreased during arid, glacial periods, infiltration was reduced. Furthermore, periglacial conditions in highland catchments have the effect of increasing runoff. Discharge is increased even though actual precipitation may be reduced (Barrows et al. 2001).

This fluvial regime was probably active as far back as the commencement of Oxygen Isotope Stage 3, when global climates became relatively more arid. The initiation of this fluvial regime may correspond to the Snowy River Glacial Advance in the southeastern highlands at 59.3 ± 5.4 ka (Barrows et al. 2001, 2002). Further glacial advances in the Late Pleistocene such as the first advance in the Kosciusko glaciation at 32 ± 2.5 ka may also have influenced the aggradation of this system with increasing aridity.

Aeolian activity of the Neds Corner Land System and activation of the fluvial Lindsay Land System: Oxygen Isotope Stage 3 to the last glacial maximum

Significant dust transport, represented by ~30 cm of parna accumulation on the floodplain surface followed the final phase of channel evolution in the Neds Corner Land System (Figure 3). This implies intensification of the arid conditions, which culminated during the last glacial maximum (~24–18 ka) (Mix and Schneider 2001). The polycyclic clay cutans and higher degree of carbonate pedogenesis in the lower part of the parna profile indicate successive depositional cycles. The first phase of aeolian mobilization is likely to have occurred late in Oxygen Isotope Stage 3 (~64 – 32 ka), coeval with a significant dust flux recorded throughout southeastern Australia (Bowler 1986b; Wasson 1989; Page and Nanson 1996).

The last major phase of parna dispersal is likely to have coincided with the last glacial maximum (Butler 1956; Bowler 1976, 1986a; Bowler et al. 1976; Wasson and Donnelly 1991; Hesse et al. 2003; Hesse and McTainsh 2003). Aluminium- and quartz-rich aeolian detritus within the Vostok ice core are indicative of the greater wind speeds and dust mobilization that occurred during the last glacial maximum (McTainsh 1989; Petit et al. 1990). Cool, arid conditions peaked at ~ 20 ka. Sea level was ~120 m below present, which increased aridity and landmass area, linking the Australian mainland with New Guinea and Tasmania (Williams et al. 1998). Continental temperatures were 6–10 °C lower than modern temperatures (Harrison and Dodson 1993, Harrison and Prentice 2003). Palynological evidence from Lake Frome, South Australia shows decreased woodland vegetation at this time, enabling strong glacial winds to mobilise sediments for aeolian transport and dune formation (Singh and Luly 1991).

The lunette lakes of the study area incise, and thus postdate, the Neds Corner fluvial channels (Figure 2). Almost all lunettes in Australia are relict, with the exception of some reactivated lunettes at Lakes Tyrrell (Teller et al. 1982) and Eyre (Dulhunty 1983). Thus the lunettes of the Neds Corner Land System provide an insight into previous hydrological conditions (Bowler 1986b). The pelletal clay compositions of the lunettes within the study area indicate that they formed during cool, arid conditions when saline water tables were near the surface (Bowler 1973; Macumber 1980; Dare-Edwards 1984). Lake Wallawalla preserves two lunette phases, which are potentially analogous to the two cycles of parna activity attributed to late Oxygen Isotope Stage 3 and the last glacial maximum.

The large meander wavelengths, channel widths and associations with source-bordering dunes suggest that, similar to the Neds Corner Land System, the Lindsay Land System consisted of laterally-migrating bedload-dominated fluvial channels with seasonally high discharge to expose point bars and provide a sediment source for source-bordering dune formation (Figure 2, Table 2). Gilgai clays on the floodplain of this land system (Figure 4a) support the notion of increased seasonal flow, with the deep cracks suggesting strong seasonal wetting and drying regimes (Retallack 1997).

Unlike the Neds Corner Land System, the upper surface of the Lindsay Land System does not preserve evidence of parna deposition. This suggests it was fluvially active during or after the last glacial maximum when the last phase of parna deposition occurred. OSL dates of 17.2 ± 4.4 ka obtained from Lindsay Land System source-bordering dunes (Table 5) indicate that activation of the Lindsay palaeochannels was roughly synchronous with lunette dune building and parna accumulation on the Neds Corner land surface. Equivalents of the Lindsay Land System fluvial phase from the Riverine Plain have been identified by Bowler (1978) (Kotupna phase) and Page et al. (1991; Page and Nanson 1996) (Yanco phase), both of which have been dated (by radiocarbon and thermoluminescence) to the last glacial maximum. These systems are characterised by large, laterally migrating meanders, source-bordering dunes and channel widths wider than the modern River Murray, but smaller than the Neds Corner palaeochannels. These channel forms imply a regime of seasonally high discharge, on a smaller scale than the Neds Corner Land System.

The last advances of the Kosciusko glaciation in the southern highlands occurred at 19 ± 1.6 ka and 16.8 ± 1.4 ka (Barrows et al. 2001, 2002). The less extensive advances of these glaciers, compared with the previous glaciations (Barrows et al. 2001), suggest lower precipitation may account for the smaller channel widths and meander wavelengths of the Lindsay Land System compared with the Neds Corner Land System.

The effect of these hyper-arid conditions on human populations is reflected in the lack of archaeological sites in the region dating to the last glacial maximum (Hope 1998; Mulvaney and Kamminga 1999). Glacial conditions would cause a reduction in the availability of food resources, shelter and raw materials. This suggests that as climatic conditions deteriorated, people were forced to retreat towards the coastal margins of Australia. It is only after the last glacial maximum that inland sites were once again colonised (Mulvaney and Kamminga 1999).

Mulcra Island Land System fluvial phase:The transition from glacial conditions
Channels of the Mulcra Island Land System cut across, and thus must postdate, those of the Lindsay Land System (Figure 2). Several billabongs, characteristic of more sinuous, suspended-load systems (Page and Nanson, 1996), have formed from meander-cutoffs within the Mulcra Island Land System (Figure 2, Table 1). The absence of source-bordering dunes suggests the return of channel-margin vegetation before the conversion to the modern system (Bowler 1978; Page et al. 1991; Page and Nanson 1996; Ogden et al. 2001). The sinuous, mixed- to suspended-load channels of the Mulcra Island Land System (Figure 2, Table 2) imply a change in hydrologic regime at the end of the Lindsay Land System during the last glacial maximum. The relatively larger channels of the Mulcra Island Land System compared with the modern Murray preserve a transitional, deglacial phase of channel evolution, no equivalents of which have been recognised on the Riverine Plain. Between ~17 and ~14.5 ka, sufficient rain in the central Murray Valley allowed eucalypt woodland to recolonise and restabilise the landscape (Singh and Luly 1991). This observation is consistent with the higher sinuosity and absence of source-bordering dunes on this system.

From Vostok ice core records, dust flux decreased and temperatures began to increase immediately following the peak of the last glacial maximum, but did not approach modern conditions until the middle Holocene (Barnola et al. 1987; Jouzel et al. 1987; McTainsh 1989). Global sea levels gradually rose following the end of the last glacial maximum, with a rapid increase at 15 ka from the melting of North American ice sheets (Chappell 1993). The final retreat of glaciers from the southeastern highlands at 16.8 ± 1.4 ka (Barrows et al. 2001, 2002, 2004), may have supplied the climatic impetus for this change in hydrologic regime. By inference, the evolution from bedload to suspended load channels was probably more gradual than has previously been assumed from studies of the Riverine Plain.

Some studies contend that substantial occupation of the central Murray Valley region did not occur until the Late Holocene, with the most intensive settlement in the last 2000 years (Ross 1981, 1982 1985; Hope 1998). However, radiocarbon ages from the surrounding regions (Edmonds 2004) show a spread of ages over a greater period, with a significant number of dates clustering around the terminal Pleistocene and early Holocene. The 15,000 cal BP dates obtained from this study as well as large shell midden accumulations on source-bordering dunes suggest substantial occupation, at least of the riparian margins, during the Mulcra Island climatic phase. At this time, the river evolved from a high-velocity bedload-dominated regime to a mixed- to suspended-load sinuous channel system. The change in fluvial regime would have increased the resource potential of the riparian margins (Mulvaney and Kaminga 1999). As the channel became narrower and more sinuous, water temperature increased and current velocity decreased, creating conditions more conducive to the support of fish and mollusc populations. Billabongs, characteristic of this land system, provided an additional resource as seasonal wetlands, attracting bird and mammal populations, which could be hunted for their meat, skins, sinew and bone for tools.

Murray Land System fluvial phase: The Holocene
The higher sinuosity, smaller channel width, and smaller meander wavelengths of the Murray Land System are a function of the current fluvial regime: a suspended-load, low-gradient flow system (Table 2). Change towards modern regimes in both the Mallee and the Riverine Plain is likely to have been gradual, occurring some time in the early Holocene. Bowler (1978) has dated this change to 9 ka BP, which is coeval with an increase in woodland vegetation at Lake Frome (Singh et al. 1981). Decreased runoff due to the presence of woodland vegetation, along with increasing summer insolation throughout the Holocene, caused increased evaporation to deplete water availability, reducing flow peaks and decreasing channel and meander size (Harrison and Dodson 1993).

During natural flow conditions in the Holocene, the Murray River dried to a series of saline water holes, which would have necessitated the exploitation of the surrounding Mallee hinterland (Pardoe 1990; Cupper 2003). Elevated regions (such as lunettes) may have been used as campsites for short term or seasonal hunting and gathering expeditions. The surface artefact deposits on dry lake floors and relict dunes in the study area may reflect such seasonal occupation. These deposits are low density, unstratified and clearly postdate the last glacial maximum active phase of the lacustrine units in the Neds Corner Land System.

A variety of food resources are found in the hinterland, including seeds from saltbush that could have been processed with grinding stones like those found at Little Lake Wallawalla, Tip Site and Lindsay Island Dune (Figure 2, Table 3). The presence of grinding implements suggests that these sites may be Holocene as previous studies have shown grinding implements to be a late Holocene innovation in arid and semi-arid Australia (Ross 1981; Ross et al. 1992).

Conclusions

This project addresses a previously unstudied area, linking fluviatile, aeolian and lacustrine systems in the central Murray Basin. A multidisciplinary approach to a complex set of environmental parameters inevitably poses more questions than it answers. However, through application of land systems mapping and archaeology, significant conclusions emerge.

Five land systems have been identified in the central Murray Valley, each reflecting different hydrological and climatic regimes. The Neds Corner, Lindsay, Mulcra Island and Murray Land Systems show a progressive evolution from large, laterally-migrating bedload-dominated regimes to the modern Murray River regime of narrow, highly sinuous, suspended-load channels. The progression of fluvio-aeolian-lacustrine environments observed within the study area demonstrates climatic evolution through successive humid-arid cycles.

Extensive archaeological evidence is preserved in the study area. OSL and radiocarbon dating show that humans were present in the landscape from at least 15,000 years ago. This was a time when the fluvial regime and climate was substantially different from that of today. The climate was shifting from the aridity caused by the last glacial maximum towards the characteristic humid Holocene climates. The recognition of fluvial and aeolian environments presents background for the interpretation of human adaptation to changing environmental regimes in this part of the Murray Basin.

Acknowledgements

Thank you to Aboriginal Affairs Victoria, particularly Mark Dugay-Grist for supporting the work and securing funding for radiocarbon dating, and to Latje Latje community for allowing this work on their land. Trust for Nature provided logistical support. Gordon Holm assisted with thin section preparation. Chris Rowley and Brian Smith identified the sub-fossil shell species. Tim Stone and Malcolm Wallace provided constructive advice throughout the project. The authors are grateful to Dr Patricia Fanning, Dr Adrian Fisher and one anonymous reviewer for their helpful reviews of an earlier draft of the manuscript.

References

Adamiec, G. and M. J. Aitken. 1998. Dose-rate conversion factors: Update. *Ancient TL* 16, 37–50.

Barnola, J. M., D. Raynaud, Y. S. Korotkevich and C. Lorius. 1987. Vostok ice core provides 160,000-year record of atmospheric CO2. *Nature* 329:408–414.

Barrows, T. T., J. O. Stone, L. K. Fifield and R. G. Cresswell. 2001. Late Pleistocene glaciation of the Kosciusko massif, Snowy Mountains, Australia. *Quaternary Research* 55(2):179–189.

Barrows, T. T., J. O. Stone, L. K. Fifield and R. G. Cresswell. 2002. The timing of the last glacial maximum in Australia. *Quaternary Science Reviews* 21(1–3):159–173.

Barrows, T. T., J. O. Stone and L. K. Fifield. 2004. Exposure ages for Pleistocene periglacial deposits in Australia. *Quaternary Science Reviews* 23:697–708.

Bullard, J. E. and G. H. McTainsh. 2003. Aeolian-fluvial interpretations in dry land environments: Examples, concepts and Australia case study. *Progress in Physical Geography* 27(4):471–501.

Butler, B. E. 1956. Parna, an aeolian clay. *Australian Journal of Science* 18(5):145–151.

Chappell, J. 1993. Late Pleistocene coasts and human migrations in the Austral region. In M Spriggs, D. E.Yen, W. Ambrose, R. Jones, A. Thorne and A. Andrews (eds), pp 13–17. *A Community of Culture*. Canberra: Australian National University Press.

Christian, C. S. and G. A. Stewart. 1952. *Summary of general report of survey of Katherine–Darwin region*.

Churchward, H. M. 1961. Soil studies at Swan Hill, Victoria, Australia. *Journal of Soil Science* 12:73–86.

Cupper, M. L. 2003. *Late Quaternary environments of playas in southwestern New South Wales*. Unpublished PhD, Melbourne: University of Melbourne.

Dare-Edwards, A. J. 1984. Aeolian clay deposits of southeastern Australia — parna or loessic clay. *Transactions of the Institute of British Geographers* 9(3):337–344.

Dulhunty, J. 1983. Lunettes of Lake Eyre North, South Australia. *Transactions of the Royal Society of South Australia* 107:219–22.

Dury, G. H. 1967. Climatic change as a geographical backdrop. *Australian Geographer* 10:231–242.

Edmonds, V. 2004. Indigenous Cultural Heritage assessment. Proposed culvert replacement, Lake Wallawalla, northwest Victoria. Buronga NSW: Archaeological Consulting Service for Sinclair Knight Merz, Melbourne.

Galbraith, R. F., R. G. Roberts, G. M. Laslett, H. Yoshida and J. M. Olley. 1999. Optical dating of single and multiple grains from Jimnium rock shelter, part 1: Experimental design and statistical methods. *Archaeometry* 41:339–364.

Gerard, A. J. 1981. *Soils and landforms: An integration of geomorphology and pedology*. London: Allen and Unwin.

Grist, M. J. 1995. *An archaeological investigation into the 'No Stone Saga' of far Northwest Victoria: A study of the Berribee quarries in the landscape*. Unpublished Honours Thesis. Canberra: Australian National University.

Harrison, S. P. and J. R. Dodson. 1993. Climates of Australia and New Guinea since 18,000 yr BP. In H. E. Wright, J. E. Kutzbach, T. Webb and W. F. Ruddiman (eds), pp 265–294. *Global climates since the last glacial maximum*. Minneapolis: University of Minnesota Press.

Harrison, S. P. and A. I. Prentice. 2003. Climate and CO2 controls on global vegetation distribution at the last glacial maximum: Analysis based on palaeovegetation data, biome modelling and palaeoclimate simulations. *Global Change Biology* 9(7):983–1004.

Hesse, P. P. and G. H. McTainsh. 2003. Australian dust deposits: Modern processes and the Quaternary record. *Quaternary Science Reviews* 22(18–19):2007–2035.

Hesse, P. P., G. S. Humphreys, B. L. Smith, J. Campbell and E. K. Peterson. 2003. Age of loess deposits in the Central Tablelands of New South Wales. *Australian Journal of Soil Research* 41(6):1115–1131.

Hope, J. 1998. *Lake Victoria: Finding the balance*. Canberra: Murray-Darling Basin Commission.

Jouzel, J., C. Lorius, J. R. Petit, C. Genthon, N. I. Barkov, V. M. Kotlyakov and V. M. Petrov. 1987. Vostok ice core: A continuous isotope temperature record over the last climatic cycle (160,000 years). *Nature* 329:403–408.

Mabbut, J. A. 1980. Some general characteristics of the aeolian landscapes. In R. R. Storrier and M. E. Stannard (eds), *Aeolian landscapes in the semi-arid zone of south-eastern Australia*, pp 1–15. Wagga Wagga: Australian Society for Soil Science, Riverina Branch.

Macumber, P. G. 1980. The influence of groundwater discharge on the Mallee landscape. In R. R. Storrier and M. E. Stannard (eds), *Aeolian landscapes in the semi-arid zone of south-eastern Australia*, pp 67–84. Wagga Wagga: Australian Society for Soil Science, Riverina Branch.

McTainsh, G. H. 1989. Quaternary aeolian dust processes and sediments in the Australian region. *Quaternary Science Reviews* 8:235–253.

Mix, A. C. and Schneider, R. 2001. Environmental processes of the ice age: Land, oceans, glaciers. *Quaternary Science Reviews* 20:627–657.

Mulvaney, D. J. and J. Kamminga. 1999. *Prehistory of Australia*. Sydney: Allen and Unwin.

Murray-Darling Basin Commission. 1996. River Murray mapping: Colour infrared orthophotography and natural resource data. Canberra: Murray-Darling Basin Commission.

O'Connell, J. F. and J. Allen. 2004. Dating the colonization of Sahul (Pleistocene Australia–New Guinea): A review of recent research. *Journal of Archaeological Science* 31:835–853.

Ogden, R., N. Spooner, M. Reid and J. Head. 2001. Sediment dates with implications for the age of the conversion from palaeochannel to modern fluvial activity on the Murray River and tributaries. *Quaternary International* 83–5:195–209.

Page, K. J., A. J. Dare-Edwards, J. W. Owens, P. S. Frazier, J. Kellett and D. M. Price. 2001. TL chronology and stratigraphy of riverine source bordering sand dunes near Wagga Wagga, New South Wales, Australia. *Quaternary International* 83:187–193.

Page, K. J., and G. C. Nanson. 1996. Stratigraphic architecture resulting from Late Quaternary evolution of the Riverine Plain, south-eastern Australia. *Sedimentology* 43(6):927–945.

Page, K. J., G. C. Nanson and D. M. Price. 1991. Thermoluminescence chronology of Late Quaternary deposition on the riverine plain of south-eastern Australia. *Australian Geographer* 22(1):14–23.

Pardoe, C. 1990. The demographic basis of human evolution in southeastern Australia. In B. Meehan (ed.), *Hunter-gatherer demography: Past and present*, pp 59–70. Sydney: University of Sydney.

Petit, J. R., L. Mournier, J. Jouzel, Y. S. Korotkevich, V. M. Kotyakov and C. Lorius. 1990. Palaeoclimatological and chronological implications of the Vostok core dust record. *Nature* 343:56–58.

Prescott, J. R. and J. T. Hutton. 1994. Cosmic ray contributions to dose rates for luminescence and ESR dating: Large depths and long-term time variations. *Radiation Measurements* 23:497–500.

Readhead, M. 1990. TL dating of sediments from Lake Mungo and Nyah West. In R. Gillespie (ed.), *Papers Presented at the Quaternary Dating Workshop*, pp 35–37. Canberra: Department of Biogeography and Geomorphology, Research School of Pacific Studies, Australian National University.

Retallack, G. J. 1997. *A colour guide to paleosols*. New York: John Wiley and Sons.

Ross, A. C. 1981. Holocene environments and prehistoric site patterning in the Victorian Mallee. *Archaeology in Oceania* 16:145–154.

Ross, A. C. 1982. Problems of disentangling human and climatic influences upon the Australian landscape. In B. G. Thom and R. J. Wasson (eds), *Holocene research in Australia*, pp 29–35. Duntroon: Royal Military College.

Ross, A. C. 1985. Archaeological evidence for population change in the middle to late Holocene in southeastern Australia. *Archaeology in Oceania* 20:81–89.

Ross, A. C., T. H. Donnelly and R. J. Wasson. 1992. The peopling of the arid zone: Human-environment interactions. In J. R. Dodson (ed.), *The naive lands*, pp 76–114. Melbourne: Longman Cheshire.

Rowan, J. N. and R. G. Downes. 1963. *A study of the land in north-western Victoria*. Victoria: Soil Conservation Authority.

Singh, G., N. D. Opdyke and J. M. Bowler. 1981. Late Cainozoic stratigraphy, paleomagnetic chronology and vegetational history from Lake George, NSW. *Journal of the Geological Society of Australia* 28(3–4):435–454.

Singh, G. and J. G. Luly. 1991. Changes in vegetational and seasonal climate since the last full glacial at Lake Frome, South Australia. *Palaeogeography, Palaeoclimatology, Palaeoecology* 84:75–86.

Stone, T., and M. L. Cupper. 2003. Last glacial maximum ages for robust humans at Kow Swamp, southern Australia. *Journal of Human Evolution* 45(2):99–111.

Teller, J. T., J. M. Bowler and P. G. Macumber. 1982. Modern sedimentation and hydrology in Lake Tyrrell, Victoria. *Journal of the Geological Society of Australia* 29:159–175.

Wasson, R. J. 1989. Landforms. In J. C. Noble and R. A. Bradstock (eds), *Mediterranean landscapes in Australia*, pp 13–34. Canberra: CSIRO.

Wasson, R. J. and T. H. Donnelly. 1991. *Palaeoclimatic reconstructions for the last 30,000 years in Australia — a contribution to prediction of future climate*. Canberra: CSIRO Division of Water Resources.

Williams, M. A. J., P. De Deckker, D. A. Adamson and M. R. Talbot. 1991. Episodic fluviatile, lacustrine and aeolian sedimentation in a Late Quaternary desert margin system, central western New South Wales. In D. F. Williams, P. De Deckker and A. P. Kershaw (eds), *The Cainozoic in Australia: A reappraisal of the evidence*, pp 258–287. Sydney: Geological Society of Australia Inc.

Williams, M. A. J., D. L. Dunkerley, P. De Deckker, A. P. Kershaw and J. M. A. Chappell. 1998. *Quaternary environments*. London: Arnold.

6

Seeing red: The use of a biological stain to identify cooked and processed/damaged starch grains in archaeological residues

Jenna Weston

Australian Museum Business Services
6 College St
Sydney NSW 2010 Australia

Abstract

Starchy plant foods are a major component in the diet of almost all peoples, particularly hunter-gatherers such as Aboriginal Australians. However, the archaeological preservation of such plants is rare, as is other direct evidence of plant use by past peoples. While analysis of starchy residues preserved on artefacts has gained acceptance as an effective method for identifying starchy plant use, it is very difficult using the standard morphological method to accurately identify starch grains that have been damaged by processing activities such as milling or cooking. Therefore, a method is presented for the identification of such damaged starch grains using the stain Congo Red, which dyes damaged (cooked or processed) but not undamaged starch. This method has been applied to identify cooking or milling activities in the subsistence of hunter-gatherers from south-east Queensland, Australia.

Keywords: Congo Red; starch grains; cooking; residue analysis; bevel-edged artefacts; south-east Queensland.

Introduction

The following paper presents the results of a feasibility study that was initially undertaken as part of an Honours thesis at the University of Queensland (Lamb 2003) under the supervision of Dr Tom Loy. This work resulted in the establishment of a new method for determining cooking in archaeological plant residues, which was published (Lamb and Loy 2005) and then presented at the 2005 Australasian Archaeometry Conference. This paper provides a summary of the methods and archaeological applications published by Lamb and Loy (2005), and expands upon the 2005 publication with new data on taphonomic controls and additional images. It is published in this forum subsequent to Dr Loy's death, with acknowledgement for his role in developing the method.

Background

As has been established by a number of studies (e.g., Brand and Cherikoff 1985; Gould 1969; Latz 1995; Latz and Griffin 1978; Lee 1965, 1968; Meehan 1989; O'Connell et al. 1983; Peterson 1978), plant foods, particularly starchy plants, are a major part of people's diets. This is especially true for hunter-gatherers, such as Aboriginal Australians (Beck et al. 1989:6; Kaberry 1935:6; Mountford 1960; Peterson 1973). However, these plants are often not preserved in whole form in the archaeological record, particularly in Australia, and direct evidence of plant use by past peoples is also rare (Hather 1991:661; Meehan 1989:14; Pearsall 2000:153; Piperno and Holst 1998:765; Therin et al. 1999:439). Residue analysis, particularly analysis of starchy residues preserved on artefacts, has now become accepted as an effective method for identifying past plant use and starch grains, which seem to be fairly prolific in archaeological plant residues, can also allow the identification of cooking in residue evidence (Babot 2003; Barton et al. 1998; Fullagar 1998; Loy 1994; Loy et al. 1992; Piperno and Holst 1998; Torrence and Barton 2006).

Starch

Starch grains have a very regular structure of amylose and amylopectin layers (Badenhuizen 1965:81; Banks and Greenwood 1975:242; Cortella and Pochettino 1994:172; Loy 1994:89; Reichert 1913:89). This structure imparts the distinctive 'cross' that they exhibit when viewed microscopically in cross-polarised light, which will rotate radially when the polariser plate is rotated. This cross effect is also displayed by faecal spherulites, ooliths, some coccoliths, some avian and reptile uric acid spheres, some plant calcium oxalates and some other biologically precipitated compounds, but each of these has a much higher refractive index than starch, so viewing residues through a low refractive index medium like water makes these compounds almost invisible owing to excessively high relief (Banks and Greenwood 1975; Canti 1998; Loy 1994; Reichert 1913).

The regular structure of starch is altered by cooking, however, a feature which allows the identification of cooking via residue analysis. This is important because it can provide additional evidence that an artefact was used in human subsistence-related activities, namely the cooking of plant foods, which is a uniquely human activity. Other evidence of the human subsistence use of an artefact may include the presence of other food residues from plants, use wear, and ethnohistoric accounts of certain stone tools being used to prepare food.

The known effects of cooking on starch include mainly swelling and loss of the regular structure of the grains, with a corresponding loss of the extinction cross (Banks and Greenwood 1975; Hall et al. 1989; Loy 1994). This is a process known as gelatinisation, and it occurs when the grain is in the presence of moisture and is heated at temperatures greater than 30°C. During this process, the hydrogen bonds linking the amylose chains are progressively broken and the grain absorbs water, which is what causes it to swell. The temperature and rate at which a grain gelatinises depend on factors such as the presence of additives like glucose or mineral salts, and the size and shape of the starch grain. This loss of the cross effect in cooked starch grains makes it very difficult to identify them accurately using the standard morphological method, because their appearance depends on the stage of gelatinisation, and at some point in this process the cross becomes completely invisible (Banks and Greenwood 1975; Loy 1994; Reichert 1913).

One way to address this problem of how to identify cooked starch lies in the effects of cooking on starch grains. The loss of the structure from the breaking of their hydrogen bonds, and the swelling caused by absorption of water, should allow cooked grains to absorb certain stains where raw and undamaged starch grains, which are hydrophobic, will not.

Congo Red

Congo Red (CI 22120, CI name 'Direct Red 28'; empirical formula $C_{32}H_{22}N_6O_6S_2Na_2$) is a water-soluble dye and a suspected carcinogen, which depends on linear hydrogen bonding for staining (Conn and

Lillie 1969). This stain has been used in other studies as a general contrast stain for cellulose, amyloid fibrils and agricultural starch products (Chou et al. 2001; Conn and Lillie 1969; Ramesh and Thranathan 1999). However, it stains proteins like amyloid fibrils only in acid or alkaline conditions, so at a neutral pH only starch and cellulose will stain (Badenhuizen 1965:86; Conn and Lillie 1969; Cortella and Pochettino 1994; Khurana et al. 2001 [pH2–4]; Loy 1994:92; Mehta and Rajput 1998 [pH3.5]; Reichert 1913). Congo Red appears to react with the amylose in starch, which is exposed when grains are damaged by structure loss and swollen with water, as when they have been cooked.

The amylose in starch is based on the same monosaccharide molecule as cellulose (Figure 1). However, as Figure 1 shows, the molecule is bonded in a different way in each. Therefore both cellulose, and starch grains that have broken bonds, will stain with Congo Red, but their appearance and structure are visibly different, so they can be differentiated under the microscope.

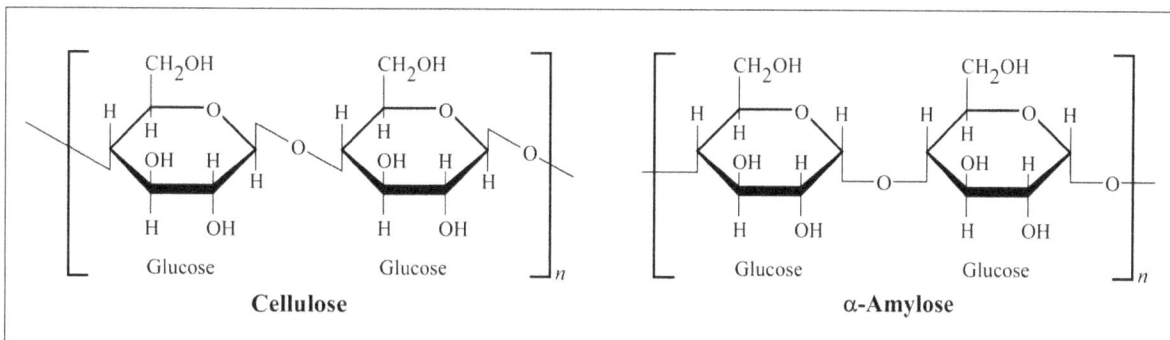

Figure 1. Primary structure of cellulose and α-amylose (from starch). n may be several thousand (after Voet and Voet 2004:365-6).

Methodology

In order to test the effects of the stain, it was necessary to use it on both experimental and archaeological residues. For the experimental phase, three plants were chosen which each had high starch content and were present in south-east Queensland, and for all of which there are ethnohistoric accounts of their having been cooked and then pounded between stones by Aboriginal people in this area. These plants were *Alocasia macrorrhiza* (elephant ear; see Brown 1893, Roth 1901, Thozet 1866), *Blechnum indicum* (Bungwahl fern; see Threlkeld 1825, Watkins 1891), and *Castanospermum australe* (Moreton Bay chestnut; see Banfield 1908, Moore [cited in Maiden1900], Roth 1901). The starchy root or seed from each of these plants was taken, and raw samples were smeared onto slides. The remaining root/seed was then cooked in an electric frypan filled with sand (to simulate roasting as if in a fire), and cooked samples were smeared onto slides. Another specimen of the three plants was then taken, and part of each was processed when raw by pounding between two stones, then another part of each was processed in the same way after having been cooked in a fire (on different days and different places in the fenced area of the University of Queensland's TARDIS excavation site, so as to avoid contamination). After the residues had dried, samples were extracted from the experimental stones and placed onto microscope slides using the procedure outlined in Lamb and Loy (2005:1435 [Table 1]).

After microscopic examination of the characteristics of each of the raw, cooked, raw and processed, and cooked and processed starch samples, each was then stained with Congo Red, using the procedure outlined in Lamb and Loy (2005:1435 [Table 1]), being very careful to take precautions against the possible carcinogenic properties of the stain. This was achieved mainly by wearing a lab coat and non-starch-powdered gloves, and sealing the slide covers once the stain was applied. The Congo Red solution used was originally made up by dissolving 50 mg of the powder into 50 ml of water; in this situation face masks and eye goggles were worn in addition to the lab coat and gloves. The solutions applied to the slides all had the low refractive index of water, to avoid confusion with the cross effects given by non-starch residues, as previously mentioned.

Archaeological residues were initially taken from three bevel-edged artefacts from the Southern Curtis Coast (Figure 2). These artefacts (Figures 3–5) were provided by Dr Sean Ulm, having been collected by him during his PhD research (Ulm 2006). A further archaeological residue was recovered from a fourth bevel-edged artefact which was excavated from a shell midden on the Gold Coast (Robins et al. 2005).

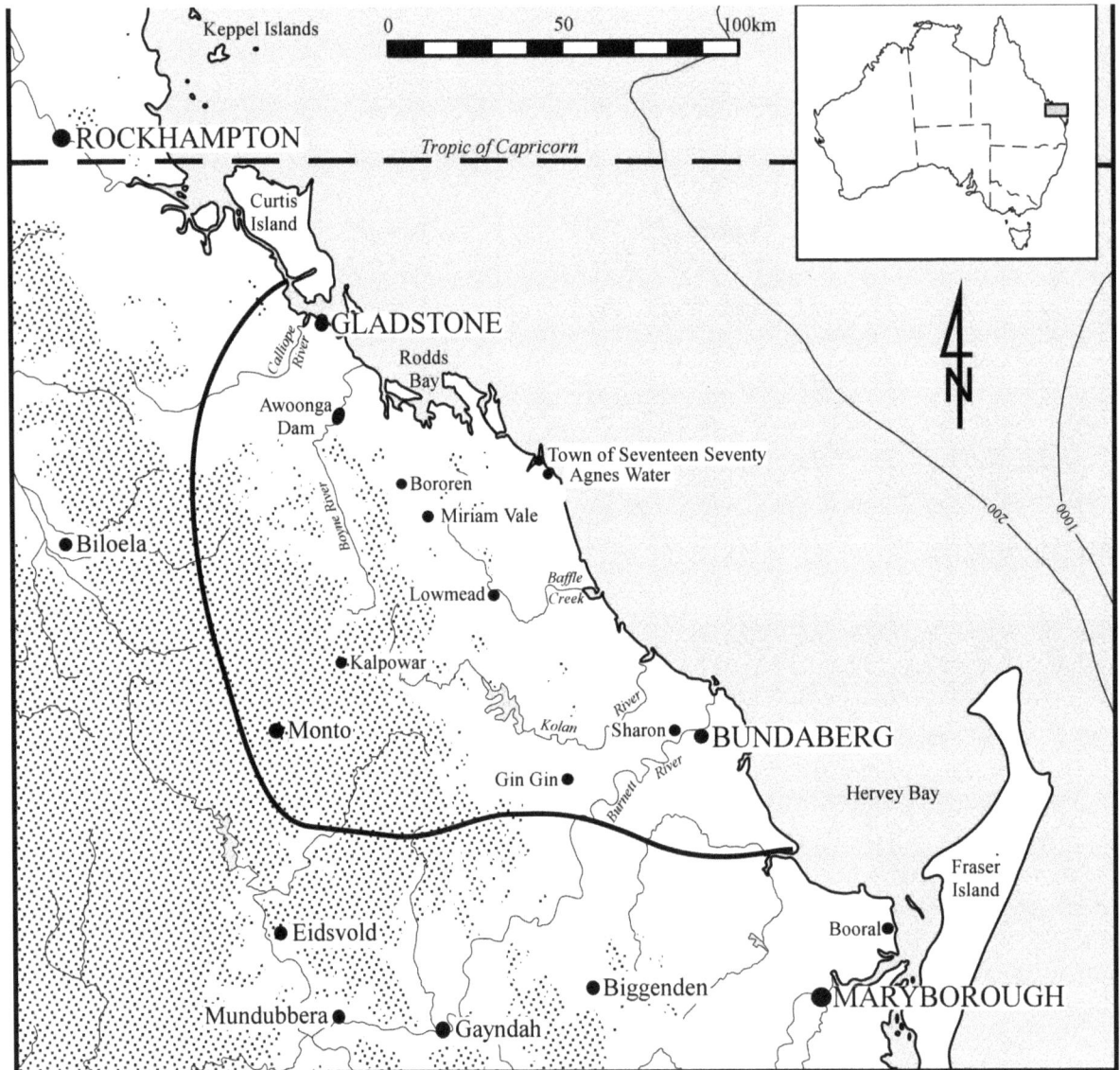

Figure 2. Location of Southern Curtis Coast from where the initial three archaeological artefacts were recovered (Ulm 2006:15).

Figure 3. Bevel-edged artefact 1, surface-collected from an eroding creek section at Eurimbula Site 1 (ES1) on the western bank of Round Hill Creek as part of the Gooreng Gooreng Cultural Heritage Project (GGCHP), on 9 March 1999. ES1 is an extensive open midden complex located in Eurimbula National Park, with deposits dating to the last 3,200 years (Ulm 2006:177).

Figure 4. Bevel-edged artefact 2, surface-collected from an eroding creek section at ES1 on the western bank of Round Hill Creek as part of the GGCHP, on 3 June 2001. Scale = 5 cm units.

Figure 5. Bevel-edged artefact 3, found 20 m from Squares O and P at the Ironbark Site Complex (ISC) as part of the GGCHP, on 13 February 1998. ISC is an extensive stone quarry/shell midden site complex located on the lower southern bank of Middle Creek estuary, with deposits dated to the last 500 years (Ulm 2006:131).

Residues were extracted from multiple positions on each artefact, mainly along the edges and on the flat, bevelled surface. All archaeological residues were extracted using the procedure outlined in Lamb and Loy (2005:1436 [Table 2]).

After examining each residue slide and noting the presence and characteristics of the starch (and other plant residues), each slide was stained using the same procedure as for the experimental residues (Lamb and Loy 2005:1435 [Table 1]).

Results

Samples of starch from the processed raw plants tended to stain light red in regular light and retained their cross effect in cross-polarised light, often as four dark points in their outline (Figure 6a and b). Partially-cooked grains may also have this appearance when stained, but would have swollen somewhat, so if the species of the plant residue is known, this distinction in activity may be determined.

6a.   6b.

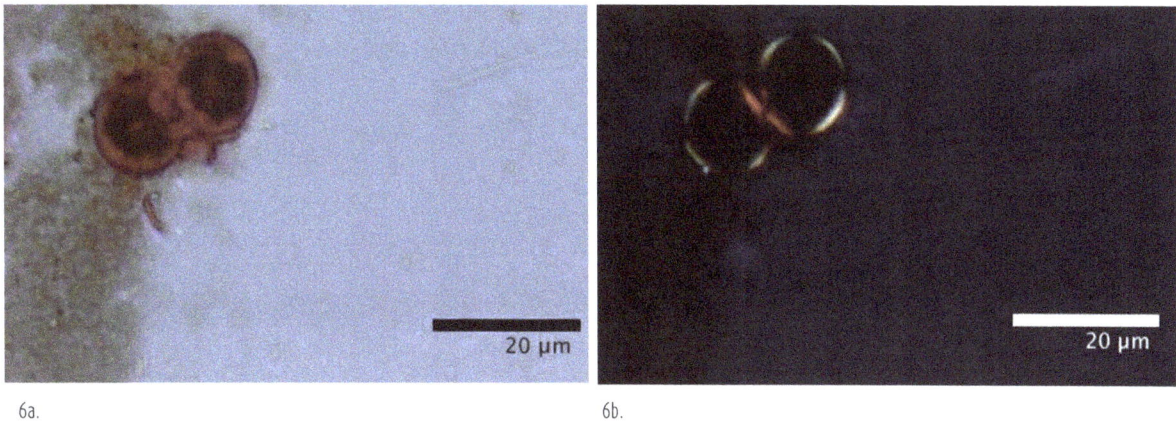

Figure 6. Damaged starch grains from raw, processed *Alocasia macrorrhiza* sample, 400x magnification, stained (a) transmitted light (b) cross-polarised transmitted light.

The raw, unprocessed starch samples did not stain, unless they had been unintentionally damaged during the experiments. However, the well-cooked, gelatinised grains stained bright red, often with either an orange-red or green-gold glow. These grains were not seen in regular light without staining, but application of Congo Red revealed that they had swollen quite dramatically. In the instance of Moreton Bay chestnut (seen in Figures 7–8), the grains had swollen to around ten times their original size. Some pitting, layering and cracking can also be seen in the swollen grains (Figure 8). Generally, cooked grains did seem to retain enough of their structure for them to maintain their primarily round or oval shape, although extensive swelling associated with heating appears to be able to change the morphology of grains somewhat (e.g. from round to oval-shaped; see Figures 7–8) and for a modified cross to appear, as four darker points in their outline, when viewed in cross-polarised light.

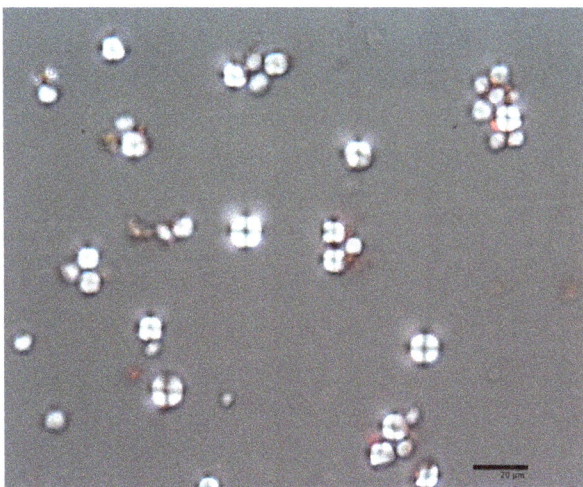

Figure 7. Starch grains from raw *Castanospermum australe* sample (cross-polarised transmitted light, 400x magnification, stained). Scale bar 20μm.

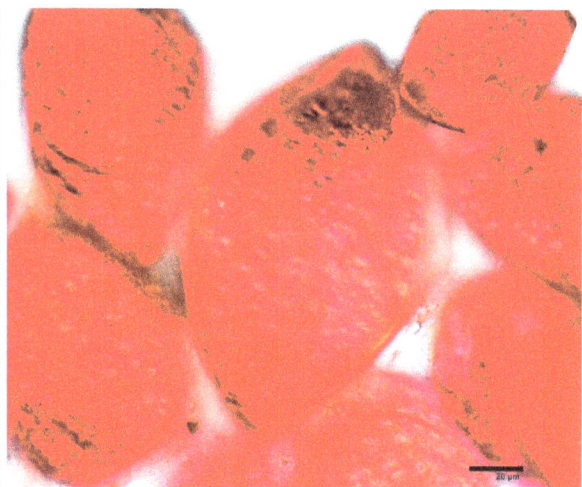

Figure 8. Starch grains from cooked *Castanospermum australe* sample (transmitted light, 400x magnification, stained). Scale bar 20μm.

Substantial quantities of both stained and unstained starch grains were seen in the archaeological residues. The unstained grains were interpreted as being raw and unprocessed, or not heavily processed. The stained grains were considered to belong to one of two categories. Figures 9 (a and b) and 10 (a and b) show some grains that are thought to have been processed to some extent but were uncooked, because they stained slightly red but retained their extinction cross. In contrast, Figures 11–15 show stained grains argued to have been cooked, because they stained bright red in regular transmitted light, and had only a modified cross as four dark points in the outline of the grain when viewed in cross-polarised light.

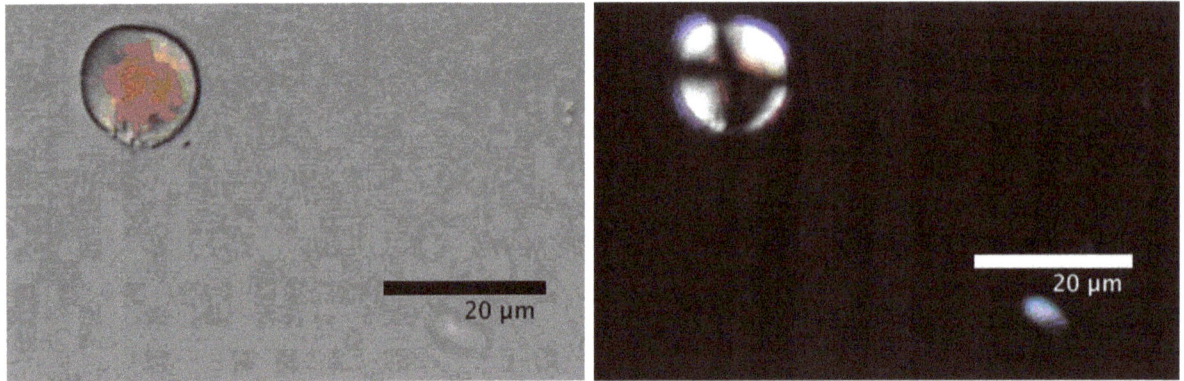

9a.   9b.

Figure 9. Damaged starch grain from bevelled face, Artefact #2, 400x magnification, stained (a) transmitted light (b) cross-polarised transmitted light.

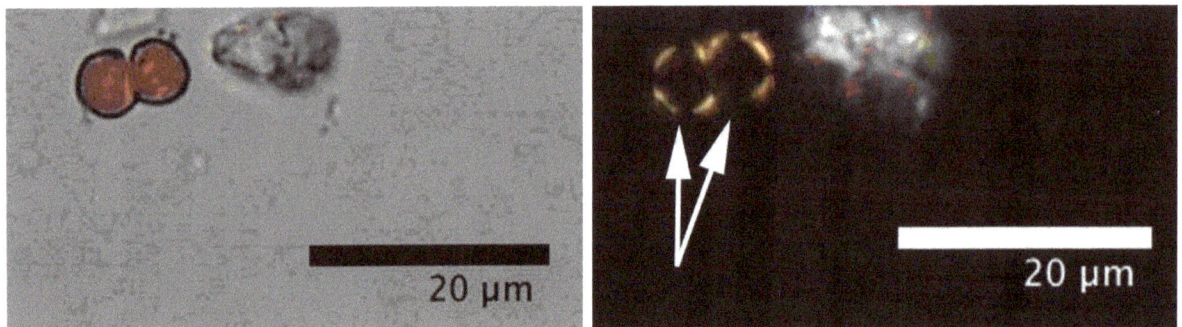

10a.   10b.

Figure 10. Starch grains (arrows in B) from bevelled face, Artefact #3, 400x magnification, stained (a) transmitted light (b) cross-polarised transmitted light.

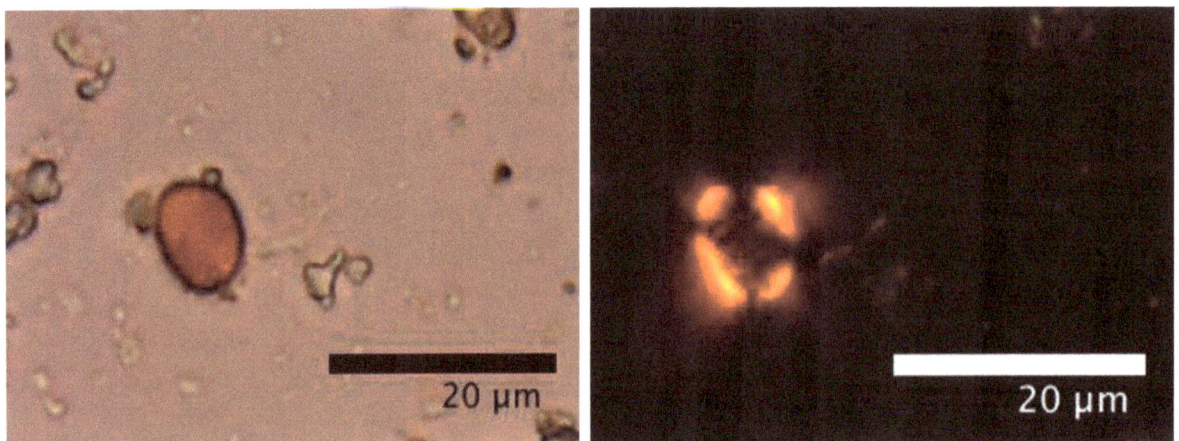

11a.   11b.

Figure 11. Starch grain from Artefact #2, 400x magnification, stained (a) transmitted light (b) cross-polarised transmitted light.

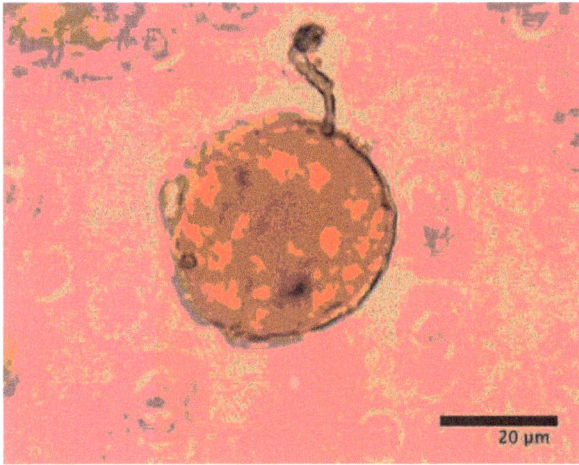

12a.

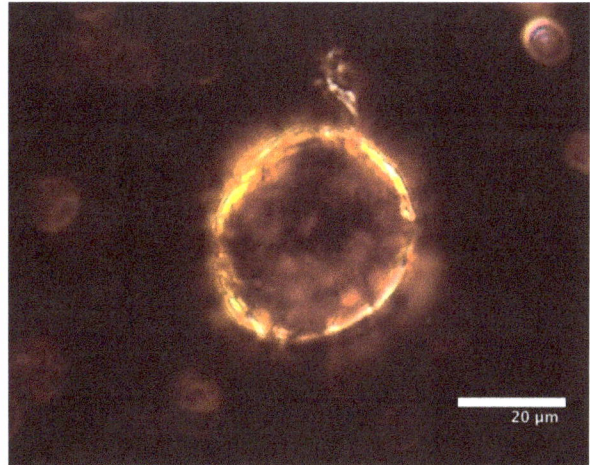

12b.

Figure 12. Gelatinised starch grain from Artefact #2, 400x magnification, stained (a) transmitted light (b) cross-polarised transmitted light.

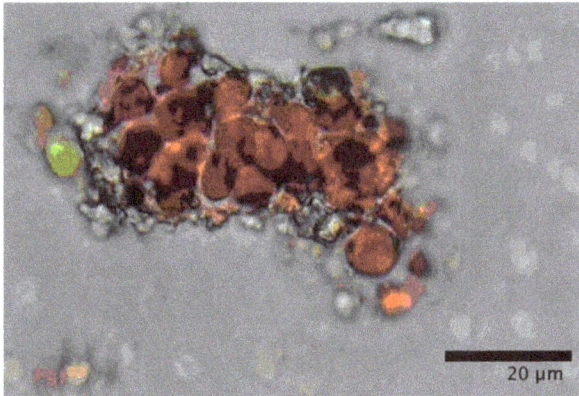

Figure 13. Group of starch grains from sloped face, Artefact #1 (transmitted light, 400x magnification, stained).

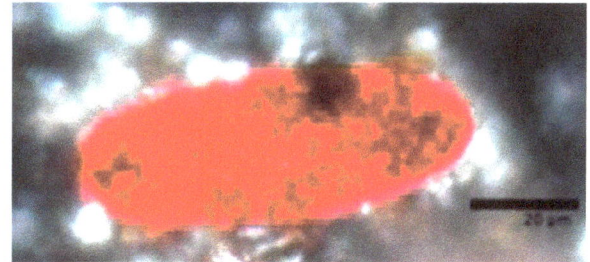

Figure 14. Gelatinised starch grain from bevelled face, Artefact #1 (cross-polarised transmitted light, 400x magnification, stained).

The fourth archaeological artefact produced plant-dominated residues, with predominantly small starch grains (approximately 10–15 microns in diameter), many of which stained and hence were interpreted as having been processed or cooked. The grain shown below in regular transmitted (Figure 16a) and cross-polarised light (Figure 16b) was assessed as having been cooked because it was stained bright red with a modified cross in the outline.

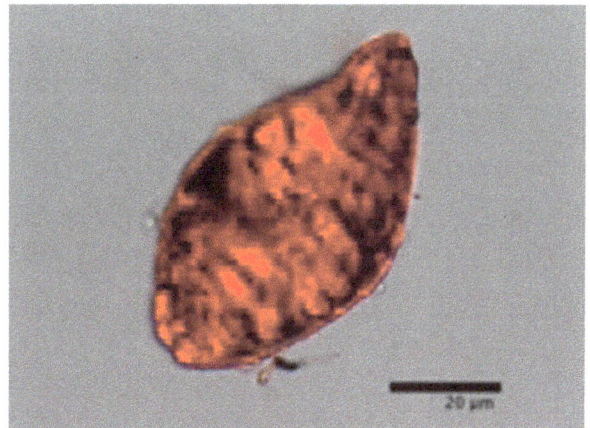

Figure 15. Gelatinised starch grain from bevelled face, Artefact #2 (transmitted light, 400x magnification, stained).

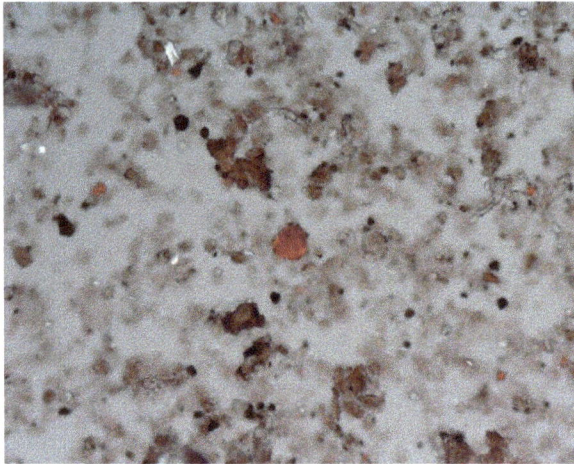
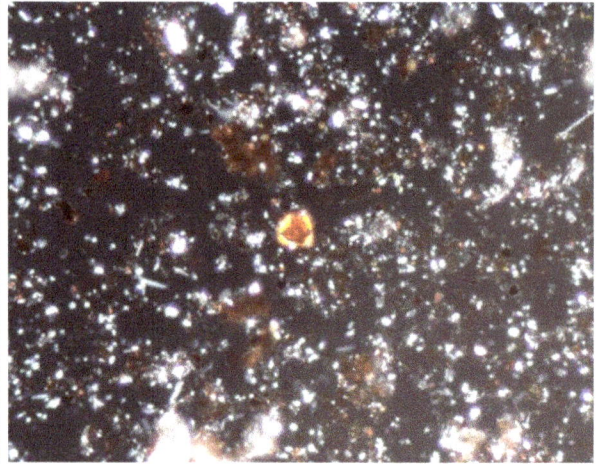

16a.   16b.

Figure 16. Damaged (12μm) starch grain, 400x magnification, stained (a) transmitted light (b) cross-polarised transmitted light.

Congo Red was observed to stain sclereids, tracheids, parenchymal tissue (Figure 17), wall thickenings, bordered pits (Figure 18) and cellulose. However, while these stained red in regular light, they usually appeared green or yellow in cross-polarised light, rather than red as was the case with the starch grains.

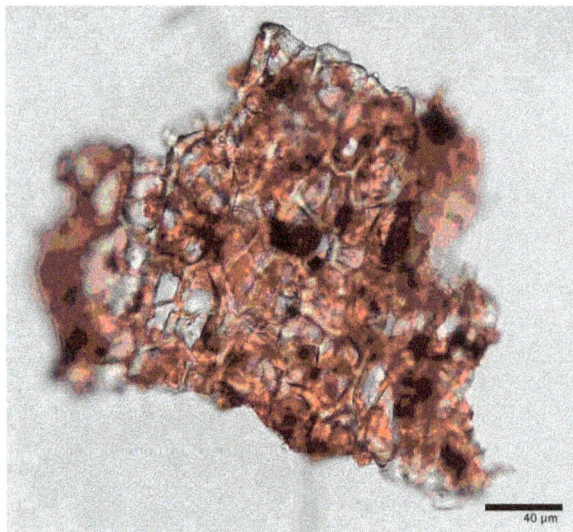
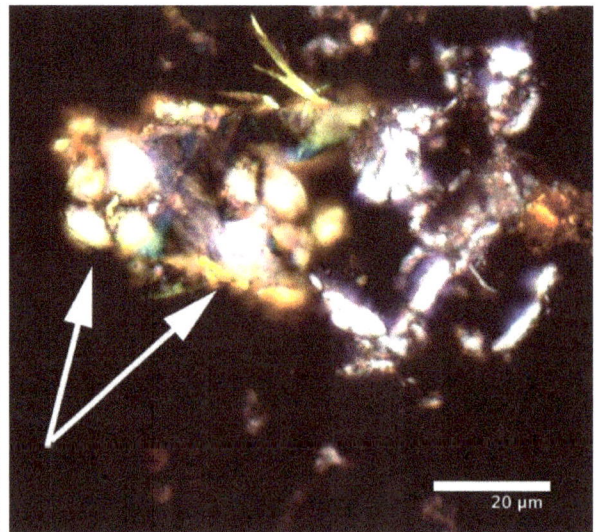

Figure 17. Parenchyma tissue from stained section, Artefact #2 (transmitted light, 200x magnification, stained).

Figure 18. Bordered pits (arrows) from Artefact #2 (cross-polarised transmitted light, 400x magnification, stained).

Discussion of Results

Nothing similar in appearance to the gelatinised starch grains was observed to stain in the same way, and the stained starch grains that were concluded to have been damaged through processing were also quite distinctive in appearance. It was therefore concluded that each artefact had most likely been used to process a starchy plant or plants, at least one of which was cooked to some extent before being processed. The described studies have therefore ascertained that Congo Red appears to be a feasible method for identifying cooking and processing in past societies.

Taphonomic Studies

Several brief taphonomic studies were also conducted to ascertain the extent of non-processing-related starch contamination on the stone artefacts. The possibility of ground contamination was tested by trampling a sterile (i.e. newly manufactured) tool into the ground where experimental processing was undertaken. Soil samples examined before processing revealed almost no starch grains; after processing starch was present in the soil and hence was a potential source of contamination if tools were trampled into the soil. Starch grains, plant material and dirt were recorded on the tools which were trampled into soil of the processing area, although only 23% of the total recorded on the processing tools was present (see Table 1, Tool Type 2).

Tool type	Number of recorded starch grains	Percentage compared with processing tools
1. Experimental processing tools	146	N/A
2. Ground contaminated tools	19	13.01
3. Touch contaminated tools	34	23.29
4. Air contamination slide	3	2.06

Table 1. Quantities of *Alocasia macrorrhiza* starch grains (raw and cooked) recovered from processing, trampling and touch-test experiments, and air contamination. Values were calculated using one unstained microscope slide for each activity.

The possibility of contamination through touch was tested by handling clean experimental tools after the user had processed each plant (Table 1, Tool Type 3). On these tools, starch grains and plant material were also seen and the quantity was about 13% of that seen on processing tools (see Table 1). All tools were subjected to the contamination mechanism for the same amount of time (ten seconds), but it is likely that more grains would have been transferred to the tools if they were subject to a longer period of exposure or handling.

Lastly, the possibility of air contamination was tested by placing a clean microscope slide in the open air near the place where processing occurred; only three starch grains were found on this slide.

Therefore, it is concluded that contamination of tools can occur through handling and trampling. Use-wear analysis provides a potential method for identifying when this is the case with archaeological tools preserving residues.

It is possible that natural attacks on starch by enzymes in biologically active soils may cause damage that would stain with Congo Red. This possibility has not been tested as part of the current research. However it seems unlikely that enzyme attack would swell the starch grains in the way that cooking does. Although it may be more difficult to distinguish processing damage from enzyme damage, use-wear analysis of tools should enable a better understanding of whether this may be the case. In both cases further experimental studies may provide useful insights into the potentially confusing effect of enzyme damage.

Archaeological Implications

The use of Congo Red when working with starchy residues from archaeological artefacts means that a cultural association may be more confidently made between the tool and its use in subsistence activity, particularly plant food preparation. This is because Congo Red only stains starch that has been damaged, whether the damage has occurred through processing or through cooking, or both, and does not stain undamaged grains. Therefore the use of Congo Red can make a significant contribution to the preponderance of evidence for human subsistence use of an artefact.

Possible Applications

Congo Red may be useful in testing the accuracy of ethnohistoric accounts of starchy plant preparation, particularly those plants that need to be cooked or processed prior to consumption because they are toxic or unpalatable raw (for example, in Australia, cycads as well as *Alocasia* and *Castanospermum australe*). It may also be used to contribute to the debate on the origins of controlled fire use, by being applied to artefacts associated with early humans such as *Homo erectus*, and detecting whether these artefacts retain the residues of cooked plants. The fact that Congo Red stains cellulose is also useful, because cellulose can often be difficult to distinguish microscopically from residues such as bone collagen fibres.

Further Work

There are a number of studies that should preferably be done to support the use of Congo Red staining of archaeological residues. Firstly, further taphonomic studies should be completed, looking specifically at whether cooked starch grains are preserved and under what circumstances, and whether their deposition for archaeological time spans changes their structure or appearance. Secondly, a reference collection should be compiled with samples of archaeological and experimental cooked starches. Experimental samples should be obtained from as many different starch species as possible, and controlled experiments should be conducted on species of different sizes, using a variety of cooking temperatures and cooking times, and different methods of cooking and processing. Thirdly, examining the use-wear on tools with residues stained with Congo Red should be undertaken to determine whether the evidence for processing of either cooked or raw starchy plants is in accordance with the conclusions drawn from residue analysis and staining.

Acknowledgements

Thanks to: the late Dr Tom Loy, formerly of the School of Social Science, University of Queensland for supervision and collaboration in developing the method; Dr Sean Ulm, Aboriginal and Torres Strait Islander Studies Unit, University of Queensland; Dr Richard Robins, Everick Heritage Consulting; Eastern Yugambeh Limited; Nathan Woolford, Aboriginal and Torres Strait Islander Studies Unit, University of Queensland; Sue Nugent, Alison Crowther, Michael Haslam, Meg Heaslop, Luke Kirkwood and Dr Gail Robinson, School of Social Science, University of Queensland; and Environmental Resources Management (Australia) Pty Ltd.

References

Babot, M. 2003. Starch grain damage as an indicator of food processing. *Terra Australis 19*. Canberra: ANU E Press.

Badenhuizen, N. P. 1965. Occurrence and development of starch in plants. In R. L. Whistler and E. F. Paschall (eds), *Starch chemistry and technology Volume 1: Fundamental aspects*, pp 65–103. New York: Academic Press.

Banfield, E. J. 1908. *The Confessions of a beachcomber*. St Lucia: University of Queensland Press.

Banks, W. and C. T. Greenwood. 1975. *Starch and its components*. Edinburgh: University Press.

Barton, H., R. Torrence and R. Fullagar. 1998. Clues to stone tool function re-examined: Comparing starch grain frequencies on used and unused obsidian artefacts. *Journal of Archaeological Science* 25:1231–1238.

Beck, W., A. Clarke and L. Head. 1989. Plants in hunter-gatherer archaeology. In W. Beck, A. Clarke and L. Head (eds), *Plants in Australian archaeology*, pp 1–13. St Lucia: Anthropology Museum, University of Queensland. *Tempus* 1.

Brand, J. C. and V. Cherikoff. 1985. The nutritional composition of Australian Aboriginal food plants of the desert regions. In G. E. Wickens, J. R. Goodin and D. V. Field (eds), *Plants for arid lands*, pp 53–69. London: George Allen and Unwin.

Brown, G. R. 1893. Notes on the uses, etc., of some native plants in the Port Macquarie district. *Agricultural Gazette of New South Wales* 4(9):680–682.

Canti, M. G. 1998. The micromorphological identification of faecal spherulites from archaeological and modern materials. *Journal of Archaeological Science* 25:432–444.

Chou, K.-S., J.-C. Tsai and C.-T. Lo. 2001. The adsorption of Congo Red and vacuum pump oil by rice hull ash. *Bioresource Technology* 78(2):217–219.

Conn, H. J. and R. D. Lillie. 1969. *H. J. Conn's biological stains: A handbook on the nature and uses of the dyes employed in the biological laboratory*, 8th edn. Baltimore: The Williams and Wilkins Company.

Cortella, A. R. and M. L. Pochettino. 1994. Starch grain analysis as a microscopic diagnostic feature in the identification of plant material. *Economic Botany* 48(2):171–181.

Fullagar, R. 1998. Use-wear, residues and lithic technology. In R. Fullagar (ed.), *A closer look: Recent Australian studies of stone tools*, pp 13–17. Sydney: Archaeological Computing Laboratory, School of Archaeology, University of Sydney. *Sydney University Archaeological Methods Series 6*.

Gould, R. A. 1969. *Yiwara: Foragers of the Australian desert*. New York: Scribner.

Hall, J., S. Higgins and R. Fullagar. 1989. Plant residues on stone tools. In W. Beck, A. Clarke and L. Head (eds), *Plants in Australian archaeology*, pp 136–160. St Lucia: Anthropology Museum, University of Queensland. Tempus 1.

Hather, J. G. 1991. The identification of charred archaeological remains of vegetative parenchymous tissue. *Journal of Archaeological Science* 18:661–675.

Kaberry, P. M. 1935. The Forrest River and Lyne River tribes of north-west Australia: A report on field work. *Oceania* 5:408–436.

Khurana, R., V. N. Uversky, L. Nielson, and A. L. Fink. 2001. Is Congo Red an amyloid-specific dye? *Journal of Biological Chemistry* 276(25):22715–22721.

Lamb, J. 2003. The raw and the cooked: A study on the effects of cooking on three Aboriginal plant foods from southeast Queensland. Unpublished BA (Hons) thesis. Department of Anthropology, Archaeology and Sociology, School of Social Science, University of Queensland, St Lucia.

Lamb, J. and T. H. Loy. 2005. Seeing red: The use of Congo Red dye to identify cooked and damaged starch grains in archaeological residues. *Journal of Archaeological Science* 32:1433–1440.

Latz, P. K. 1995. *Bushfires and bushtucker: Aboriginal plant use in Central Australia*. Alice Springs: IAD Press.

Latz, P. K. and G. F. Griffin. 1978. Changes in Aboriginal land management in relation to fire and to food plants in central Australia. In B. S. Hetzel and H. J. Frith (eds), *The nutrition of Aborigines in relation to the ecosystem of Central Australia*, pp 77–85. Melbourne: CSIRO.

Lee, R. B. 1965. *Subsistence ecology of the !Kung Bushmen*. Ann Arbor: University Microfilms.

Lee, R. B. 1968. What hunters do for a living, or, how to make out on scarce resources. In R. B. Lee and I. De Vore (eds), *Man the hunter*, pp 30–48. Chicago: Aldine.

Loy, T. H. 1994. Methods in the analysis of starch residues on prehistoric stone tools. In J. G. Hather (ed.), *Tropical archaeobotany*: Applications and new developments, pp 86–114. London: Routledge.

Loy, T. H., M. Spriggs and S. Wickler. 1992. Direct evidence for human use of plants 28,000 years ago: Starch residues on stone artefacts from the northern Solomon Islands. *Antiquity* 66:898–912.

Maiden, J. H. 1900. Native food-plants. *Agricultural Gazette of New South Wales 10* (4):279–290.

Meehan, B. 1989. Plant use in a contemporary Aboriginal community and prehistoric implications. In W. Beck, A. Clarke and L. Head (eds), *Plants in Australian archaeology*, pp 14–30. St Lucia: Anthropology Museum, University of Queensland. Tempus 1.

Mehta, S. and Y. S. Rajput. 1998. A method for staining of proteins in nitrocellulose membrane and acrylamide gel using Congo Red dye. *Analytical Biochemistry* 263(2):248–251.

Mountford, C. P. 1960. *Records of the American-Australian scientific expedition to Arnhem Land. Anthropology and Nutrition, Volume 2*. Melbourne: Melbourne University Press.

O'Connell, J. F., P. K. Latz and P. Barnett. 1983. Traditional and modern plant use among the Alyawara of central Australia. *Economic Botany* 37(1):80–109.

Pearsall, D. M. 2000. *Paleoethnobotany: A handbook of procedures*, 2nd edn. Sydney: Academic Press.

Peterson, N. 1973. Camp site location among Australian hunter-gatherers: Archaeological and ethnographic evidence for a key determinant. *Archaeology and Physical Anthropology in Oceania* 8(3):173–193.

Peterson, N. 1978. The traditional pattern of subsistence to 1975. In B. S. Hetzel and H. J. Frith (eds), *The Nutrition of Aborigines in relation to the ecosystem of Central Australia*, pp 25–35. Melbourne: CSIRO.

Piperno, D. R. and I. Holst. 1998. The presence of starch grains on prehistoric stone tools from the humid neotropics: Indications of early tuber use and agriculture in Panama. *Journal of Archaeological Science* 25:765–776.

Ramesh, H. P. and R. N. Tharanathan. 1999. Water-extracted polysaccharides of selected cereals and influence of temperature on the extractability of polysaccharides in sorghum. *Food Chemistry* 64(3):345–350.

Reichert, E. T. 1913. *The differentiation and specificity of starches in relation to genera, species, etc.: Stereochemistry applied to protoplasmic processes and products, as a strictly scientific basis for classification of plants and animals*. Washington: Carnegie Institute of Washington.

Robins, R., J. Lamb and N. Woolford. 2005. *Between swamp and sea: Archaeological investigations of an Aboriginal midden at Allisee Residential Development, Hollywell, Gold Coast Queensland*. Consultancy report for Stocklands Pty Ltd for Eastern Yugambeh Ltd.

Roth, W. E. 1901. Food: Its search, capture and preparation. *North Queensland Ethnology Bulletin* 3:1–31.

Therin, M., R. Fullagar and R. Torrence. 1999. Starch in sediments: A new approach to the study of subsistence and land use in Papua New Guinea. In C. Gosden and J. G. Hather (eds), *The prehistory of food: Appetites for change*, pp 438–462. London: Routledge.

Thozet, A. 1866. *Notes on some of the roots, tubers, bulbs, and fruits, used as vegetable food by the Aboriginals of Northern Queensland, Australia*. Rockhampton: W. H. Buzacott, Bulletin Office, Denham Street.

Threlkeld, L. E. 1825. Reminiscences of the Aborigines of New South Wales. In N. Gunson (ed) *Australian reminiscences and papers of L. E. Threlkeld, missionary to the Aborigines 1824–1859*, pp 41–82. Canberra: Australian Institute of Aboriginal Studies

Torrence, R. and H. Barton. 2006. *Ancient starch research*. California: Left Coast Press, Inc.

Ulm, S. 2006. Coastal themes: An archaeology of the Southern Curtis Coast, Queensland. *Terra Australis 24*. Canberra: ANU E Press.

Voet, D. and J. G. Voet. 2004. Biochemistry: *Biomolecules, mechanisms of enzyme action, and metabolism, Volume 1*, third edn. New York: John Wiley & Sons.

Watkins, G. 1891. Note on the Aborigines of Stradbroke and Moreton Islands. *Proceedings of the Royal Society of Queensland 8*:40–50.

7

Initial tests on the three-dimensional movement of starch in sediments

Michael Haslam

Postdoctoral Fellow in Palaeolithic Archaeology
Leverhulme Centre for Human Evolutionary Studies
University of Cambridge
Cambridge CB2 1QH
United Kingdom
mah66@cam.ac.uk

Abstract

Archaeological studies of microbotanical remains must consider taphonomic factors likely to have affected a recovered assemblage. One of the most fundamental taphonomic considerations is the movement of objects within a soil profile under the influence of groundwater. To this end, this study reports on initial tests involving the movement of starch granules in three dimensions within a constructed sand matrix. The results indicate that both lateral and upward as well as significant downward movement occurs under the influence of groundwater, with movement rates of up to 4mm/ day for some starch granules greater than 10μm in size.

Keywords: starch, taphonomy, post-depositional movement, sediment, microscopy

Introduction

Analyses of starch granules from archaeological soils complement other, longer-established microbotanical techniques including pollen and phytolith studies. As in these other fields, starch analysts are increasingly recognising the influence of taphonomic factors on constructing and biasing the recovered microfloral suite (Babot 2003; Barton et al. 1998; Haslam 2004). Of critical taphonomic importance is the estimation of starch movement through a soil profile following deposition. Pioneering work in this regard was undertaken in the early 1990s by Fullagar, Loy and Cox (Fullagar et al. 1994), and later developed and elaborated on by Therin (1998, see also 1994) in the first publication on the topic directed towards archaeologists. To date Therin's study remains the sole published starch movement study, with subsequent ancient starch studies often citing his finding that very little starch appears to move in soils (e.g. Fullagar et al. 1998:56; Lentfer et al. 2002:687; Therin et al. 1999:450), or omitting discussion of starch movement altogether (e.g. Barton 2005; Parr and Carter 2003). Therin's findings were based on downward percolation of starch of varying sizes through sterile sands, under the influence of significant levels of water simulating heavy rainfalls. His main conclusions were that small starch granules will move more often and faster than large granules, and that larger granules will move more slowly but further than smaller granules once mobile. Therin also suggested that future studies should investigate lateral and upward movement of starch, in addition to other tests. Following this advice, here I report the results of a recent experiment into the three-dimensional (3D) movement of starch in sediments.

Methods

One of the goals of the current experiment was to produce results which complement previous starch movement studies, and the methods for assessing starch movement were therefore deliberately chosen to build on and augment Therin's (1994, 1998) experiments. Certain parameters (including the amount of starch added to the experiment, and the rainfall rate simulation) were mirrored as precisely as practicable, while other parameters (including extraction procedures and the makeup of the sedimentary matrix) differed out of necessity.

The experiment

The 3D experimental design required a starch sample to be placed in the centre of a constructed sediment matrix with the freedom to move up, down and sideways under the influence of water. To this end, an apparatus was constructed of four 10cm sections of 10cm diameter PVC piping, cut in half lengthwise and joined to create an open cruciform structure (see Figure 1). Three of the four 'arms' of the cross were blocked with semi-circular wooden stops, which had several holes drilled in them to permit water flow. A piece of synthetic scouring pad was placed against the inner surface of each of the wooden stops to prevent loss of sediment. Four mm thick transparent perspex was cut to match the cruciform shape, and each of the exposed walls of the PVC pipes was lined with a thin foam/rubber window seal strip, with the result that the perspex could be pressed against the open apparatus to create a waterproof seal, and then removed again with ease. The transparent perspex enabled continuous monitoring of the sediment matrix for the development of cracks or other events that may have influenced starch movement. A metal bracket was attached to the back of the cross arm which did not have a wooden stop; this arm formed the 'top' of the apparatus during the experiment.

Sand was chosen as the sediment matrix for the experiment, following Therin (1998). 'Medium' sand was obtained from a commercial supplier, and dry-sieved through a series of Endecott test sieves of mesh sizes varying from 125μm to 1mm. The 250 – 500μm fraction was selected for the experiment as this could be expected to leave a matrix

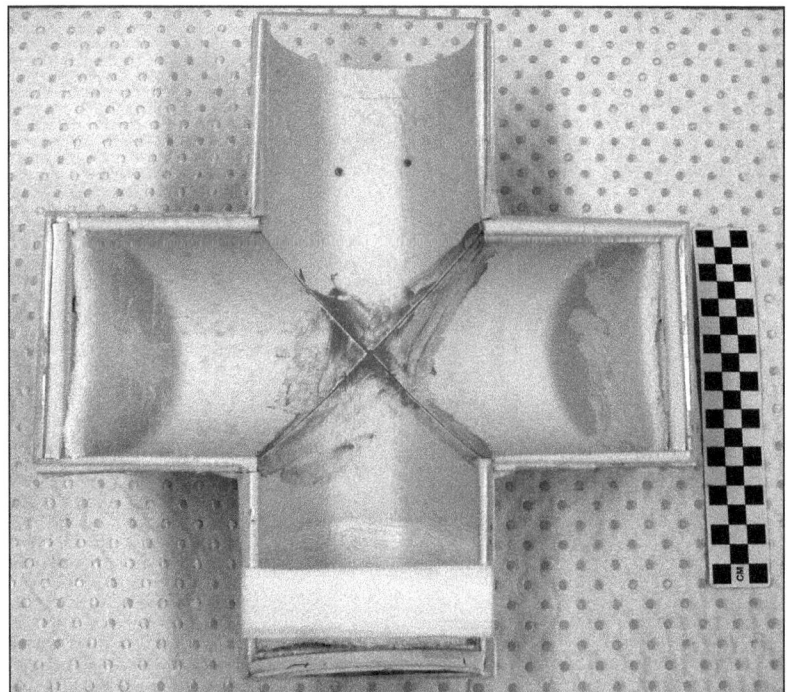

Figure 1. 3D starch movement apparatus with perspex cover.

permeable and porous enough to permit starch movement. As the sand was dry-sieved, some grains <250 µm remained in the 250–500µm fraction through aggregation. It was not desirable to construct a matrix that excluded the possibility of starch movement (for example a dense clay), or one that allowed free movement to all granules (for example loose-packed coarse sand), since this would be self-defeating. As the present study was an initial test of 3D starch movement, it was decided to err on the side of allowing movement rather than restricting it, while still constructing a plausible matrix. The sand was autoclaved for 30mins at 121°C and 105kPa prior to use in the experiment to eliminate starch-degrading fungi and bacteria, and a sample run through the starch extraction procedure described in the following section to check for contaminants.

With the cruciform apparatus upright (that is, with the wooden stops at the base and side arms of the cross, and the top arm open) and the perspex firmly attached with masking tape, dry packed sand was added to half-fill the apparatus. Using a funnel and a cut-down section of a plastic pipette, 0.1g of commercial corn starch was added to the centre of the sand (Figure 2). The corn starch chosen has starch granules which range in size from 2–27µm (average 9.07 µm), based on measurements taken from more than 2000 granules. Dry packed sand was then added to fill the apparatus to a level 1cm below the top.

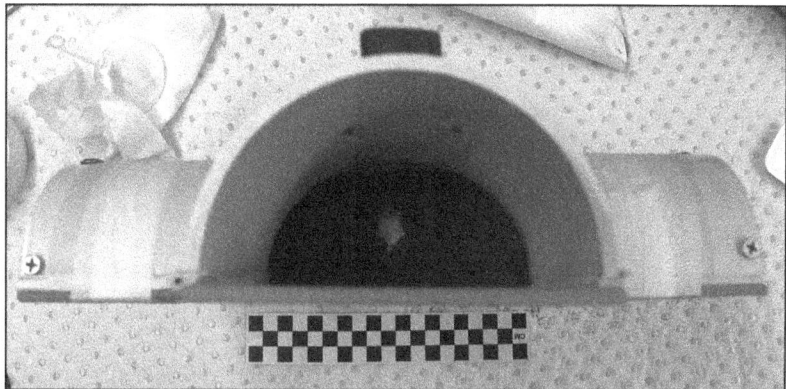

Figure 2. Starch added to experiment.

The whole experiment was hung from a retort stand via the metal bracket attached to the back (Figure 3), allowing it to hang freely vertical. 500ml of ultrapure water was gradually added to the experiment over the period of a day to initially dampen the sand.

The choice of water volume to be added to the experiment was again based on Therin (1994, 1998) to allow comparability. The irrigation of three of Therin's four starch movement columns corresponded to 4777mm of rainfall over two months (Therin 1994:53; incorrectly recorded as 4777ml in Therin 1998:63). The present experiment ran over a period of one month (30 days) and aimed therefore to simulate approximately 2400mm rainfall. With a cross-

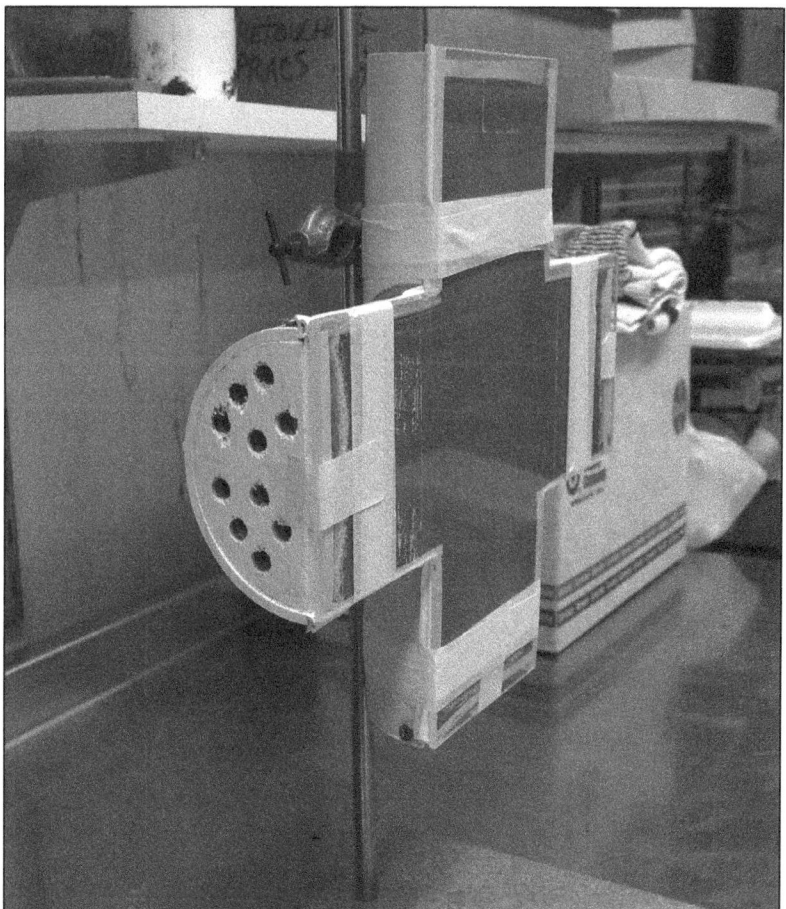

Figure 3. Experiment attached to retort stand. Note holes in wooden stop at end of arm to permit water flow.

section of 40cm2 for the current experiment 2400mm corresponds to a total irrigation of 9600ml over 30 days, or 320ml/day. This irrigation rate was achieved by adding 160ml of ultrapure water twice a day, in the morning and evening. As one of Therin's conclusions was that a high irrigation rate probably washed a large number of starch granules entirely out of his experimental column (Therin 1998:71), all water passing through the 3D experiment was collected in a separate petrie dish each day. Daily 20μl slides were made fr om the basal waste water, and additional slides were made at regular intervals of the water passing through the side arms, although the volume of water in these cases was less than that passing downwards through the basal arm. When not being irrigated, the top of the apparatus was covered with cling film to prevent evaporation, and the whole experiment kept in a cool area away from direct sunlight.

Sampling

Following the end of irrigation, the experiment was laid flat and the perspex removed. Sediment samples were taken at 2cm intervals along each arm of the cross, starting from the initial central starch addition point, using a section of plastic pipette tube with a 6mm inner diameter to 'core' into the sand matrix. The tube was cleaned thoroughly with ethanol between samples, and clean non-starched gloves were worn at all times. Twenty-five samples were collected from along the main cruciform axes, and an additional four samples were collected at diagonals to the axes for a total of 29 samples (Figure 4). Collected samples were transferred to 10ml conical-base plastic tubes with screw-top caps.

Starch extraction

Starch extraction involved first deflocculation then heavy liquid separation steps, based in part on Horrocks (2005). The procedure is summarised in Table 1. The use of sterile sand as a matrix obviated the use of more time-consuming

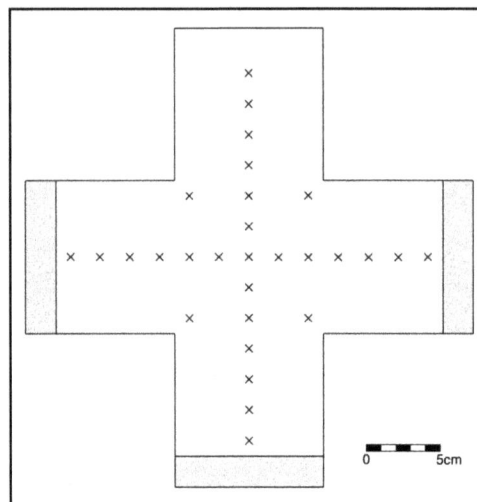

Figure 4. Sediment sample points.

extraction protocols designed to remove organics and other unwanted soil components. Important differences between this protocol and that employed by Therin (1994) include the use of non-toxic sodium polytungstate as opposed to caesium chloride for density separation, and lower centrifuge speeds owing to the equipment available at the University of Queensland Department of Earth Sciences sample preparation laboratory.

1. Place 2g sediment in a 10ml conical-base plastic sample tube.
2. Add 6ml 5% sodium hexametaphosphate solution to each 10ml tube for deflocculation. Leave samples overnight, shake vigorously occasionally to assist sediment breakup.
3. Pour sample into 50ml beaker, swirl and decant back into 10ml tube. Add 3ml water to beaker and repeat. Make tube volume up to 10ml using water. Dispose of settled fraction in beaker.
4. Centrifuge 10ml tube at 3000rpm for 3 min, decant and dispose of supernatant.
5. Add water to 10ml in tube, vortex to disaggregate pellet, centrifuge at 3000rpm for 3 min, then decant. Repeat until supernatant is clear.
6. Add 3ml sodium polytungstate solution (1.7 specific gravity), vortex briefly, then centrifuge at 1500rpm for 3 min. Pipette separated top layer into a second tube, add another 1ml sodium polytungstate to the first tube, and repeat centrifugation and removal of separated layer.
7. Add water to 10ml in second tube. Vortex, then centrifuge at 3000rpm for 6 min and decant supernatant (NB: keep the supernatant from steps 6 and 7 for sodium polytungstate recovery).
8. Add water to 10ml, vortex briefly, centrifuge at 3000rpm for 3 min. Discard supernatant, leaving ca. 100μl water with starch pellet in tube.
9. Stir sample, transfer 20μl to microscope slide.

Table 1. Summary of starch extraction procedure.

Analysis

Analysis was conducted using an Olympus BX50 light microscope fitted with polarising filters, rotating stage and 100–1000x magnification. All slides made from the waste water collected each day and the starch extractions were initially scanned at 100x in cross-polarised light. Examination of possible starch granules and size/shape recording were carried out at 400x. Every starch granule on each slide was recorded using a transect approach to ensure complete slide coverage.

Results

Porosity calculations

That the sand matrix permitted the flow of water attests to its permeability, although this attribute was not further quantified. For porosity calculations, an approximation can be obtained by taking the lower limit of sieved particle size (excluding minor attached smaller particles) of 250µm, and using the equations:

1. Porosity = 100 x (1-[bulk density/particle density])
2. Average pore space = [porosity(%)/soil particle volume(%)] x average particle diameter.

Bulk density in (1) is a measure of the weight of the sediment per unit volume (i.e. g/cm3). In the present experiment, the bulk density may be calculated from the dry weight of the sand added to the apparatus (2957g) and the volume taken up by that sand (calculated at 1688cm3), leaving a bulk density of 1.7518g/cm3. Particle density in this instance can be taken as 2.65, the specific gravity of quartz. Using equation (1), this produces a porosity of just under 34%, which is typical for packed sand with little grain-size variation (Tolman 1937:115). The average pore space for 250µm sand can then be calculated as 128.16µm using equation (2), given that the percentages of porosity and soil particle volume must add to 100%. Clearly this allows for movement of particles above the size range of the corn starch in the experiment, although the presence of some particles below 250µm which were not removed during sieving would reduce the available pore space and permeability of the matrix somewhat.

Starch waste water slides

A total of 38 slides were collected from the waste water passing through the experiment. None of these contained starch granules, of any size. The waste water did contain sedimentary particles up to approximately 5µm diameter, however, indicating that any starch granules of this size or smaller which had migrated through the sand matrix would have been able to exit the experiment. The complete lack of starch suggests that either small starch granules were not able to move beyond 13cm (the distance from the starch addition point to the water exit point) under the sedimentary and watering parameters of this experiment, or they did so in undetectably low numbers.

Sediment extractions

Of the 29 samples removed from the experiment, 27 were run through the extraction procedure outlined in Table 1, along with the control autoclaved sand sample. The two samples not analysed were those closest to the point of water addition at the top of the apparatus. Starch was recovered from nine of the 27 samples, while no starch was recovered from the control sample. Figure 5 shows the frequency of starch granules recovered from these nine locations. The highest number of starch granules (n=237) was, as expected, recovered from the sediment sample at the starch addition point at the centre of the experiment. The next highest frequency (n=44) was 2cm below the starch addition point, and the third highest frequency (n=6) 2cm to one side of the addition point, although the latter number is obviously quite low. Only very low numbers of starch granules were recovered from any other location, although three granules were located at a lateral distance of 12cm, indicating the potential for this kind of movement under the influence of groundwater. One starch granule was recovered from the sediment 2cm above the starch addition point.

The average size of the starch granules recovered from the starch addition point is 13.34 µm, while those 2cm below is 11.82µm. All starch granules recovered from the other seven locations were >10µm in maximum dimension, and the three starch granules displaced laterally by 12cm ranged from 11–26µm. No starch granules below 6µm in size were recovere d from any of the extractions, including the sample at the starch addition point, indicating that these granules were lost during the extraction process. While this is regrettable, archaeological starch analyses frequently focus on starches over 5µm as the storage starches important in human diets are often (though not always) greater than this size (e.g. Torrence et al. 2004). The results of this initial 3D test therefore

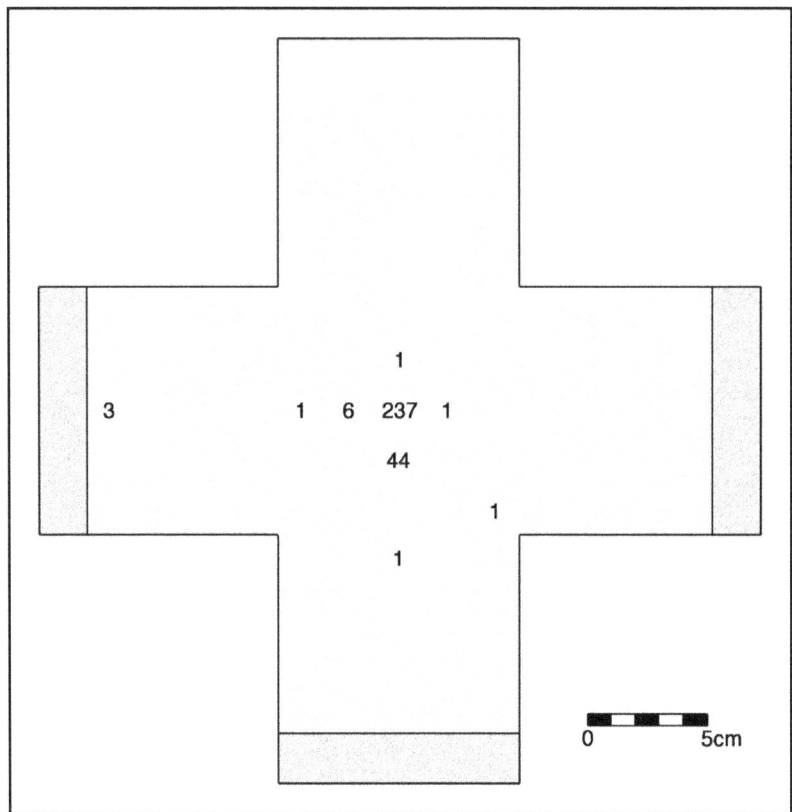

Figure 5. Recovered starch frequencies.

carry useful information for ancient starch research, and for assessment of the movement of small starch granules the reader is directed to Therin (1998). The loss of small granules during the extraction process does not affect the monitoring of the experiment waste water, as this was applied directly to each slide without further processing.

Discussion

Two constructive conclusions can be drawn from this experiment. First, the findings of this study are in concordance with those of Therin (1994, 1998) in that only a limited percentage of starch moved any distance within the sediment matrix. Second, the study has shown that lateral and upward movement of starch granules does occur under the influence of groundwater, albeit to a lesser extent than for granules moving with the direction of main water flow.

To draw out these conclusions, Therin's (1998:66) study found that in his experimental column with sand particle sizes and rainfall simulation closest to those of the current study (designated column 1 by Therin), approximately 0.03% of starch moved downward when compared to starch frequency at the addition point. The percentage in the current experiment is much higher, at almost 16%. To some extent, the difference may be due to the lack of starch granules below 6µm in size counted in the current study, although Therin's starch size breakdown for section B of column 1 (Therin 1998:Figure 5), which is 2cm below his starch addition point and therefore comparable to the second highest starch frequency in the 3D study, shows that over 90% of the starch granules in this section were >5µm. The difference therefore more likely results from the greater porosity afforded by the larger minimum sediment granule size in the current study, and emphasises the necessity of assessing sediment compaction and particle size when interpreting the stability or movement (i.e. provenance) of starches in archaeological sediments. Similarly, it is worth repeating Therin's (1998:71) recognition that finer sediments such as clays and silts will restrict percolating starch movement (while remaining susceptible to cracking and other turbation, as noted below).

Even though no starch <6μm was recovered following the 3D experiment extraction procedure, no starch of any size was recovered from the waste water exiting the base or arms of the experiment, showing that very few, and possibly no, small starch granules travelled the necessary 13cm. While the 3D experiment was set up to permit starch movement if possible (although not excessively), the results do show that higher numbers of larger starch granules may move than suggested by Therin's initial tests. The movement of even three granules greater than 10μm in size over a distance of 12cm in 30 days (a rate of 4mm/day) is certainly intriguing, and shows that while mobility may be low in relative terms, it is very real and a factor to consider in interpreting low-frequency archaeological starch data.

It is important to stress that this experiment ran for only 30 days, and the results represent only the beginning of starch movement activity for this sample. There is no reason at this stage to suspect that starch granules will remain stationary following initial movement under the influence of water, and with low recovery rates typical for ancient starch from sediments or artefacts (e.g. Parr and Carter 2003; Perry 2004; see also Haslam 2006 Appendix 2), the possibility that the original starch source was in fact ten or more centimetres away can become significant, particularly in sites where a vertical distance of that magnitude equates to a period of hundreds or thousands of years. It might be tempting to say that the current results imply that most archaeological starch recovered from sediments was collected in its primary context, however consideration of both these and Therin's (1998) results show that this is often likely to be an unwarranted assumption which blurs the record of activity at a site. While artefact residue studies may rely on use-wear corroboration to determine if starches are associated with use (and calculations of artefact movement must of course then still be made), studies of starches directly extracted from archaeological sediments may have to turn more often to starch assemblage composition studies if claims that recovered starch was in its primary context are to be supported (see also palynological discussion below). Other taphonomic issues will however come into play at that point, including differential decay of starches from different species and of different sizes (Haslam 2004).

The second main conclusion and perhaps the most important contribution of this study is that lateral and upward movement of starch granules does occur, although it is clear that this is less significant than downward movement (or more strictly, movement with the main direction of water flow). The implications of this finding for ancient starch research centre mainly on the study of activity areas, although upward movement raises the same issues of incorrect dating or strata association as the downwards movement discussed previously. Two published spatial analyses use starch to identify ancient activity areas, conducted at Petzkes Cave in Australia (Balme and Beck 2002) and at Tor Faraj in Jordan (Rosen 2003; see also Henry et al. 2004). The Australian study identified clear areas of starch concentration in the top 3cm of sediment at Petzkes Cave, with all dated hearths from the same layer dating to the last few hundred years (Balme and Beck 2002:163). The authors did consider moisture at the site (based on wet/dry observations following rain), but this did not correlate with the starch pattern. The relatively young age of the site may have played a part in maintaining distinct starch-rich zones, although the portions of the site with the highest starch loadings do possess a halo of decreasing starch content (Balme and Beck 2002: Figure 1) which may have at least partially been influenced by lateral starch movement.

Rosen (2003) sampled a Middle Paleolithic living floor at Tor Faraj, with 0–177 starch granules recovered from each of 30 analysed samples. Four of the samples have >100 granules each, while the remainder have <60. The three highest recovery sites are recognised as 'concentrations' and translated as spatially discrete areas (Henry et al. 2004:24). Again, these starch-rich clusters possess what may be viewed as a lateral halo of decreasing starch content. Henry et al. (2004:21) report that 'the deposit is in primary context with only minor post-depositional disturbance from low energy sheetwash', although the effect of this lateral water movement is not discussed in relation to the recovered starches. Interestingly, the sediment matrix of the Tor Faraj living floor is comprised of aeolian sand (Henry et al. 2004:20), providing a potentially useful parallel to the current study. As these two studies demonstrate, if ancient starch research continues at its current rate it is likely that increasing numbers of spatial/activity area studies will be conducted that will need to consider the presence and rate of lateral starch movement.

Movement is not confined to that initiated by groundwater, of course, and processes such as physical turbation of sediments, cracking clays and human activity will likely play a part in any starch movement. Whether or not the conditions within a rockshelter or cave differ in the effect of starch movement when compared with more open site remains to be seen, although the lesser exposure to rainfall in a shelter could be expected to lead to a reduction in water-induced movement. Sediment starch studies from shelter sites would then need to pay close attention to soil micromorphological evidence where this was available (e.g. McConnell and Magee 1993). As an example, a recent study revealed micromorphological evidence of episodic wetting/drying, water movement and bioturbation molds around 4mm in size at Niah Cave, Sarawak (Stephens et al. 2005). A report of the starch analysis of sediments from the site (Barton 2005), however, makes no mention of the possible effects of these (or any other) processes on post-depositional starch movement. Development of closer ties between starch researchers and other soil analysts is one means by which explanations of any presumed mobility or immobility of starch within a sediment profile may be strengthened.

As a final point of comparison, the notion of starches moving through archaeological sediments may be considered in light of previous research into pollen grain movement. Pollen movement through sediments under the influence of water (or percolation) is well studied (Dimbleby 1985; Horrocks and D'Costa 2003; Kelso 1993, 1994a, 1994b; Kelso et al. 1995; Kelso et al. 2000), and relevant to the present study thanks to overlap in grain sizes with starch. Dimbleby (1985) produced detailed theoretical models of pollen movement and preservation at archaeological sites, and his conclusions broadly suggest that the lower one looks in a sediment profile, the higher the percentage of older pollen, and the higher the percentage of degraded pollen grains, one should find. Samples taken from higher up in a profile should contain mixtures of washed-down pollen of various ages, trending towards a high percentage and frequency of recently-deposited pollen grains at the surface. In addition, because of the interplay between percolation and degradation (including differential decomposition), the absolute numbers of pollen grains recovered from a given sample will show a more complex pattern than that calculated via percentages (Dimbleby 1985:5). Subsequent empirical research supports Dimbleby's results (e.g. Horrocks and D'Costa 2003; Kelso 1994a, 1994b; Russell 1993), and emphasises the necessity of not just recording the presence of particular taxa within a given sample, but evaluating the entire microfossil assemblage to accurately identify the influence of movement and degradation. Such results have obvious implications for the conduct of sediment-starch analysis. From archaeological data, calculations of the average rate of downward pollen percolation fall within a range of approximately 1cm per 1–30 years (Dimbleby 1985; Horn et al. 1998; Horrocks and D'Costa 2003; Kelso 1994a, 1994b; Kelso et al. 1995), with generally faster rates in tropical and non-forested areas.

Significantly, Dimbleby (1985:2) and others suggest that much of the pollen present in sediments may be trapped within soil aggregates and is not free to move with groundwater. A similar situation has been suggested as a mechanism of starch survival, including the possible triggering of aggregate formation by the starch itself (Haslam 2004:1725). In such cases, it is the breakdown of these aggregates, or their wholesale movement, which is necessary to transport microfossils within the profile (Russell 1993). Both scenarios are frequently accomplished through the agency of macro-fauna. Both pollen and phytolith (e.g. Hart 2003) studies have noted the influence of earthworms, ants and termites in translocating either individual microfossils or soil aggregates. Earthworms in particular may homogenise a considerable volume of soil (Armour-Chelu and Andrews 1994; Canti 2003), although pH and vegetation characteristics will influence the number and species of worms present in a given setting. In a multi-analytical approach, Grave and Kealhofer (1999) used phytoliths to demonstrate that even where there is 'strong evidence of insect activity and percolation' in a Thailand site, microfossils may avoid much of the homogenisation because of their small size (<50μm). Such investigations have yet to be replicated ated by starch analysts, however they indicate that the potential complexity of starch movement should not be underestimated for any site. Starch-specific sediment deposition patterns not seen for other microfossils (e.g. in situ underground decay of starchy organs; see Zarillo and Kooyman 2006:485) also require further study to properly adapt pollen and phytolith results to ancient starch research.

Conclusion

All available evidence suggests that starch granules do move through sediments under the influence of groundwater, just as pollen grains, phytoliths, and in fact any sedimentary particle with sufficient space and the appropriate influences will do. The initial testing of three-dimensional movement of starch has shown that upward and particularly lateral movement occurs, although to a lesser degree than downward motion. This movement is not confined to starch granules below 5μm in size (those often considered by archaeologists not to be representative of human economic activity), and starch granules well over 10μm have been shown to move at a rate of up to 4mm/day laterally. The volume of water added to the experiment over a period of one month suggests two conflicting caveats when interpreting these results — firstly that rainfall rates in many parts of the world will not be contributing as much water to the soil per day as in this experiment (although certain events, for example floods, sheetwash or irrigation certainly would), and second that the much greater time scale of archaeological sediments is expected to permit much greater movement of mobile starch granules than seen in the current experiment. Future studies will need to vary sediments, water input and time periods according to local conditions of interest. Continued research into the taphonomy of starch granules remains an important component of archaeological starch research, and movement studies employing varying environmental conditions and starch types will form an integral component of such research.

Acknowledgements

This experiment benefited from discussions with Tom Loy, who also helped build the apparatus and donated his garage and toolkit to that end. I thank Sandra Shearer, Peter Colls, the Department of Earth Sciences, University of Queensland, and Alison Crowther for assistance with various aspects of the project, and those present at the session organised by Judith Field at the Australasian Archaeometry Conference in Canberra, December 2005, where a partial version of this paper was presented, for their ideas. I also thank the anonymous reviewer. The Department of Archaeology and Natural History at the Australian National University assisted with funding to attend the conference. Ian Lilley commented on an earlier draft of this paper.

References

Armour-Chelu, M. and P. Andrews. 1994. Some effects of bioturbation by earthworms (Oligochaeta) on archaeological sites. *Journal of Archaeological Science* 21:433–443.

Babot, M. d. P. 2003. Starch grain damage as an indicator of food processing. In D. M. Hart and L. A. Wallis (eds), *Phytolith and starch research in the Australian-Pacific-Asian regions: The state of the art*, pp 69–81. Canberra: Pandanus Books.

Balme, J. and W. Beck. 2002. Starch and charcoal: Useful measures of activity areas in archaeological rockshelters. *Journal of Archaeological Science* 29:157–166.

Barton, H. 2005. The case for rainforest foragers: The starch record at Niah Cave, Sarawak. *Asian Perspectives* 44(1):56–72.

Barton, H., R. Torrence and R. Fullagar. 1998. Clues to stone tool function re-examined: Comparing starch grain frequencies on used and unused obsidian artefacts. *Journal of Archaeological Science* 25:1231–1238.

Canti, M. G. 2003. Earthworm activity and archaeological stratigraphy: A review of products and processes. *Journal of Archaeological Science* 30:135–148.

Dimbleby. G. W. 1985. *The palynology of archaeological sites. Studies in archaeological science*. New York: Academic Press.

Fullagar, R., T. Loy and S. Cox. 1994. Starch grains in archaeological sediments: Implications for vegetation history and residue studies. Paper presented at the 15th Congress of the Indo-Pacific Prehistory Association, Chiang Mai, Thailand.

Fullagar, R., T. Loy and S. Cox. 1998. Starch grains, sediments and stone tool function: Evidence from Bitokara, Papua New Guinea. In R. Fullagar (ed.), *A closer look: Recent Australian studies of stone tools*, pp 49–60. Sydney: Archaeological Computing Laboratory, University of Sydney.

Grave, P. and L. Kealhofer. 1999. Assessing bioturbation in archaeological sediments using soil morphology and phytolith analysis. *Journal of Archaeological Science* 26:1239–1248.

Hart, D. M. 2003. The influence of soil fauna on phytolith distribution in an Australian soil. In D. M. Hart and L. A. Wallis (eds), *Phytolith and starch research in the Australian-Pacific-Asian regions: The state of the art*, pp 83–91. Canberra: Pandanus Books.

Haslam, M. 2004. The decomposition of starch grains in soils: Implications for archaeological residue analyses. *Journal of Archaeological Science* 31:1715–1734.

Haslam, M. 2006. *Archaeological residue and starch analysis: Interpretation and taphonomy*. Unpublished PhD thesis, School of Social Science, The University of Queensland.

Henry, D. O., H. J. Hietala, A. M. Rosen, Y. E. Demidenko, V. I. Usik, and T. L. Armagan. 2004. Human behavioural organization in the Middle Paleolithic: Were Neanderthals different? *American Anthropologist* 106 (1):17–31.

Horn, S., K. H. Orvis, J. C. Rodgers and L. A. Northrop. 1998. Recent land use and vegetation history from soil pollen analysis: Testing the potential in the lowland humid tropics. *Palynology* 22:167–180.

Horrocks, M. 2005. A combined procedure for recovering phytoliths and starch residues from soils, sedimentary deposits and similar materials. *Journal of Archaeological Science* 32:1169–1175.

Horrocks, M. and D. M. D'Costa. 2003. Stratigraphic palynology in porous soils in humid climates: An example from Pouerua, northern New Zealand. *Palynology* 27:27–37.

Kelso, G. K. 1993. Pollen-record formation processes, interdisciplinary archaeology, and land use by mill workers and managers: The Boott Mills Corporation, Lowell, Massachusetts, 1836–1942. *Historical Archaeology* 27(1):70–94.

Kelso, G. K. 1994a. Pollen percolation rates in Euroamerican-era cultural deposits in the northeastern United States. *Journal of Archaeological Science* 21:481–488.

Kelso, G. K. 1994b. Palynology in historical rural-landscape studies: Great Meadows, Pennsylvania. *American Antiquity* 59(2):359–372.

Kelso, G. K., S. A. Mrozowski, D. Currie, A. C. Edwards, M. R. Brown III, A. J. Horning, G. J. Brown and J. R. Dandoy. 1995. Differential pollen preservation in a seventeenth-century refuse pit, Jamestown Island, Virginia. *Historical Archaeology* 29(2):43–54.

Kelso, G. K., D. Ritchie and N. Misso. 2000. Pollen record preservation processes in the Salem Neck Sewage Plant Shell Midden (19-ES-471), Salem, Massachusetts, U.S.A. *Journal of Archaeological Science* 27:235–240.

Lentfer, C., M. Therin and R. Torrence. 2002. Starch grains and environmental reconstruction: A modern test case from West New Britain, Papua New Guinea. *Journal of Archaeological Science* 29:687–698.

McConnell, A. and J. W. Magee. 1993. The contribution of microscopic analysis of archaeological sediments to the reconstruction of the human past in Australasia. In B. L. Fankhauser and J. R. Bird (eds), *Archaeometry: Current Australasian research*, pp 131–140. Canberra: Department of Prehistory, Research School of Pacific Studies, The Australian National University.

Parr, J. F. and M. Carter. 2003. Phytolith and starch analysis of sediment samples from two archaeological sites on Dauar Island, Torres Strait, northeastern Australia. *Vegetation History and Archaeobotany* 12:131–141.

Perry, L. 2004. Starch analyses reveal the relationship between tool type and function: An example from the Orinoco valley of Venezuela. *Journal of Archaeological Science* 31:1069–1081.

Rosen, A.M. 2003. Middle Paleolithic plant exploitation: The microbotanical evidence. In D. O. Henry (ed.) *Neanderthals in the Levant: Behavioral organization and the beginnings of human modernity*, pp 156–171. London: Continuum.

Russell, E. W. B. 1993. Early stages of secondary succession recorded in soil pollen on the North Carolina piedmont. *American Midland Naturalist* 129(2):384–396.

Stephens, M., J. Rose, D. Gilbertson and M. G. Canti. 2005. Micromorphology of cave sediments in the humid tropics: Niah Cave, Sarawak. *Asian Perspectives* 44(1):42–55.

Therin, M. 1994. *Subsistence through Starch: The examination of subsistence changes on Garua Island, West New Britain, Papua New Guinea, through the extraction and identification of starch in sediments*. Unpublished BA (Hons) thesis. Sydney: Department of Prehistoric and Historical Archaeology, University of Sydney.

Therin, M. 1998. The movement of starch grains in sediments. In R. Fullagar (ed.), *A closer look: Recent Australian studies of stone tools*, pp 61–72. Sydney: Archaeological Computing Laboratory, University of Sydney.

Therin, M., R. Fullagar and R. Torrence. 1999. Starch in sediments: A new approach to the study of subsistence and land use in Papua New Guinea. In C. Gosden and J. Hather (eds), *The Prehistory of Food: Appetites for change*, pp 438–462. London: Routledge.

Tolman, C. F. 1937. *Ground Water*. 1st ed. New York: McGraw Hill.

Torrence, R., R. Wright and R. Conway. 2004. Identification of starch granules using image analysis and multivariate techniques. *Journal of Archaeological Science* 31:519–532.

Zarillo, S. and B. Kooyman. 2006. Evidence for berry and maize processing on the Canadian plains from starch grain analysis. *American Antiquity* 71(3):473-500.

8

Re-viewing raphides: Issues with the identification and interpretation of calcium oxalate crystals in microfossil assemblages

Alison Crowther

School of Social Science
The University of Queensland
St Lucia Qld 4072, Australia
alison.crowther@gmail.com

Abstract

Needle-shaped calcium oxalate crystals called raphides are increasingly being identified in microfossil assemblages from archaeological sites, including in sediments and as residues on artefacts. This paper reviews issues associated with the identification and interpretation of raphides in the archaeological record of the Pacific. It is argued that, with a better understanding of the occurrence, distribution, and morphological variability of raphides across higher plant families, more accurate and meaningful interpretations can be made of these archaeological microfossils. Problems with differentiating raphides from naturally occurring calcite crystals of similar morphology and optical properties are briefly raised and illustrated using the analysis of microscopic residues on Lapita pottery from sites in Melanesia.

Keywords: Calcium oxalate crystals, raphides, microfossils, residue analysis, Araceae, Pacific

Introduction

The analysis of plant microfossils such as starch, phytoliths and pollen is gaining momentum in Pacific archaeology. Such studies have made important contributions to palaeoenvironmental reconstructions and our understanding of past human-plant interactions. This paper focuses on the analysis of calcium oxalate crystals, which have been reported in a number of archaeological microfossil assemblages worldwide, including as residues on stone tools (e.g. Bernard-Shaw 1984 cited in Bernard-Shaw 1990:193; Fullagar 1993b; Fullagar et al. 2006; Hardy 2004; Hardy et al. 2001; Loy et al. 1992; Shafer and Holloway 1979; Sobolik 1994) and pottery surfaces (Crowther 2005; Horrocks and Bedford 2005), in coprolites (Danielson and Reinhard 1998; Jones and Bryant 1992; Reinhard and Danielson 2005), and in sediments from hearths and roasting pits (Bernard-Shaw 1990; Koumouzelis et al. 2001), burials (Lalueza Fox et al.1996) and wetland ditch systems (Horrocks and Barber 2005). These crystals, which are sometimes referred to in the literature as calcium oxalate phytoliths, are the most commonly formed biomineral in higher plants (Webb et al. 1995).

Calcium oxalate crystals are generally considered to be of lesser archaeological importance than other microfossils owing to their more restricted range of forms, which are rarely diagnostic to family (Lalueza Fox et al. 1996:106; cf. Jones and Bryant 1992). In the Pacific region, however, needle-shaped calcium oxalate crystals known as raphides have been drawn upon as an important indicator of aroids (a group that includes the Pacific food staples *Colocasia esculenta* [taro], *Alocasia macrorrhiza*, and *Cyrtosperma chamissonis* [giant swamp taro]) in microfossil assemblages. The aroids are a subgroup of the Araceae, which is a family of mainly herbaceous plants that are known to produce raphides in abundance. Raphides have long been associated with the acridity of the aroids, meaning that when eaten raw they cause swelling of the lips, mouth and throat. It is now understood, however, that raphides only act as a conduit for the injection of an associated chemical irritant into the skin. Detoxification via cooking, pounding or leaching neutralises the chemical, hence making the aroids edible, but does not destroy or degrade the raphides (Bradbury and Nixon 1998; Johns and Kubo 1988; Paull et al. 1999; Sakai 1979). In the Pacific region, raphides alongside other microfossil types have been drawn on as evidence for Pleistocene processing of starchy plants (Loy et al. 1992), the development of agriculture (Denham et al. 2003; Fullagar et al. 2006), the timing of plant introductions, and the integration of horticulture into prehistoric economies (Horrocks and Bedford 2005).

Given the increasing incorporation of raphides into microfossil studies, this paper aims to review issues associated with their identification and interpretation. First, it examines how microfossil researchers in the past have interpreted raphides in archaeological assemblages from the Pacific, and how these interpretations might be improved with a better understanding of both the widespread occurrence of raphides in higher plant families, and higher resolution analyses of raphides using scanning electron microscopy (SEM). Second, it examines identification problems with raphides, in particular the possibility that naturally occurring calcite crystals of similar morphologies and optical properties might be misidentified as raphides in microfossil assemblages. Both these issues are non-trivial considering the importance of raphide-bearing plants in the Pacific economy, and the growing use of these microfossils to identify such plants in the archaeological record.

Occurrence of calcium oxalate crystals in higher plants

Calcium oxalate crystals occur in more than 215 higher plant families, as well as the algae, lichen and fungi, in the form of whewellite ($CaC_2O_4.H_2O$) or weddelite ($CaC_2O_4.2H_2O$) (Bouropoulos et al. 2001; Horner and Wagner 1995:54). They can form in any organ or tissue within plants, including in stems, leaves, roots, tubers, and seeds, and have a variety of functions including calcium storage, defence and providing structural strength (Franceschi and Horner 1980; Franceschi and Nakata 2005; Nakata 2003). Unlike phytoliths, which vary considerably in size and shape across families, calcium oxalate crystals are generally restricted to five basic morphological types: needle-shaped raphides, rectangular or pencil-shaped styolids, mace-head-shaped aggregates called druses, block-shaped aggregates called crystal sand, and variously shaped prisms (Horner and Wagner 1995) (Figure 1). Given that the size and appearance of these crystals can differ within and between families, and that these morphological characteristics are probably under genetic control (e.g. a plant species always produces the same form of crystal in the same tissue [Bouropoulos et al. 2001]), calcium oxalate crystals may have taxonomic potential for both botanists and archaeologists (Horner and Wagner 1995).

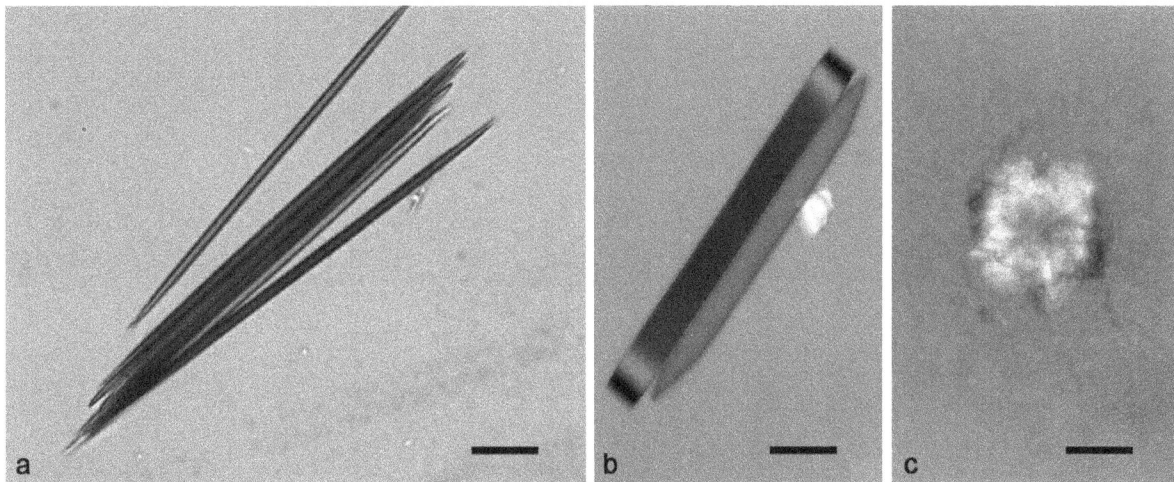

Figure 1. Examples of calcium oxalate crystal forms most commonly found in monocotyledons:

(a) needle-shape raphides, (*Pandanus* sp. leaf; brightfield transmitted light, scale bar equals 10 μm);
(b) rectangular and pencil-shaped (pointed) styloids (*Cordyline* sp. leaf; brightfield transmitted light, scale bar equals 20 μm); and
(c) druse (*Colocasia esculenta* corm; brightfield transmitted light, scale bar equals 20 μm).

Raphides most commonly occur in bundles of tens to hundreds of crystals in specialised cells called idioblasts (Figure 2). Although less common, they are also known to form in extracellular bundles (e.g. Barabé et al. 2004) and there has been one report of raphides forming within starch granules (Okoli and Green 1987). Idioblast cells are structurally modified to accommodate crystals and are therefore markedly different in form, structure, and contents from other cells in the same tissue (Gallaher 1975:320; Sunell and Healey 1981:235). A variety of idioblast cell forms have been identified (see Keating 2004), which can be grouped broadly into those having defensive or non-defensive functions (Sunell and Healey 1985). The latter form crystals in relatively loose arrangement compared to defensive cells, which are usually smaller in size and suspended in the intercellular airspaces from where they are easily dislodged (Sunell and Healey 1985). When a defensive idioblast is disrupted, the tightly packed raphides are ejected through thin-walled papillae at its ends (Sunell and Healey 1981:236), a feature likened to an 'automatic microscopic blowgun' (Middendorf 1983:9) (Figure 3). This mechanism probably developed as a defence against herbivores.

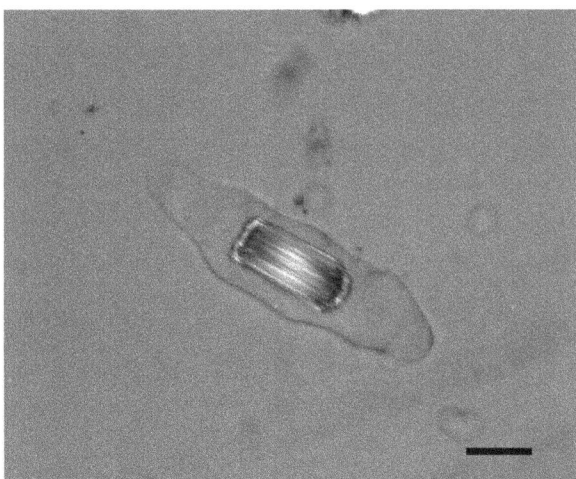

Figure 2. Bundle of   sp. (tuber) raphides in idioblast cell (part-polarised brightfield transmitted light, scale bar equals 100 μm).

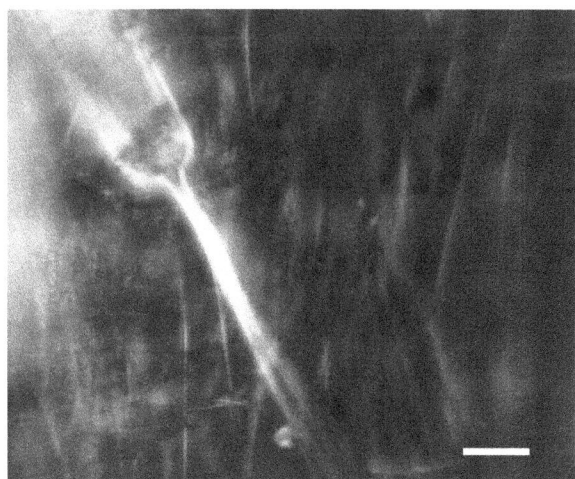

Figure 3. Spindle-shaped defensive idioblast cell from *Colocasia esculenta* corm. Raphides are partially ejected through the papilla at its end (cross-polarised brightfield transmitted light, scale bar equals 20 μm).

Raphides and other crystals of calcium oxalate are relatively stable minerals. They are only slightly water soluble (Graustein et al. 1977:199; Webb 1999:752), and are insoluble in alkalis (i.e. bases; high pH conditions) and organic acids (e.g. carboxylic acids such as acetic acid) (Gallaher 1975:315). They are heat resistant and can therefore survive cooking. As Monje and Baran (2002:711) summarise, 'the oxalates have the particular advantage to be very resistant water-insoluble plant products and, thus, they could be found where other plant residues are no longer evident'. Raphides are not entirely indestructible, however, as they are susceptible to damage from mechanical action as well as dissolution in acidic environments (Mulholland and Rapp 1992:4).

Background to the identification and interpretation of raphides in the Pacific archaeological record

The presence and potential importance of raphides in the archaeological record of the Pacific region was first highlighted by Loy, Spriggs and Wickler (1992; see also Loy 1994), who found these crystals in association with starch residues on the edges of stone tools as old as 28,000 years from the Solomon Islands. Following analysis of limited modern reference material from aroids, yams, and other economic species from the region, Loy and his colleagues made two key findings regarding raphide identification and interpretation. First, the abundance of raphides in aroids and near absence from other plants analysed in their study could broadly support the identification of starch residues from aroids when the two microfossil types are associated in use-residues. Notably, Loy et al. (1992) relied on starch granule size and shape as a more significant taxonomic indicator than the presence of raphides. Second, the size and morphology of raphides from the aroids was sufficiently varied to potentially differentiate them at species level, although these differences were only briefly described (Loy et al. 1992:900, 905–906). Loy et al.'s (1992) study has formed the basis of raphide identifications in Pacific microfossil studies, but their findings have been subject to little further review.

Raphides are known to occur in at least 49 monocotyledon and 27 dicotyledon families worldwide (Dahlgren and Clifford 1982:92–93; McNair 1932; Metcalfe and Chalk 1979:217; Prychid and Rudall 1999), including many that are commonly associated with archaeological sites in the Pacific region or had probable prehistoric economic importance. These include palms (Arecaceae) (Tomlinson 1971), pandanus (Pandanaceae), cordyline (Laxmanniaceae), bananas (Musaceae) (Osuji and Ndukuwu 2005; Osuji et al. 2000) and yams (Dioscoreaceae) (Ayensu 1972; see also Al-Rais et al. 1971; Okoli and Green 1987). For the last family, it is reported that raphides occur in at least 51 species worldwide (Ayensu 1972), even though Loy et al. (1992) only observed them in one of seven species (*Dioscorea esculenta*) in their study, possibly as a result of sampling error (cf. Loy et al. 1992:910). Hence, the presence of raphides in microfossil assemblages from the Pacific region could potentially represent a number of economically important taxa. Of the published studies in which raphides have been identified in microfossil assemblages from this region (see Table 1), all have identified the observed crystals to the Araceae, although Fullagar (1993a:335, 1993b:25) acknowledges the possibility that raphides on a stemmed tool from West New Britain could also be from yam. It is therefore important to examine and evaluate the reasoning underlying these identifications, and identify avenues by which they can be refined and strengthened.

The Araceae are known to produce abundant quantities of raphides in all or most organs – for example, a single *Colocasia esculenta* corm may contain up to 120000 calcium oxalate raphides and druses per cubic centimetre (Sunell and Healey 1979:1031; see also Bradbury and Nixon 1998; Genua and Hillson 1985; Keating 2004; Sakai 1979). This abundance is one of the most commonly used characteristics to distinguish Araceae raphides in microfossil studies (Crowther 2005; Horrocks and Bedford 2005; Horrocks and Weisler 2006; Loy et al. 1992), even though quantitative comparative analyses of raphide concentrations in economic plants from the Pacific have yet to be undertaken. It is suggested, for

Location	Sample type	Taxonomic identification of raphides	Reference
Kilu Cave, Buka, Solomon Islands	Stone tools	**Araceae** (*Alocasia macrorrhiza* and *Colocasia esculenta* – direct association with starch)	Loy et al. 1992
Bitokara, West New Britain, PNG	Stone tools	**Araceae** and/or **Dioscoraceae** (probably *C. esculenta* and/or *Dioscorea* sp. – direct association with starch)	Fullagar 1993b (see also Fullagar 1993a)
Anir, PNG	Sediments & potsherds	**Araceae** (probably *Colocasia esculenta* – possibly associated with starch)	Crowther 2005
New Zealand	Sediments	**Araceae** (*C. esculenta* – indirect association with starch in 2 of 6 samples)	Horrocks and Barber 2005
Uripiv, Vanuatu	Sediments & potsherds	**Araceae** (non-*Colocasia* – indirect association with starch)	Horrocks and Bedford 2005
Henderson and Marshall Islands.	Sediments	**Araceae** (non-*Colocasia* – indirect association with starch)	Horrocks and Weisler 2006
Kuk Swamp, PNG	Stone tools	**Araceae** (*Colocasia esculenta* – direct association with starch)	Fullagar et al. 2006 (see also Denham et al. 2003)

Table 1. Summary of studies that identify calcium oxalate raphides in archaeological sites in the Pacific region.

example, that raphides are also both abundant and widespread in the palms (Tomlinson 1961), suggesting that this feature is not unique to the Araceae. Quantitative analyses also need to account for the possibility that one idioblast cell alone can contain hundreds of raphides that, in a single event, could be deposited in the archaeological record in a concentrated mass. Cultural processes potentially leading to the build up of raphides in sediments, as well as taphonomic processes that might differentially affect archaeological raphide assemblages and subsequently their quantitative analysis, must also be assessed. Within our current state of knowledge, it can be argued that the quantity of raphides present in a microfossil assemblage is essentially meaningless in relation to their taxonomic identification.

More reliable raphide identifications may be possible in cases where they co-occur with other taxonomically distinctive microfossils, such as starch granules or phytoliths. The strength of such identifications, however, relies in part on their degree of association in the archaeological record. For example, it can be argued that raphides and starch granules observed in direct association on the used edge of a stone tool are likely to be related and therefore together they may enable stronger taxonomic identifications of the residue, based on a wider range of attributes (Fullagar 1993b; Fullagar et al. 2006; Loy et al. 1992). Where similar identifications have been made of raphides associated with starch in sediments, however, the relationship between the two is arguably relatively indirect given that, when analysed for pollen and phytoliths, the same samples are found to contain residues of a number of plant taxa (e.g. Horrocks and Barber 2005; Horrocks and Bedford 2005). As it cannot be assumed that the microfossils in sediment samples are taxonomically related, the evidence for each microfossil type must be independently as strong as possible.

The final line of reasoning by which analysts have taxonomically identified raphides in Pacific microfossil assemblages has been on the basis of morphometric characteristics, following Loy et al.'s (1992) descriptions of aroid raphides. These attributes, however, have yet to be evaluated based on more comprehensive reference collection analyses and have been only minimally incorporated into archaeological raphide analyses since. Fullagar et al. (2006), for example, mention the square cross-section of a raphide found in association with *Colocasia esculenta*-type starch granules on a stone tool from Kuk Swamp in the central highlands of Papua New Guinea. Other studies simply describe microfossil raphides as typical of or similar to those from the aroids without presenting detailed morphometric data of both the archaeological samples and modern reference specimens (including non-Araceae) to lend weight to their identifications (Crowther 2005; Fullagar 1993b:25; Horrocks and Barber 2005; Horrocks and Bedford 2005). It is insufficient for taxonomic purposes to note that a crystal or broken crystal segment fits within the size and shape range of a particular genus or family without discussing other possible matches. Combining this point with that concerning microfossil co-occurrence, cases where raphides alone have been used to identify taxa present in sediments, even to species level (e.g. Horrocks and Barber 2005:Figure 6), are clearly problematic. As outlined below, the taxonomic significance of raphide morphology may have greater value than realised in these studies, particularly for the Araceae, and may reduce reliance on the potentially ambiguous criteria discussed above.

Micromorphological identification of raphides using SEM

Anatomical and tissue culture studies indicate that raphide morphology and formation in specific locations within a plant are genetically controlled (Horner and Wagner 1995). Key attributes for differentiating raphides include size, cross-section and termination morphology, all of which appear to vary to differing degrees depending on taxa of origin. Fragmentation and degradation of archaeological raphides, however, limits the potential for taxonomic differentiation using some of these attributes (see Crowther 2005:64–65; Horrocks and Bedford 2005:70–71). For example, morphologically distinctive features such as terminations may not be present and size can be difficult to estimate from broken crystals. While cross-section has been noted in some studies using light microscopy (e.g. Fullagar et al. 2006:605; Loy et al. 1992), personal experience shows that unambiguous analysis of raphide sectional shape is often beyond the resolution limits of the light microscope. Scanning electron microscopic analysis of raphide cross-sectional shape and terminations (where present), however, may have greater taxonomic potential.

SEM studies of raphides from a number of higher plants have described four general morphological types, referred to here as Types I–IV (Figure 4) (Cody and Horner 1983:328).

Type I is the most common raphide form and consists of four-sided single crystals that have two symmetrical pointed ends (Figure 5).

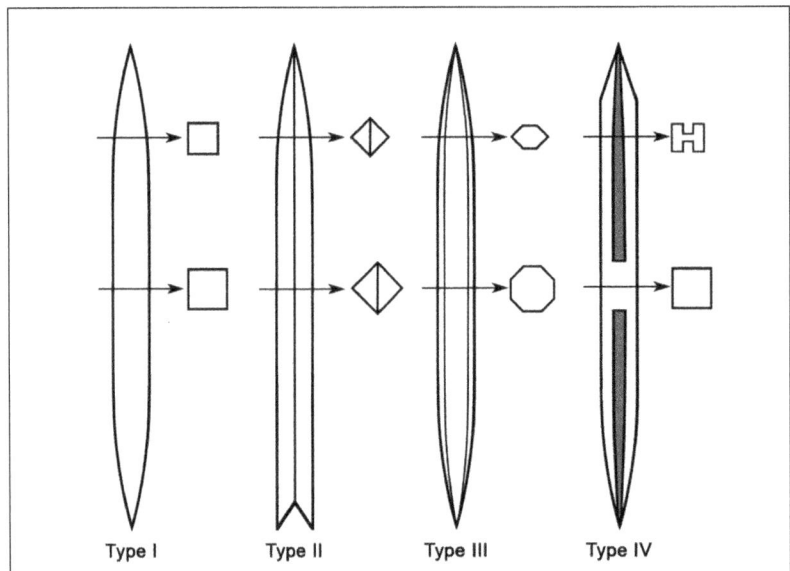

Figure 4. Diagram showing the four basic types of raphide cross-sections. End forms are described in text (redrawn from Horner and Wagner 1995: 58).

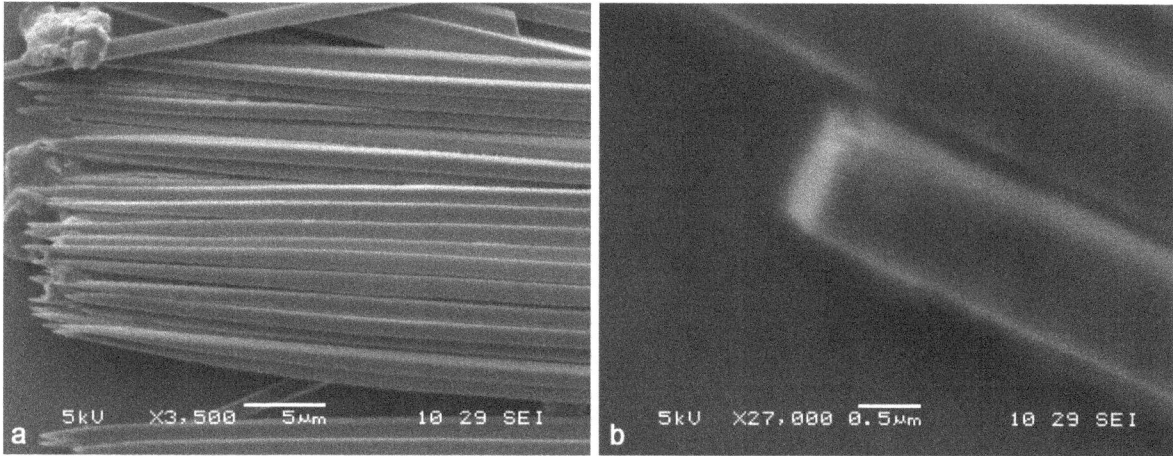

Figure 5. SEM micrographs of *Pandanus* sp. (leaf) raphides showing (a) part of bundle of four-sided crystals, and (b) broken raphide with square cross-section.

Type II raphides, which are also four-sided, have one pointed and one bidentate or forked end (Prychid and Rudall 1999:726). This type of raphide has so far only been recorded in a few families such as the Vitaceae (Figure 6) (Cody and Horner 1983; Webb 1999). The bidentate end is formed by crystal twinning (Arnott and Webb 2000:133).

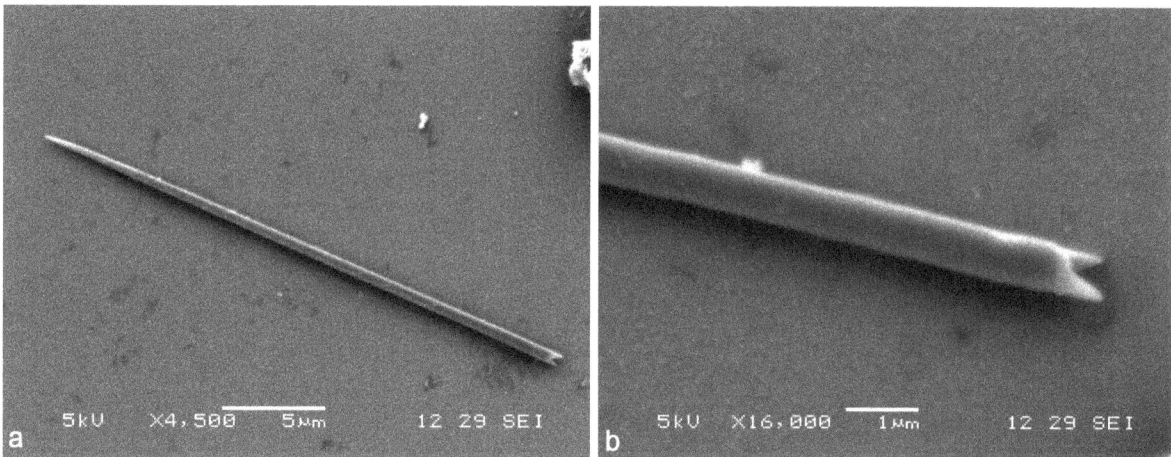

Figure 6. (a) SEM micrograph of Type II raphide from *Vitis* sp. (stem). (b) Detail of bidentate end.

The third form (Type III) are crystals with six to eight sides and symmetrical pointed ends. This raphide type is known to occur in the Agavaceae (Wattendorf 1976), Typhaceae (Horner et al. 1981), and Dioscoreaceae (Figure 7).

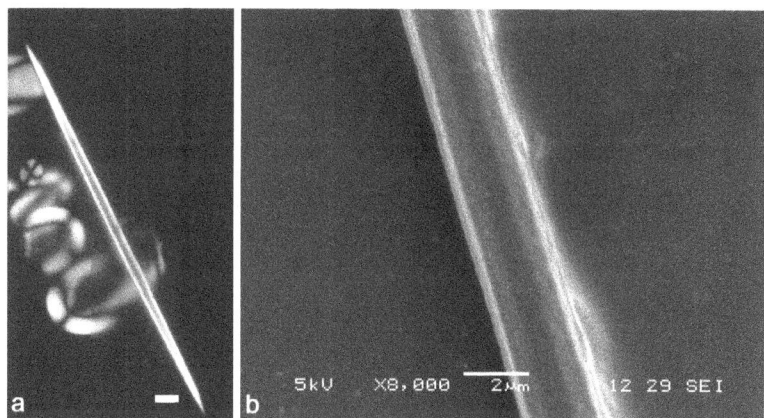

Figure 7. (a) Type III raphide from *Dioscorea alata* (tuber) (brightfield transmitted light, composite photograph, scale bar equals 10 µm) and (b) SEM micrograph of mid-section of *Dioscorea* sp. (tuber) raphide showing faceted faces.

Crystals of this morphological type are sometimes described as appearing elliptical or circular in cross-section as viewed under a light microscope (Loy et al. 1992:900; Prychid and Rudall 1999:726).

The fourth raphide form (Type IV) comprises twinned crystals with H-shaped cross-sections and asymmetrical ends (one wedge-shaped and the other sharply pointed) (Figure 8a) (Bradbury and Nixon 1998; Kostman and Franceschi 2000). A medial groove along two of the crystal's opposite faces gives this type its overall H-shaped cross-section, although the solid mid-point of the raphide is 4-sided and sometimes visible under a polarised light microscope (Figure 8b–c). Type IV raphides also develop barbs on their pointed ends as they mature (Bradbury and Nixon 1998:614; Prychid and Rudall 1999:726; Sakai et al. 1972). Significantly for microfossil researchers, the H-shaped form is thought to be unique to raphides from the Araceae and therefore offers great potential for taxonomic identification (Prychid and Rudall 1999). Barbs, on the other hand, have been noted in other taxa (Cody and Horner 1983:319).

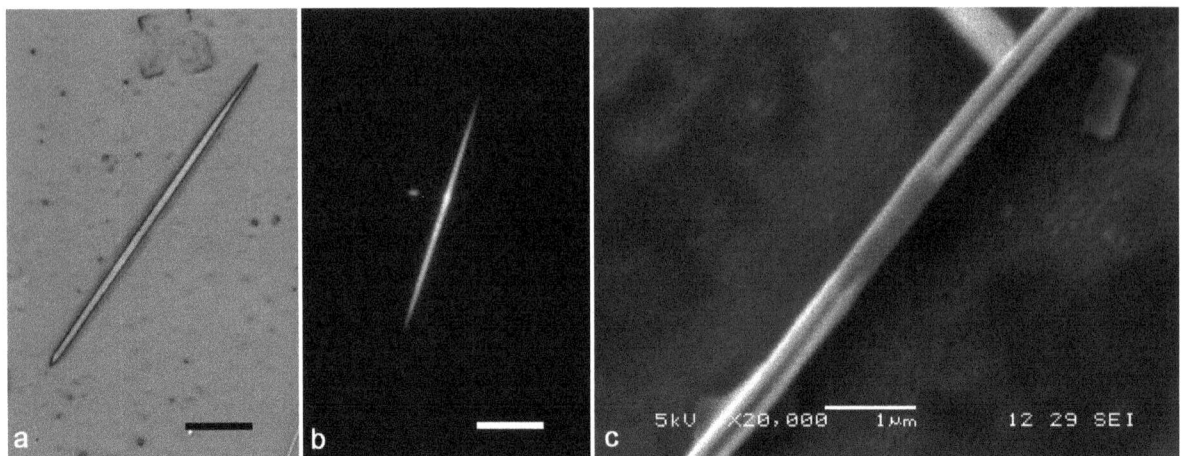

Figure 8. Type IV raphides from   (leaf) showing (a) asymmetrical end terminations (brightfield incident light, scale bar equals 10 μm), (b) thickened central region as visible under cross-polarised light (brightfield transmitted light, scale bar equals 20 μm), and (c) groove and thickened region as observed under SEM.

A final point on using morphometric characteristics to identify calcium oxalate crystals relates to size. The relationship between crystal size and taxa of origin is complicated by the fact that size can vary within a species, depending partly on intrinsic (i.e. genetic) variables, such as the function of the cell in which the crystal is formed, and partly on environmental factors including the amount of calcium available during crystal formation (see Franceschi and Nakata 2005). The degree of variation within species therefore needs to be assessed before size is accepted as a reliable indicator of taxa of origin.

Based on this review, it is likely that more reliable identifications of microfossil raphides may be achieved by incorporating higher resolution examinations of micro-morphological features into the analyses. This potential can only be fully assessed, however, through comprehensive reference collection analysis.

Needle-fibre calcite: occurrence and implications for raphide analyses

A second major issue in the analysis of raphides from Pacific microfossil assemblages relates to their differentiation from other crystals with similar morphologies and optical properties. All of the published cases that have identified raphides in the archaeological record have done so based solely on light microscopy. Using this technique, raphides have usually been identified as such by their acicular or needle-like morphology and optical properties such as pleochroism (colours that shift with varying degrees of polarisation) and birefringence (double refraction of light, which is evident by the bright appearance of the crystals in cross-polarised light) (see Figure 8b). During recent examination of Lapita potsherds from sites in Island Melanesia for surface residues, however, it was found that calcite crystals which had precipitated on the pottery can have forms and optical properties similar to calcium oxalate raphides (Figure 9). The archaeological crystals were identified as calcite, based on morphological descriptions provided by Phillips and Self (1987), Borsato et al. (2000) and Verrecchia and Verrecchia (1994), and chemical tests, as described below.

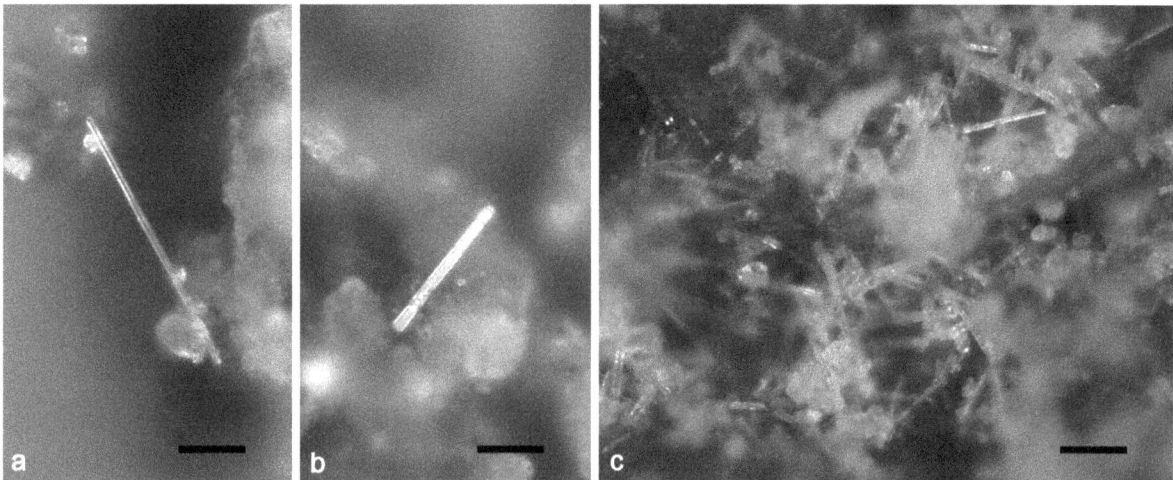

Figure 9. Examples of individual (a, b) and clustered (c) calcite crystals *in situ* on archaeological potsherds from the Mussau Islands, Papua New Guinea. Note the crystals' birefringence (bright appearance) and acicular forms, although ends are square rather than pointed. (All images: brightfield cross-polarised incident light, scale bar equals (a) 10 μm, (b) 10 μm, (c) 20 μm).

Needle-shaped calcite crystals are described in the literature as whisker crystals, acicular calcite, rock-milk, moonmilk, and lublinite, but most commonly as needle-fibre calcite. They form as smooth, single microrods that can fuse together in pairs of two, giving more complex morphologies with either H- or X-shaped cross-sections (Figure 10; cf. Type IV raphides) (Phillips and Self 1987; Verrecchia and Verrecchia 1994). Although some calcite crystals have small, tapered tips, most have blunt, square ends, while raphides generally have at least one sharp point. In highly fragmented archaeological assemblages, however, this distinction may not be as apparent (cf. Figure 11). Needle-fibre calcite can occur in bundles of sub-parallel rods, somewhat similar to the arrangement of raphides in an idioblast cell, or in randomly oriented meshes of individual crystals. Although the most common form of calcite crystals is straight-sided, some can also be curved (Philips and Self 1987:435), a trait shared with some calcium oxalate raphides (see Loy et al. [1992:900] on 'whisker' raphides). In terms of size, both raphides and calcite crystals can have a similar range, with average widths between 1–2 μm and lengths up to c. 150–200 μm, although some calcite crystals can be up to 1 mm in length (Bradbury and Nixon 1998; Phillips and Self 1987). The high degree of similarity in size, shape, optical properties and arrangement between needle-fibre calcite and calcium oxalate raphides raises the possibility that the two could be confused when observed in microfossil assemblages.

Needle-fibre calcite precipitation can be related to either biological (e.g. produced as coatings on fungal mycelia) or inorganic processes (e.g. crystallisation from supersaturated calcium solutions) (Borsato et al. 2000; Verrecchia and Verrecchia 1994). They have been reported from a variety of environments exhibiting active pedogenic processes, including calcareous soils, calcrete, aeolianite, weathered limestone and limestone caves (Phillips and Self 1987:429). Finding an abundance of these crystals on the Lapita potsherds is not surprising, therefore, given that those examined are all from former beach-side locations. The occurrence of calcite crystals in the environments listed above will also have implications for studies of artefacts and sediments from limestone caves, upon which a great deal of Pacific archaeology relies. Not only might it be possible that naturally occurring needle-fibre calcite crystals be present in sites from these depositional environments, but if calcium oxalate raphides associated with plant-processing are also present, it may be difficult to differentiate them based on their morphological and optical properties alone.

Figure 10. SEM micrographs of needle-fibre calcite crystals *in situ* on Lapita potsherds from the Mussau Islands, showing their grooved morphology similar to Type IV raphides from the Araceae.

Figure 11. SEM micrographs at the same magnification comparing (a) a dense cluster of broken *Alocasia macrorrhiza* (leaf) raphides with (b) a concentration of *in situ* needle-fibre calcite crystals on a Lapita potsherd from the Reef Islands, southeast Solomon Islands. The fragmented raphides are virtually indistinguishable in size and form from the calcite crystals.

Where suspected raphides are found, a simple acid test could resolve their identification. For example, when a sample of crystals removed from one of the Lapita potsherds was treated with a small amount of weak (5%) acetic acid, which is known to dissolve calcium carbonate but not calcium oxalate, all crystals rapidly dissolved. If the reaction is observed under the microscope, it may also be possible to see bubbles of carbon dioxide gas produced by the calcite as it dissolves, although in the experiments performed here the reaction occurred too quickly and vigorously to observe this phenomenon. By comparison, a sample of *Alocasia macrorrhiza* raphides treated with the same solution was not affected. This simple test supported an SEM analysis which suggested that the archaeological crystals were needle-fibre calcite rather than raphides. The use of histological stains that differentially stain calcium oxalate and calcium carbonate, such as silver nitrate ($AgNO_3$) (Pizzolato 1964; Yasue 1969) or Alizarin red S (Macnish et al. 2003; Proia and Brinn 1985), is another avenue currently under investigation. In any case, a combination of methods, including light microscopy, scanning electron microscopy and chemical tests, should be used to confirm the identification of raphides as such when found in archaeological sites, particularly from environments where needle-shaped crystals of other types might occur. Previous studies can also be strengthened by reanalysis of recovered assemblages using these techniques.

Conclusion

The major implication of this review is that microfossil and residue studies published to date, in which needle-like crystals have been identified through light microscopy as raphides from Pacific food plants, must treated with caution until more robust arguments for their identifications are presented. Many of the criteria upon which acicular crystals have been identified as raphides and then assigned a taxon of origin have yet to be shown to be reliable. Incorporation of higher resolution analyses of morphometric characteristics for taxonomic identification, and chemical tests, are necessary further steps for raphide analysis. This finding requires that a reexamination of previous and ongoing studies of raphides from Pacific archaeological sites be conducted before secure conclusions based on these crystals can be drawn. This is a non-trivial issue, as these conclusions influence our understanding of early horticulture and the timing of plant introductions in the region, especially concerning the use of important aroids such as taro.

With discussion of archaeological Pacific raphides increasing, it is timely to assess what is known about these crystals, how they are interpreted, and how they can be identified more accurately to taxa of origin. Previous archaeological studies of raphides have been restricted in the scope of their interpretation, which has focused on an association between raphides and the Araceae, as well as their identification, for which there has been a reliance on light microscopy rather than SEM, which is more suitable for examining taxonomically significant features. Before even attempting to answer questions about what type of raphides are involved, however, analysts need to establish that the crystals found in microfossil assemblages are in fact calcium oxalate raphides and not naturally occurring crystals of similar appearance. By resolving these issues, the analysis of calcium oxalate crystals, alongside starch granules, phytoliths, pollen and other microbotanical remains, can lead us to a firmer and more holistic understanding of plant use in Pacific prehistory.

Acknowledgements

Pottery samples were provided by Roger Green and Peter Sheppard (University of Auckland) and Patrick Kirch (University of California, Berkeley). This research was funded by the School of Social Science, The University of Queensland. Funding to attend the 6th Australasian Archaeometry Conference, 2005 was provided by The Department of Archaeology and Natural History, The Australian National University. I thank Michael Haslam, Ian Lilley and the reviewers for their valuable comments on earlier drafts of this paper, and Sandy Shearer (School of Social Science) and staff at the Centre for Microscopy and Microanalysis, The University of Queensland, for assistance with the SEM. CRC Press gave permission for the reproduction in modified form of Horner and Wagner's (1995:58) diagram as Figure 4.

References

Al-Rais, A. H., A. Myers and L. Watson. 1971. The isolation and properties of oxalate crystals from plants. *Annals of Botany* 35:1213–1218.

Arnott, H. J. and M. A. Webb. 2000. Twinned raphides of calcium oxalate in grape (vitis): Implications for crystal stability and function. *International Journal of Plant Science* 161:133–142.

Ayensu, E. S. 1972. *Anatomy of the Monocotyledons VI: Dioscoreales*. Oxford: Clarendon Press.

Barabé, D., C. Lacroix, M. Chouteau and M. Gibernau. 2004. On the presence of extracellular calcium oxalate crystals on the inflorescences of Araceae. *Botanical Journal of the Linnean Society* 146:181–190.

Bernard-Shaw, M. 1990. Experimental agave fiber extraction. In M. Bernard-Shaw and F. Huntington (eds), *Rincon Phase seasonal occupation of the Northern Tucson Basin*, pp 181–196. *Technical Report* 90-2. Tucson: Centre for Desert Archaeology.

Borsato, A., S. Frisia, B. Jones, and K. Van Der Borg. 2000. Calcite moonmilk: Crystal morphology and environment of formation in caves in the Italian Alps. *Journal of Sedimentary Research* 70:1171–1182.

Bouropoulos, N., S. Weiner, and L. Addadi. 2001. Calcium oxalate crystals in tomato and tobacco plants: Morphology and in vitro interactions of crystal-associated macromolecules. *Chemistry* 7:1881–1888.

Bradbury, J. H. and R. W. Nixon. 1998. The acridity of raphides from the edible aroids. *Journal of the Science of Food and Agriculture* 76:608–616.

Cody, A. M. and H. T. Horner, Jr. 1983. Twin raphides in the Vitaceae and Araceae and a model for their growth. *Botanical Gazette* 144:318–330.

Crowther, A. 2005. Starch residues on undecorated Lapita pottery from Anir, New Ireland. *Archaeology in Oceania* 40:62–66.

Dahlgren, R. M. T. and H. T. Clifford. 1982. *The Monocotyledons: A comparative study*. London: Academic Press.

Danielson, D. R. and K. J. Reinhard. 1998. Human dental microwear caused by calcium oxalate phytoliths in prehistoric diet of the lower Pecos Region, Texas. *American Journal of Physical Anthropology* 107:297–304.

Denham, T. P., S. Haberle, C. J. Lentfer, R. L. K. Fullagar, J. Field, M. Therin, N. Porch and B. Winsborough. 2003. Origins of agriculture at Kuk Swamp in the Highlands of New Guinea. *Science* 301:189–193.

Franceschi, V. R. and H. T. Horner, Jr. 1980. Calcium oxalate crystals in plants. *The Botanical Review* 46:361–427.

Franceschi, V. R. and P. A. Nakata. 2005. Calcium oxalate in plants: Formation and function. *Annual Review of Plant Biology* 56:41–71.

Fullagar, R. L. K. 1993a. Flaked stone tools and plant food production: A preliminary report on obsidian tools from Talasea, West New Britain, PNG. In P. C. Anderson, S. Beyries, M. Otte and H. Plisson (eds), *Traces et Fonction: Les Gestes Retrouvés, pp 331–337. Colloque International de Liége, Éditions ERAUL 50. Liége*: ERAUL.

Fullagar, R. L. K. 1993b. Taphonomy and tool-use: A role for phytoliths in use-wear and residues analysis. In B. Fankhauser and R. Bird (eds), *Archaeometry: Current Australasian research*, pp 21–27. Canberra: Department of Prehistory, Research School of Pacific Studies, The Australian National University.

Fullagar, R. L. K., J. Field, T. P. Denham and C. J. Lentfer. 2006. Early and mid Holocene tool-use and processing of taro (*Colocasia esulenta*), yam (*Dioscorea sp.*) and other plants at Kuk Swamp in the highlands of Papua New Guinea. *Journal of Archaeological Science* 33:595–614.

Gallaher, R. N. 1975. The occurrence of calcium in plant tissue as crystals of calcium oxalate. *Communications in Soil Science and Plant Analysis* 6:315–330.

Genua, J. M. and C. J. Hillson. 1985. The occurrence, type and location of calcium oxalate crystals in the leaves of fourteen species of Araceae. *Annals of Botany* 56:351–361.

Graustein, W. C., K. Cromack, Jr. and P. Sollins. 1977. Calcium oxalate: Occurrence in soils and effect on nutrient and geochemical cycles. *Science* 198:1252–1254.

Hardy, B. L. 2004. Neanderthal behaviour and stone tool function at the Middle Palaeolithic site of La Quina, France. *Antiquity* 78:547–565.

Hardy, B. L., M. Kay, A. E. Marks and K. Monigal. 2001. Stone tool function at the Paleolithic sites of Starosele and Buran Kaya III, Crimea: Behavioral implications. *Proceedings of the National Academy of Sciences* 98:10972–10977.

Horner, H. T., Jr, A. P. Kausch and B. L. Wagner. 1981. Growth and change in shape of raphide and druse calcium oxalate crystals as a function of intracellular development in *Typha angustifolia* L. (Typhaceae) and *Capsicum annuum* L. (Solanaceae). *Scanning Electron Microscopy* 1981:251–62.

Horner, H. T., Jr. and B. L. Wagner. 1995. Calcium oxalate formation in higher plants. In S. R. Khan (ed.), *Calcium Oxalate in Biological Systems*, pp 53–72. Boca Raton: CRC Press.

Horrocks, M. and I. Barber. 2005. Microfossils of introduced starch cultigens from an early wetland ditch site in New Zealand. *Archaeology in Oceania* 40:106–114.

Horrocks, M. and S. Bedford. 2005. Microfossil analysis of Lapita deposits in Vanuatu reveal introduced Araceae. *Archaeology in Oceania* 40:67–74.

Horrocks, M. and M. I. Weisler. 2006. Analysis of plant microfossils in archaeological deposits from two remote archipelagos: The Marshall Islands, Eastern Micronesia, and the Pitcairn Group, Southeast Polynesia. *Pacific Science* 60:261–280.

Johns, T. and I. Kubo 1988. A survey of traditional methods employed for the detoxification of plant foods. *Journal of Ethnobiology* 8(1):81–129.

Jones, J. G. and V. M. Bryant. 1992. Phytolith taxonomy in selected species of Texas cacti. In G. Rapp, Jr. and S. C. Mulholland (eds), *Phytolith systematics: Emerging issues*, pp 215–238. New York: Plenum Press.

Keating, R. C. 2004. Systematic occurrence of raphide crystals in Araceae. *Annals of the Missouri Botanical Garden* 91:495–504.

Kostman, T. A. and V. R. Franceschi. 2000. Cell and calcium oxalate crystal growth is coordinated to achieve high-capacity calcium regulation in plants. *Protoplasma* 214:166–179.

Koumouzelis, M., B. Ginter and J. K. Kozlowski. 2001. The early Upper Palaeolithic in Greece: The excavations in Klisoura Cave. *Journal of Archaeological Science* 28:515–539.

Lalueza Fox, C., J. Juan and R. M. Albert. 1996. Phytolith analysis on dental calculus, enamel surface, and burial soil: Information about diet and paleoenvironment. *American Journal of Physical Anthropology* 101:101–113.

Loy, T. H. 1994. Methods in the analysis of starch residues on prehistoric stone tools. In J. G. Hather (ed.), *Tropical archaeobotany: Applications and new developments*, pp 86–114. London: Routledge.

Loy, T. H., M. Spriggs and S. Wickler. 1992. Direct evidence for human use of plants 28,000 years ago: Starch residues on stone artefacts from the northern Solomon Islands. *Antiquity* 66:898–912.

Macnish, A. J., D. E. Irving, D. C. Joyce, V. Vithanage, A. H. Wearing, R. I. Webb and R.L. Frost. 2003. Identification of intracellular calcium oxalate crystals in *Chamelaucium uncinatum* (Myrtaceae). *Australian Journal of Botany* 51:565–572.

McNair, J. B. 1932. The interrelation between substances in plants: Essential oils and resins, cyanogen and oxalate. *American Journal of Botany* 19:255–72.

Metcalfe, C. R. and L. Chalk. 1979. *Anatomy of the Dicotyledons II: Wood structure and conclusion of the general introduction*. Oxford: Clarendon Press.

Middendorf, E. A. 1983. The remarkable shooting idioblasts. *Aroideana* 6:9–11.

Monje, P. V. and E. J. Baran. 2002. Characterization of calcium oxalates generated as biominerals in cacti. *Plant Physiology* 128:707–713.

Mulholland, S. C. and G. Rapp, Jr. 1992. Phytolith systematics: An introduction. In G. Rapp, Jr. and S. C. Mulholland (eds), *Phytolith systematics: Emerging issues*, pp 1–13. New York: Plenum Press.

Nakata, P. A. 2003. Advances in our understanding of calcium oxalate crystal formation and function in plants. *Plant Science* 164:901–909.

Okoli, B. E. and B. O. Green. 1987. Histochemical localization of calcium oxalate crystals in starch grains of yams (*Dioscorea*). Annals of Botany 60:491–394.

Osuji, J. O. and B. C. Ndukuwu. 2005. Probable functions and remobilisation of calcium oxalates in *Musa* L. *African Journal of Biotechnology* 4:1139–1141.

Osuji, J. O., B. E. Okoli and R. Ortiz. 2000. Taxonomic value of calcium oxalate crystals in *Musa* germplasm. *Acta Horticulturae* 540:137–146.

Paull, R. E., C.-S. Tang, K. Gross, and G. Uru. 1999. The nature of the taro acridity factor. *Postharvest Biology and Technology* 16:71–78.

Phillips, S. E. and P. G. Self. 1987. Morphology, crystallography and origin of needle-fibre calcite in Quaternary pedogenic calcretes of South Australia. *Australia Journal of Soil Research* 25:429–44.

Pizzolato, P. 1964. Histochemical recognition of calcium oxalate. *Journal of Histochemistry and Cytochemistry* 12:333–336.

Proia, A. D. and N. T. Brinn. 1985. Identification of calcium oxalate crystals using Alizarin red S stain. *Archives of Pathology and Laboratory Medicine* 109:186–189.

Prychid, C. J. and P. A. Rudall. 1999. Calcium oxalate crystals in monocotyledons: A review of their structure and systematics. *Annals of Botany* 84:725–739.

Reinhard, K. J. and D. R. Danielson. 2005. Pervasiveness of phytoliths in prehistoric southwestern diet and implications for regional and temporal trends for dental microwear. *Journal of Archaeological Science* 32:981–988.

Sakai, W. S. 1979. Aroid root crops, acridity and raphides. *Tropical Foods* 1:265–278.

Sakai, W. S., M. Hanson and R. C. Jones. 1972. Raphides with barbs and grooves in *Xanthosoma sagittifolium* (Araceae). Science 178:314–15.

Shafer, H. J. and R. G. Holloway. 1979. Organic residue analysis in determining stone tool function. In B. Hayden (ed.), *Lithic Use-Wear Analysis*, pp 385–399. New York: Academic Press.

Sobolik, K. D. 1994. Lithic organic residue analysis: An example from the southwestern Archaic. *Journal of Field Archaeology* 23:461–469.

Sunell, L.A. and P.L. Healey. 1979. Distribution of calcium oxalate crystal idioblasts in corms of taro (*Colocasia esculenta*). *American Journal of Botany* 66:1029–32.

Sunell, L. A. and P. L. Healey. 1981. Scanning electron microscopy and energy dispersive x-ray analysis of raphide crystal idioblasts in taro. *Scanning Electron Microscopy* 1981:235–244.

Sunell, L. A. and P. L. Healey. 1985. Distribution of calcium oxalate crystal idioblasts in leaves of taro (*Colocasia esculenta*). *American Journal of Botany* 72:1854–1860.

Tomlinson, P. B. 1961. *Anatomy of the Monocotyledons* II: Palmae. Oxford: Clarendon Press.

Verrecchia, E. P. and K. E. Verrecchia. 1994. Needle-fibre calcite: A critical review and a proposed classification. *Journal of Sedimentary Research* A64:650–664.

Wattendorff, J. 1976. A third type of raphide crystal in the plant kingdom: Six-sided raphides with laminated sheaths in *Agave americana* L. *Planta* 130:303–311.

Webb, M. A. 1999. Cell-mediated crystallization of calcium oxalate in plants. *Plant Cell* 11:751–761.

Webb, M. A., J. M. Cavaletto, N. C. Carpita, L. E. Lopez and H. J. Arnott. 1995. The intravacuolar organic matrix associated with calcium oxalate crystals in leaves of *Vitis*. *The Plant Journal* 7:633–648.

Yasue, T. 1969. Histochemical identification of calcium oxalate. *Acta Histochemica et Cytochemica* 2:83–95.

9

Archaeobotany of Sos Höyük, northeast Turkey

Catherine Longford[1]
Andrew Drinnan[2]
Antonio Sagona[3]

[1]Department of Archaeology
The University of Sheffield
S1 4ET, UK
c.longford@sheffield.ac.uk

[2]School of Botany
The University of Melbourne
3010 Australia

[3]School of Historical Studies, Centre for Classics and Archaeology
The University of Melbourne
3010 Australia

Abstract

Sos Höyük, located in the Pasinler Valley of northeast Turkey, is a multiperiod site occupied from the Late Chalcolithic to the Medieval Period. No major archaeobotanical analysis has been conducted on the carbonised plant remains from Sos Höyük. Samples from three time periods, the Late Chalcolithic, Middle Bronze and Iron Age were selected for analysis. Examination of both seeds and wood charcoals has indicated a complex and changing pattern of plant use over time. From wood charcoals, four vegetation zones (riparian woodland, open oak woodland, mountain pine forest and birch forest) were identified in the Pasinler Valley during antiquity and exploited by the occupants of Sos Höyük. Change in wood procurement patterns suggests the depletion of the open oak woodland in the Pasinler Valley between the Middle Bronze Age and Iron Age. A reduction in wood resources is supported by the use of dung fuel as a firewood supplement from the Middle Bronze Age. These findings correlate with the pattern of deforestation across the Near East and suggest a combination of human and environmental causes for vegetation change in the Bronze Age. A second deforestation event is proposed for the modern period.

Keywords: Vegetation history; deforestation; climate change; archaeobotany; Turkey

Introduction

Sos Höyük is an archaeological site located in the modern village of Yiğittaşı in Erzurum Province, northeast Turkey (Figure 1). The site is situated at an altitude of 1800m in the narrow Pasinler Valley flanked by the Karapazarı Mountains to the north and the Palandöken Range to the south, both of which reach elevations in excess of 3000 m. As is the case today, in antiquity the Pasinler Valley lay on one of the main routes through the mountains of Eastern Anatolia linking Western Turkey to Iran and the Caucasus. Members of the University of Melbourne's Northeastern Anatolia project, led by Antonio and Claudia Sagona, excavated the mound of Sos Höyük from 1994 to 2000 (see Sagona et al. 1995, 1996, 1997, 1998; Sagona and Sagona 2000. Study seasons, accompanied by a field survey of the surrounding region, were conducted from 2001 to 2003.

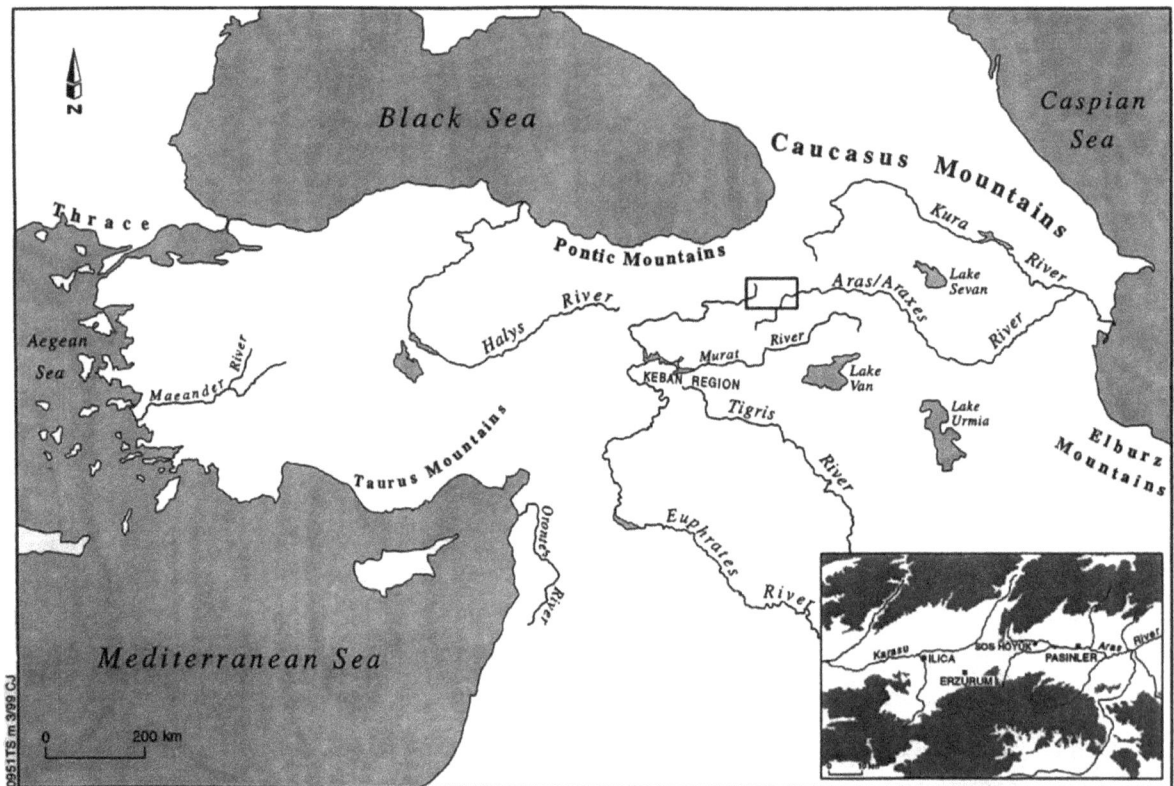

Figure 1. Map of Ancient Anatolia, Transcaucasia, and Northwest Iran showing the location of Sos Höyük (Sagona 2000)

Excavations at the site have revealed that Sos Höyük was initially used in the Late Chalcolithic (late 4th millennium BC) by agro-pastoralists who were part of the Early Transcaucasian (or Kura-Araxes) cultural complex. Communities of this tradition continued to occupy Sos Höyük throughout the greater part of the Bronze Age. Judging from architectural remains and analysis of some of the faunal data, these early villagers incorporated a degree of transhumance into their lifestyle, though the exact nature of this economic activity and how it related to sedentary agricultural practices has yet to be determined. During the late second millennium BC (Late Bronze Age) occupation appears to have been seasonal, though this may reflect the limited area of this period that was excavated; more permanent buildings may still lie unexcavated at the southern end of the mound. In the Iron Age, the site was home to a farming community on the edge of the Urartian Empire and later within the 18th Satrapy of the Persian Empire. The Iron Age plant remains are characterised by an excellent level of preservation with carbonised rush matting, basketry, rope, furniture and a pair of charred sandals recovered from a burnt house. Sos Höyük was abandoned in the second century BC and reoccupied briefly after the Seljuk conquest of Anatolia in the eleventh century AD before the present village was established some time within the modern era (Table 1) (Sagona and Sagona 2000).

Period	Date
Late Chalcolithic	3500–3000 BC
Bronze Age	
Early Bronze Age	3000–2200 BC
Middle Bronze Age	2200–1500 BC
Late Bronze Age	1500–1000 BC
Iron Age	1000–300 BC
Post-Achaemenid	300–200 BC
Medieval	1100–1300 AD

Table 1. Sos Höyük chronology (Sagona 2000).

Post-excavation research at Sos Höyük has included zooarchaeology (Howell-Muers, 2001), ethnoarchaeology (Hopkins 2000), osteoarchaeology (Parr et al. 1999) and palynology (Connor n.d.). While wood charcoal is plentiful throughout the sequence, little analysis, other than radiocarbon dating, has so far been undertaken on these samples. As part a broad environmental study of the Bronze Age at Sos Höyük, Newton (2004) performed a limited study on the wood charcoal remains, recommending that further intensive analysis be performed. The closest archaeobotanical studies to Sos Höyük have been conducted beyond the southern Palandoken Mountains in south-eastern Anatolia at Arslantepe (Frangipane et al. 2001), Aşvan (Willcox 1974) and Kurban Höyük (Miller 1986) and to the east at Patnos (Dönmez 2003). No archaeobotanical analysis has been completed in northeast Anatolia.

In the Near East most archaeobotanical work has focused on the early domestication of crop plants in the Neolithic and relatively little research has been undertaken into the plant based subsistence economy of the Chalcolithic, Bronze and Iron Ages (Miller 1991, Reihl and Nesbitt 2003). Analysis of the plant remains from Sos Höyük would help to fill this void by providing information on subsistence in Eastern Turkey from the Late Chalcolithic to Medieval Period. Sos Höyük is located in the Eastern Highlands of Anatolia and provides a rare opportunity to investigate the subsistence strategies of a mountain village in comparison to the lowland settlements already documented to the south and west of the Palandöken Range.

Similarly, Sos Höyük is ideally placed to examine the vegetation history of a mountain zone. Tod ay the Pasinler Valley is practically treeless, dominated by alpine wildflowers with riparian trees restricted to riverbanks, and pollarded poplars along roadsides and around villages. The surrounding mountains are devoid of trees except in steep gullies near the plateau of the Karapazarı range where isola ted pockets of birches (Betula sp.) pines (Pinus sylvestris) and stunted oaks (Quercus macranthera) remain (Newton 2004; personal observation). In contrast to the current treeless steppe, Xenophon, writing of his retreat through the Pasinler Valley from Persia to the Black Sea in the fourth century BC, recorded 'wood in abundance' in the Erzurum region (Xenophon IV.5.2, Sagona 2004). Thus, in ancient times, the Pasinler Valley and nearby mountain ranges were likely to have been wooded. A pollen core from Bulemaç, south east of Sos Höyük on the Pasinler Plain, suggests that the area was dominated by a Pinus woodland for m ost of the Holocene (Collins et al. 2005). Prehistoric deforestation has been identified via wood charcoal analysis across southwest Asia (Willcox 1974, Miller 1985). The present vegetation mosaic of isolated trees amid herbaceous steppe found across Central and Eastern Turkey is thought to have resulted from deforestation in antiquity (McNeill 1992, Woldring and Cappers 2001).

The research described here was a pilot study that aimed to bring a highland perspective to archaeobotanical research in the Near East and begin much needed work in northeast Anatolia. In particular, the project aimed to study material from the Late Chalcolithic, Middle Bronze and Iron Ages to reconstruct the vegetation history of the Pasinler Valley and relate this to environmental change across the Near East. Here we present the preliminary findings of the study, concentrating primarily on the ecological changes in the Pasinler Valley during that period.

Method

Samples for archaeobotanical analysis were taken from recognised floor levels rich in organic material (up to 10 cm above the floor), but not from the mixed fill between layers. A minimum of 60 litres of soil was collected randomly from the floor level of each targeted excavation locus, with hearths preferentially sampled. The soil was processed by bucket flotation to collect the charred plant remains, with the heavy residue wet sieved and manually sorted to collect non-floating organic remains. After being air dried in cloth bags the material was stored in Yiğittası until 2003 when it was transported to Australia for laboratory analysis.

After initial examination, three loci were selected for analysis based on the amount of carbonised material and period of deposition. The Late Chalcolithic sample was from the surface of a burnt plaster floor of a Transcaucasian roundhouse dated to 3100–2600 cal BC (Beta-135363, calibrated using OxCal v3.10). The Middle Bronze Age (2580–2340 cal BC (OZF943)) sample was from a secondary deposit in a plastered rubbish pit. The Iron Age remains were from the destruction debris of a burnt house from 1260–890 cal BC (Beta-95214). Detailed explanation of each context is found in Sagona and Sagona (2000). Each sample was sieved into three fractions: >2 mm, >0.5 mm and <0.5 mm, as suggested by Pearsall (2000).

For seed analysis, whole seeds and seed fragments were collected from the >2 mm sample fractions and only whole seeds from the >0.5 mm portion, following Miller (1998). These were identified using an Olympus SZ60 stereomicroscope with the aid of illustrative archaeobotanical texts by van Zeist and Bakker-Heeres (1985, 1986), Zohary and Hopf (2000), and seed atlases (Berggren 1969, 1981; Anderberg 1994). Cereal grain identification was assisted by reference to modern seed samples of wild and domesticated cereals sent from the Australian Winter Cereal Collection in Tamworth. Where possible, taxa were identified to species level, otherwise to genus. Nomenclature follows Zohary and Hopf (2000) for cultivated species and for wild taxa follows the *Flora of Turkey* by Davis (1965–1985).

Wood charcoal fragments were sorted from the >2 mm sample fractions. For this preliminary study an initial subsample size of 50 charcoal pieces per sample was selected as a compromise between the five of Miller (2004) and 200 of Asouti and Hather (2001). To check the appropriateness of examining 50 pieces per unit studied, a variant of Smart and Hoffman's (1988) taxon-number curve was used. In each instance the taxon number curve plateaued prior to identification of all 50 fragments, indicating that a 50 fragment count provided an adequate sub-sample. The 50 charcoal fragments were randomly selected for examination under an Olympus SZH stereomicroscope. Each piece was cut with a razor blade to expose the transverse, tangential longitudinal and radial longitudinal planes. The wood was identified using keys for European and Near Eastern wood (Greguss 1955; Brazier and Franklin 1961; Fahn et al. 1986; Schoch et al. 2004) and through reference to a modern comparative collection created as part of this research. Fragments that could not be identified under the stereomicroscope were identified under the Phillips XL30 Field Emission Scanning Electron Microscope (FESEM), in the School of Botany at the University of Melbourne. Confirmation of identifications made under the stereomicroscope was conducted by randomly selecting a charcoal piece from each taxon group identified, for FESEM examination and photographing as a quality control. Nomenclature follows the *Flora of Turkey* (Davis 1965–1985). Identifications were made to genus or, if possible, to species. For Salicaceae, identification was made to family level since, by wood anatomy, the two genera *Salix* and *Populus* are indistinguishable except at high magnification (Schoch et al. 2004). With the majority of identifications made with the stereomicroscope, *Salix* and *Populus* could not be differentiated and were recorded at the family level.

Results

Seeds

A summary of seed abundance data for the Late Chalcolithic, Middle Bronze and Iron Age levels is presented in Table 2. In the Late Chalcolithic sample, only wheat (*Triticum* sp.) and barley (*Hordeum vulgare*) grains were identified with naked wheats (*Triticum aestivum* ssp. *vulgare/T. turgidum* ssp. *durum*) being most abundant. The recovery of a *T. aestivum* ssp. *vulgare* rachis indicates hexaploid bread wheat may have been present. A greater variety of plant taxa were discovered in the Middle Bronze Age pit. Wheat and barley were found with hulled barley (*Hordeum vulgare*) the most common taxa. A diverse wild seed flora with *Galium* sp., *Asperula* sp., *Polygonum* sp., *Lolium* sp, and members of the Boraginaceae, Caryophylaceae, Convolvulaceae, Fabaceae and Poaceae families were present. Remains of sheep/goat dung were also found. In the Iron Age sample, naked wheat was the most dominant grain type, with seeds of millet (*Setaria italica*), grape (*Vitis vinifera*), lentil (*Lens culinaris*), and five wild taxa also found.

Taxon (English name)	Plant Part	Period		
		Late Chalcolithic	Middle Bronze Age	Iron Age
Crops				
Triticum aestivum L. ssp. *vulgare* /T. turgidum L. conv. *durum* (bread/durum wheat)	grain	84	16	715
Triticum sp. (wheat)	grain	35	4	71
Triticum aestivum ssp. *vulgare* (bread wheat)	rachis	2	-	-
Hordeum vulgare L. (hulled barley)	grain	46	175	14
H. vulgare L var. *nudum* (naked 6-row barley)	grain	-	1	-
Cerealia (cereal)	grain frags.	-	-	1
Setaria italica L. P. Beauv. (foxtail millet)	grain			
Lens culinaris L. (lentil)	seed	-	-	1
Vitis vinifera L. (grape)	seed	-	-	1
Wild plants				
Galium L. sp.	seed	-	2	2
Asperula L. sp.	seed	-	1	1
Polygonum L. sp.	seed	-	3	-
Lolium L. sp.	seed	-	1	2
Ranunculus arvensis L.	seed	-	-	1
Boraginaceae	seed	-	1	-
Caryophylaceae	seed	-	1	-
Convolvulaceae	seed		1	
Fabaceae	seed	-	7	2
Poaceae	seed	4	2	-
Unidentified	seed	-	3	6
Pinus sp. cone	cone	-	-	1
Other				
Sheep/goat faecal pellets		-	2	-

Table 2. Summary of seed and chaff data for Sos Höyük for the Late Chalcolithic, Middle Bronze and Iron Age levels examined. Data is recorded as sum individual seeds/units with weight (g) recorded for Cerealia fragments. Nomenclature follows the modern system described in Zohary and Hopf (2000) for cereal taxa and the *Flora of Turkey* (Davis 1965–1985) for wild taxa.

These preliminary findings from Sos Höyük are an interesting contrast to the rest of the Near East where hulled wheats (emmer and einkorn) are known to have decreased in dominance over the Bronze Age with naked and free-threshing wheats becoming the main cereal crops by the Iron Age (Zohary and Hopf 2000). In Eastern Turkey this transition is thought to have occurred in the Early Bronze Age (Riehl and Nesbitt 2003). Preliminary results from Sos Höyük, showing the prevalence of naked wheat from the earliest period studied may indicate that the shift from hulled wheat to naked wheat may have occurred earlier in the northeastern highlands of Anatolia than in the rest of the region. This could be because this region in the Late Chalcolithic was part of the Early Transcaucasian cultural complex. Genetic evidence suggests bread wheat was first domesticated on the southern flanks of the Caucasus (Dvorak et al. 1998), and in many settlements from the Neolithic onwards it is the most dominant wheat type identified (Wasylikowa et al. 1991). The presence of bread wheat rachis segments and quantity of naked wheat in the Late Chalcolithic may be because *T. aestivum* ssp. *vulgare* was the main wheat of the Early Transcaucasian crop assemblage.

Wood

Seven taxa were identified in the wood charcoal assemblages (Table 3, Figure 2). In the Late Chalcolithic *Pinus* cf. *sylvestris* (Scots Pine), *Quercus sp.*(oak) and Salicaceae (willow/poplar) were found with Salicaceae being the most common taxa present. *P.* cf. *sylvestris* was the most abundant taxon in a more diverse assemblage from the Middle Bronze Age pit, which also contained *Betula* sp. (birch), *Quercus* sp., *Ulmus* sp. (elm), *Acer* sp. (maple) and Salicaceae. In the Iron Age *P.* cf. *sylvestris*, Salicaceae, *Betula* sp. and *Lonicera* sp. (honeysuckle) were identified with the riverine Salicaceae being the most dominant charcoal type.

Period	Late Chalcolithic		Middle Bronze Age		Iron Age	
Sample numbers	L17b 4299 142		L16c 4035 60		J14 1299 122	
Taxon	Count	%	Count	%	Count	%
Pinus cf. *sylvestris* L.	10	20	21	42	17	34
Salicaceae	38	76	9	18	19	38
Betula sp.	-	-	1	2	9	18
Lonicera sp.	-	-	-	-	1	2
Quercus sp.	2	4	12	24	-	-
Ulmus sp.	-	-	1	2	-	-
Acer sp.	-	-	3	6	-	-
Indeterminate	-	-	3	6	4	8
Total	50	100	50	100	50	100

Table 3. List of the taxa present in the wood charcoal assemblage from Sos Höyük in the Late Chalcolithic, Middle Bronze and Iron Age samples, recorded as absolute fragment counts (Count) and relative percentages (%).

The Environmental History of Sos Höyük

While this study is based on only a small number of samples, the preliminary archaeobotanical record from Sos Höyük preserves a diverse charcoal flora for the Pasinler Valley. Although very few trees are present in the vicinity of Sos Höyük today, mainly along roadsides, rivers and near villages, the past environment can be inferred based on the species' modern habitat preferences as recorded in Zohary (1973),

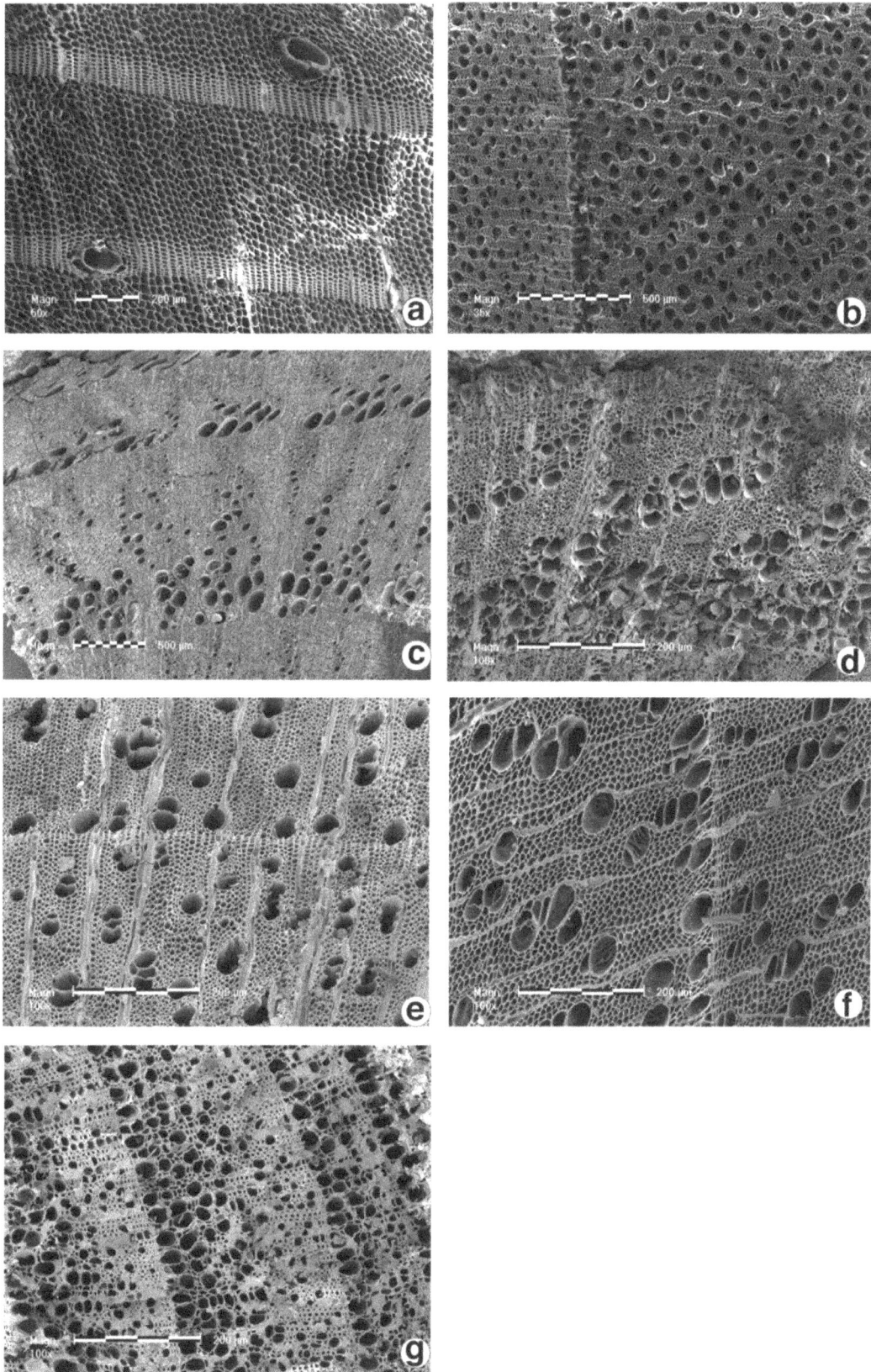

Figure 2. Representative transverse sections of taxa identified in the wood charcoals studied from Sos Höyük photographed under the FESEM. Scale bars are in increments of 50 μm. a) *Pinus* cf. *sylvestris;* b) Salicaceae; c) *Quercus* sp.; d) *Ulmus* sp.; e) *Acer* sp.; f) *Betula* sp.; g) *Lonicera* sp.

Fairbairn et al. (2002), Newton (2004) and Davis (1965–1985). When integrated, the charcoal taxa form five vegetation types that have distinct distributions in the site's locale (Table 4, Figure 3). The environmental reconstruction (Figure 3) incorporates the northern half of the Pasinler Valley, in which Sos Höyük is located, and the Karapazarı Mountains, since the Aras River to the south of the site forms a socio-geographical boundary in the Valley which may have marked the edge of village territory in ancient times (Sagona 2004). Along the rivers and alluvial flats, riparian woodlands dominated by willows and poplars were probably present. Disturbed land, including cultivated fields near villages and the village itself would have been home to wild and weedy annuals and perennials. On the Pasinler Plain, the severe winter frosts and winds depress the tree line, limiting tree growth to scattered stands (Stern 1983). This Bronze Age steppe savannah was probably an open oak-dominated woodland with maple, elm, almond, and shrubby Rosaceae, and juniper as is still present in some parts of Anatolia. The oak woodland would have spread across the wild subalpine flower-covered valley and into the foothills. Further into the Karpazarı Range, the open oak woodland would have graded into the Pinus sylvestris forest that may have had an oak and maple understorey at lower altitudes. At the upper tree line beneath the Karapazarı plateau, birch would have grown, particularly dense in sheltered north-facing niches, together with stunted pines, Rhododendron and juniper as is found in Northern Turkey today (Connor pers. comm.). This reconstruction matches the habitat requirements of wild animals found at Sos Höyük in the Early Bronze Age with brown bears in coniferous forests, wild sheep in the wooded mountains, red deer in woodlands and golden eagles preferring the open plain with scattered trees (Howell-Meurs 2001; Newton 2004; Sagona et al. 1997).

Location	Vegetation type	Taxa identified in assemblage	Possible taxa in association
river banks/alluvial flats	riparian	willow, poplar (Salicaceae), elm (*Ulmus*), *Polygonum*	ash (*Fraxinus*) alder (*Alnus*), *Carex*, *Juncus*,
plain	disturbed land	*Ranunculus*, *Galium*, *Lolium*, *Asperula*, *Triticum*, *Hordeum*, *Lens*, *Cucumis*, *Convolvulaceae*	*Bromus*, *Phalaris*, *Rumex*, *Trifolium*
plain and foothills	oak steppe-woodland	oak (*Quercus*), maple (*Acer*), honeysuckle (*Lonicera*), elm (*Ulmus*)	hawthorn (*Crataegus*), pears (*Pyrus*) cherry (*Prunus*), terebinth (*Pistachia*) juniper (*Juniperus*), rose (*Rosa*) almond (*Amygdalus*)
lower mountain slope	pine forest	pine (*Pinus sylvestris*), honeysuckle (*Lonicera*), maple (*Acer*), oak (*Quercus*),	juniper (*Juniperus*)

Table 4. Summary of vegetation reconstruction for the Pasinler Valley, recording vegetation type, the taxa identified in the charcoal assemblage and the possible taxa in association for habitat location. Reconstruction based on plant ecological preferences (from data in Zohary 1973, Fairbairn et al. 2002, Newton 2004 and Davis 1965–1985).

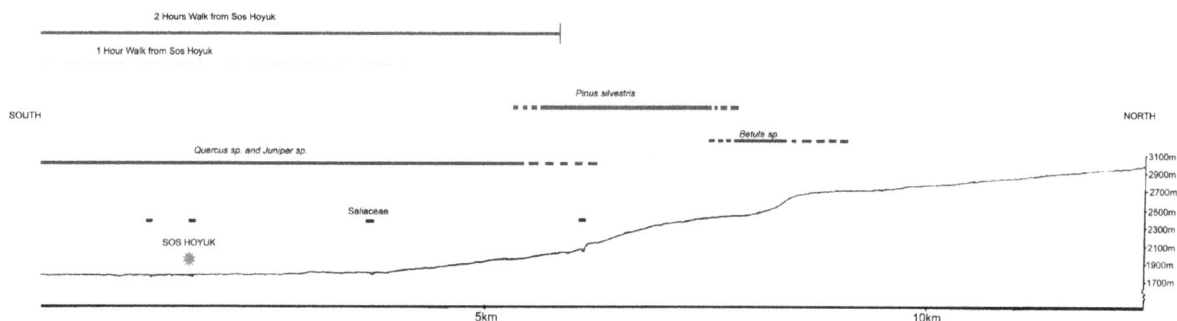

Figure 3. Possible distribution across the northern Pasinler Valley into the Karapazarı mountain range of the four vegetation types identified in charcoal from Sos Höyük. Riparian woodland along the rivers and in moist places, open oak woodland on the valley plain and into the foothills, mountain pine on the slopes of the Karapazarı mountains and birch forest at the upper treeline. Elevation and walking distance from Sos Höyük derived from topographical maps of Newton (2004).

All the wood charcoal fragments identified are elements of the Euro-Siberian Euxine phytogeographical province (Davis 1965–1985). This is in accordance with Zohary (1973) and Davis (1965) who view the East Anatolian Highlands as being a mix of relict Euro-Siberian forests and Irano-Turanian steppe. As the Euro-Siberian forests were depleted over time, the Irano-Turanian steppe and shrubbery became the dominant vegetation type present in the region today. In comparison to Collins et al. (2005), this model suggests a more complex vegetation history for the Pasinler Valley with several different ecotypes present together with the Pinus woodland identified in the Bulemaç pollen core.

Resource Depletion

The depletion of the Euro-Siberian woodland is visible in the charcoal assemblage examined from Sos H öyük when tabulated in terms of vegetation types used over time (Table 5). From the burnt plaster floor of the Late Chalcolithic round house, Salicaceae, pine and oak were found, with Salicaceae charcoal constituting the majority of the sample. This room had a portable hearth in the form of a twin-horned andiron and is part of a succession of plaster floor levels suggesting that the occupation at this time was transient, by semi-nomadic agro-pastoralists (Sagona and Sagona 2000). Wood gathering appears to have occurred in the riparian woodland along the river close to the site with some collection of oak and pine extending wood procurement into the foothills. This use of poor-burning Salicaceae wood matches the predictive model of nomadic pastoralist behaviour proposed by Asouti and Austin (2005), in which wood selection is driven more by the proximity and quantity of fuel required for mass milk processing than the calorific quality, and hence burning properties, of the wood itself.

Vegetation Type	Late Chalcolithic	Middle Bronze Age	Iron Age
riparian woodland	xxx	xx	xxx
open oak woodland	x	xxx	
pine forest	xx	xxx	xxx
birch forest		x	xx

Table 5. Summary of vegetation zones exploited in the Late Chalcolithic, Middle Bronze and Iron Age. Zone composition detailed in Table 4. x = minor exploitation. xx = some exploitation. xxx = heavy exploitation. Intensity of use based on proportion of wood charcoal from each zone in Table 3.

In the Middle Bronze Age pit, a greater diversity of charcoal taxa is present from all four recognised arboreal associations. More species from the plain, including maple and elm, and from the mountain slopes, birch, are found in the sample, indicating exploitation of vegetation communities beyond the settlement periphery, perhaps due to depletion of the riparian arboreal flora. This matches findings from Neolithic Çatalhöyük where most wood in the early phase of settlement was gathered from the local wetlands. Over time the resource base was broadened to include more ecological niches, including intensive harvesting in the foothills 10km to the south, which at Sos Höyük seems to have occurred by the Middle Bronze Age (Fairbairn et al. 2002).

Charcoal from the Iron Age house includes taxa from the riparian woodland and mountain forests, but lacks taxa from the open oak woodland. While honeysuckle is present, its use as an ornamental in some Anatolian villages impinges on its reliability as a woodland indicator (Ertuğ 2000). This scarcity of fragments from the steppe woodland, particularly the lack of oak, may be an artificial product of sampling but could equally represent the decline of the open oak stands. Asouti and Austin (2005) also suggest that the function of different archaeological contexts leads to botanically different charcoal assemblages. Thus the difference in context between the Late Chalcolithic and Middle Bronze Age samples, which are likely to be hearth remains, and the mixed house debris from the Iron Age could be the reason for the lack of oak woodland charcoal in the Iron Age. However, this is unlikely, since a household context would usually contain a representative flora of most taxa near to the site due to the greater diversity of wood forms that are burnt. A house fire would include a mix of wood from building construction, furniture, tools, domestic objects, and firewood stores. Hearth remains primarily contain taxa that were intentionally gathered from nearby vegetation and burnt as firewood (Willcox 2002). A household context is likely to contain a wider range of woods selected for a variety of purposes and objects and thus is representative of the surrounding vegetation. Therefore the lack of taxa from the oak woodland in the Iron Age probably reflects a depletion of the trees from the plain that forced the inhabitants of Sos Höyük to travel further into the Karapazarı Mountains to collect wood.

The seed results from Sos Höyük also are indicative of woodland reduction (Table 2). In the Middle Bronze and Iron Age samples the presence of a diverse weed seed flora together with sheep faecal pellets may indicate the use of dung fuel as a wood supplement, similar to findings from modern and ancient Maylan in Iran (Miller and Smart 1984). Indeed, having examined a range of samples from throughout the Bronze Age, Mark Nesbitt (cited in Newton 2004) found that at Sos Höyük weed seeds only entered the charcoal record in the Middle Bronze Age and concluded that this was suggestive of the introduction of dung fuel. In modern Yiğittaşı, dung bricks (tezek) are still used in the earth stove (tandoor) and as an occasional supplement to the coal and wood bought in Pasinler (Hopkins 2000). At Sos Höyük, the remains of dung fuel were found in the Late Bronze Age industrial sector preserved as a highly alkaline chocolate-brown layer and were probably incorporated as fuel supplement into a lime production pit (Sagona et al. 1997). Thus, as the oak woodland resources were coming under pressure in the Middle Bronze Age, dung fuel may have been used as a fuel supplement.

This pattern of woodland depletion is comparable to findings from Malyan in Iran, Aşvan in south eastern Anatolia and pollen cores from the region. During the Middle Bronze Age at Malyan, juniper was exhausted from the pistachio-almond-maple-juniper woodland and was replaced as fuel by oak from mountain slopes 10 kms away (Miller 1985). At Aşvan, charcoal analysis revealed that riverine taxa were severely depleted over the Late Chalcolithic and Early Bronze Age following initial settlement along the river (Willcox 1974). Sometime after the Early Bronze Age and before the first century BC, deforestation of the oak woodland led to the cultivation of poplar and finally the importation of timber in the Medieval Period. In Central Anatolia, the oak woodland around Eski Acıgöl is seen to decline sharply in pollen records from the Bronze Age, as the pine increases (Roberts et al. 2001). This decline of oak woodland is confirmed by a pollen core from Lake Imera in the highlands of Eastern Georgia, climatically similar to the Pasinler Valley, where the *Quercus* pollen decreases as the *Pinus* pollen increases in the second millennium BC (Kvavadze and Connor 2005; Connor et al. n.d.). The same pattern is present in several cores from Lake Van in southeastern Anatolia where *Quercus* pollen starts decreasing after 4000 BP and

Pinus pollen begins expanding (Wick et al. 2003:673; Zeist van and Woldring 1978:270). Deforestation in the highlands of eastern Turkey during the Late Bronze Age is apparent through increased sedimentation along the Euphrates River Valley, when land clearing and climate desiccation caused increased erosion (Kuzucuoğlu 2003).

Human Impact

Human impact, either direct, through tree felling and fires, or indirect, through animal grazing, is the probable cause of deforestation in the Pasinler Valley. At some point between the Middle Bronze and Iron Age at Sos Höyük, arboreal diversity decreased considerably. In the Late Bronze Age, the entire northern sector of the site was an industrial quarter, with signs of intensive bone processing for marrow extraction and a complex of pits for plaster manufacture (Sagona et al. 1998). In these pits limestone was burnt to produce lime plaster, used in antiquity for floors, bowls, statues and other artefacts (Sagona et al. 1997, Miller 1990a). Lime manufacture required high temperatures and much fuel, one ton of lime plaster requiring two tons of limestone and two tons of wood fuel for burning (Miller 1990a). Fresh wood or charcoal can be used to convert limestone into plaster. Charcoal has a higher calorific value but its production is very inefficient and requires sapling timber (Miller 1985). Work at Bronze Age Maylan and surrounds found that certain villages were centres of lime production that serviced the nearby villages that had no kiln (Miller 1990b). It could be that Sos Höyük was one such centre of lime manufacture for some of the villages in the Pasinler Valley in the Late Bronze Age. The quantities of fuel required for plaster production in combination with land clearing for agriculture and the usual wood procurement for hearths and housing may have placed excessive pressure on the vegetation of the Pasinler Valley and contributed to the decline of the oak woodland. In particular, oak has the characteristics necessary for good charcoal manufacture (Horne 1982). Selection of saplings for lime-making would have impinged on oak proliferation and over time reduced the extent of the oak woodland. The advent of iron working in the Pasinler Valley may also have affected the vegetation cover.

In tandem with wood fuel collection, livestock grazing may have diminished the plant resources of the Pasinler Valley. Sheep and goats were the most prevalent animals slaughtered in the Early Bronze and Iron Ages (Howell-Meurs 2001:90) and were probably the primary herd type throughout the occupation of Sos Höyük, just as they are today (Sagona et al. 1997). These animals graze on soft-leaved vegetation, and are noted to prefer oak (Riehl 1999). As flocks of sheep and goats range across the land they constantly graze, killing freshly germinated seedlings and defoliating saplings. Over time, this would prevent the growth of the new generations of trees and eventually stop regeneration of certain species. In Central Anatolia, most woodland remnants are found on isolated rock outcrops where they are protected from grazing (Woldring and Cappers 2001). Grazing promotes the proliferation of anti-herbivorous taxa with protective spines and unpalatable leaves and selects against soft-leaved taxa. In the Pasinler Valley, Crataegus and Rosa shrubs, both bearing woody spines, are present, whereas unprotected trees are rare, located in marshes and sheltered mountain niches (Newton 2004). Furthermore, shepherds and graziers of north-eastern Anatolia often burn the subalpine meadows in the autumn to promote grass growth after winter (Connor et al. 2004). For the mesophilic open oak woodland, intensive grazing in the Pasinler Valley together with increased fuel procurement in the Late Bronze Age, may have led to its depletion.

Climatic Aridity

Overgrazing in the Late Bronze Age may itself have been a function of the increasing climatic aridity in this period. From the beginning of the Holocene, humidity increased in the Near East which, with the improved climatic conditions, enabled the oak woodlands to first expand along the Mediterranean coasts and then, after a 3000-year delay, expand in the montane interiors of Anatolia as found at Eski Acıgöl and Lake Van around 8200BP (Roberts et al. 2001; Wick et al. 2003). This climatic optimum was maintained through the Bronze Age until 4000 BP when humidity levels began to decrease, which lowered the water levels at Lake Van (Wick et al. 2003; Lemcke and Sturm 1997) and Eski Acıgöl (Roberts et al. 2001). At

Lake Van the oak woodlands were depleted and the drought intolerant trees from Eski Acıgöl were considerably reduced (Roberts et al. 2001; Wick et al. 2003). Similarly in Eastern and Southern Georgia, rainfall and temperature kept increasing until 4500 BP, permitting oak forest expansion and the raising of the tree line before conditions began to deteriorate around 3500 BP (Kvavadze and Connor 2005). The Pasinler Valley in the second millennium BC, located between Lake Van and Georgia, was probably similarly affected, with the treeline lowering and decrease of oak woodland in an already marginal environment, leaving pine as the dominant tree by the Iron Age, as occurred at Lake Imera (Kvavadze and Connor 2005). At Lake Van (Wick et al. 2003) and Gravgaz in Southwest Anatolia (Vermoere et al. 2000) the oak woodland's decline is interpreted as a product of Holocene desiccation reducing the oak woodland's climatically favourable distribution. In contrast, at Eski Acıgöl after the loss of the drought intolerant species at the onset of the Holocene arid phase, the oak pollen has a delayed but marked fall which Roberts et al. (2001:733) interpreted as an anthropogenic change. Around Sos Höyük, as at Eski Acıgöl, the growing climatic aridity from the Middle Bronze Age onward, in tandem with human agricultural and settlement practices, may have led to the depletion of the open oak woodland.

Indeed over several years, increased summer aridity may have hastened the death of pasture plants early in the summer, leaving the flocks of goats and sheep to graze on the surrounding trees for the remainder of the season (Newton 2004). These trees, in the Late Bronze Age were already in decline due to decreased rains in the spring growing season (Wick et al. 2003). Today semi-nomadic herders from southern Turkey walk their flocks over the Taurus Mountains into Erzurum to graze on the Karapazarı plateau for the summer months (personal observation). This is a tradition that may have occurred for thousands of years (A. Sagona personal communication). With increased aridity across the Near East during the Late Bronze Age, the amount of migratory herders travelling to the relatively rich pastures of the Pasinler Valley in summer may have increased. As a modern analogue, the influx of migratory cattle herds to Lake Victoria in Tanzania during a recent arid period caused substantial deforestation and promotion of cattle-tolerant species (Hongo and Masikini 2003).

Therefore, these preliminary findings suggests that the depletion of the open oak woodland in the Pasinler Valley was initiated by the onset of climatic aridity in the Middle Bronze Age and was exacerbated by human activities. It should be remembered that Sos Höyük is not the only known settlement in the valley. A survey of the Pasinler region located 22 Middle Bronze Age sites scattered across the valley floor and 45 Iron Age sites mainly in the foothills zone (NEAAP). By the time Xenophon marched through the Pasinler Valley in the fourth century BC the abundant woods he passed were probably composed primarily of pine and birch. The timing of the final deforestation of the Pasinler Valley which eliminated the pine and birch forests is uncertain.

A pollen record from north of Sos Höyük, dating back 700 years (620±60 BP), indicates that since then the arboreal vegetation on the Karapazarı Mountain slope has been sparse, with only a few isolated pines and birches (Connor personal communication). To the south, on the northern slopes of the Palandöken Mountains, pine and birch may have been more common. Today, the mountain slopes are practically treeless. This final deforestation probably occurred in the last 500 years, when a sizable nomadic population of Eastern Anatolia settled to form village communities (McNeill 1992). Yiğittası, the village of Sos Höyük, may have been resettled at this time. At Van the oak woodland was cleared within the last 600 years (Wick et al. 2003), coinciding with village proliferation. Indeed Erzurum recorded a two thousand-fold increase in population in the 16th Century AD (Jennings 1976). The effect of this population shift was seen by the French botanist Joseph Pitton de Tournefort in 1702. Travelling through Erzurum he commented on 'the scarcity and dearness of Wood. Nothing but Pine-wood is known there, and that too they fetch two or three days Journey from the Town; all the rest of the Country is quite naked.' (Tournefort 1718:194). As Tournefort passed through the Pasinler Valley he noted 'There is not a Tree to be seen in all this part of the Country, which otherwise is flat, well cultivated and water'd as abundantly as the Fields of Erzeron [Erzurum]' (Tournefort 1718:212). By the 18th century AD the forests of the Pasinler Valley were gone, perhaps felled only a few centuries before.

Conclusions

From these preliminary findings, a picture of deforestation in the Pasinler Valley emerges from the macrobotanical remains of the Late Chalcolithic, Middle Bronze and Iron Ages. The modern alpine meadows of the Pasinler Valley are a cultural landscape, created by two distinct deforestation events. While this study was limited to only three loci, a number of initial observations can be made. In the Late Chalcolithic, initial occupants of the site gathered their wood from nearby in the riparian woodland, oak woodland on the plain and from the mountain pine forest. In the Middle Bronze Age the inhabitants were exploiting the full range of vegetation types hypothesised for the Pasinler Valley, collecting timber from the nearby riparian woodland, the open oak woodland and the pine and birch forests on the southern slopes of the Karapazarı Mountains. By the Iron Age only the riparian woodland and the mountain pine and birch forests were used for wood gathering. Representatives of the oak woodland on the plain were absent from the Iron Age sample. This suggests that sometime between the Middle Bronze and Iron Age the oak woodland was depleted which is in accordance with findings from Lake Van (Wick et al. 2003), Eski Acıgöl (Roberts et al. 2001), and Eastern Georgia (Kvavadze and Connor 2005). The use of dung fuel as a supplement for firewood may indicate that the deforestation of the Pasinler Valley was a gradual process beginning in the Middle Bronze Age. Increasing climatic aridity together with goat grazing, land clearing and intensified industrial activities in settlements may have contributed to the decline of the oak woodland in the late second millennium BC and caused the inhabitants of Sos Höyük to change their wood procurement strategies. The final deforestation of the Pasinler Valley, the felling of the pine and birch forests, occurred prior to 1700 AD perhaps as a result of the settling of the nomadic population of Eastern Anatolia. This two-tiered deforestation model may be the pattern for vegetation change across the highland regions of the Near East. Due to the limited nature of this study further analysis of the charcoals from Sos Höyük is needed to confirm these trends and to help clarify the timing of these events. Future investigations into the archaeobotantical material of Sos Höyük and neighbouring sites may modify these initial conclusions.

Acknowledgements

Thanks to: Simon Connor and Andy Fairbairn for many discussions on the material from Sos Höyük; Michael Mackay from the Australian Winter Cereal Collection in Tamworth who sent seed samples to aid with analysis; Jenny Newton for initiating work on this project.

References

Anderberg, A. 1994. *Atlas of seeds and small fruits of northwest-European plant species with morphological descriptions* Part 4. Stockholm: Swedish Museum of Natural History.

Asouti, E. and P. Austin. 2005. Reconstructing woodland vegetation and its exploitation by past societies, based on the analysis and interpretation of archaeological wood charcoal macro-remains. *Environmental Archaeology* 10(1):1–18.

Asouti, E. and J. Hather. 2001. Charcoal analysis and the reconstruction of ancient woodland vegetation in the Konya Basin, south-central Anatolia, Turkey: Results from the Neolithic site of Çatalhöyük East. *Vegetation History and Archaeobotany* 10:23–32.

Berggren, G. 1969. *Atlas of seeds and small fruits of northwest-European plant species with morphological descriptions Part 2*. Stockholm: Swedish Museum of Natural History.

Berggren, G. 1981. *Atlas of seeds and small fruits of northwest-European plant species with morphological descriptions Part 3*. Stockholm: Swedish Museum of Natural History.

Brazier, J. and G. Franklin. 1961. *Identification of hardwoods — A microscope key*. London: Her Majesty's Stationery Office.

Collins, P. E. F., D. J. Rust, M. Salih Bayraktutan and S. D. Turner. 2005. Fluvial stratigraphy and palaeoenvironments in the Pasinler Basin, eastern Turkey. *Quaternary International* 140–141: 121–134.

Connor, S. n.d. *A Holocene palaeoenvironmental reconstruction of the South Caucasus*. PhD thesis (in progress). Melbourne: School of Anthropology, Geography and Environmental Studies, The University of Melbourne.

Connor, S. E., C. Longford, J. Newton, and A. Sagona. n.d. A history of highland vegetation and prehistoric human activity in the Pasinler Valley, eastern Anatolia (in progress).

Connor, S. E., I. Thomas, E. V. Kvavadze, G. J. Arabuli, G. S. Avakov and A. Sagona. 2004. A survey of modern pollen and vegetation along an altitudinal transect in southern Georgia, Caucasus region. *Review of Palaeobotany and Palynology* 129:229–250.

Davis, P. (ed). 1965–1988. *Flora of Turkey and East Aegean Islands* Vols 1–9. Edinburgh: Edinburgh University Press.

Dönmez, E. 2003. Urartian crop plant remains from Patnos (Ağri) eastern Turkey. *Anatolian Studies* 53:89–95.

Dvorak, J., M.-C. Luo, Z.-L. Yang and H.-B. Zhang. 1998. The structure of Aegilops tauschii genepool and the evolution of hexaploid wheat. *Theoretical Applied Genetics* 97:657–670.

Ertuğ, F. 2000. An ethnobotanical study in Central Anatolia (Turkey). Economic Botany 54:155–182.

Fahn, A., E. Werker and P. Baas. 1986. *Wood anatomy and identification of trees and shrubs from Israel and adjacent regions*. Jerusalem: Israel Academy of Science and Humanities.

Fairbairn, A., E. Asouti, J. Near and D. Martinoli. 2002. Macro-botanical evidence for plant use at Neolithic Çatalhöyük, south-central Anatolia, Turkey. *Vegetation History and Archaeobotany* 11:41–54.

Frangipane, M., G. M. Di Nocera, A. Hauptmann, P. Morbidelli, A. Palmieri, L. Sadori, M. Schultz and T. Schmidt-Schultz. 2001. New symbols of a new power in a 'royal' tomb from 3,000 BC Arslantepe, Malatya (Turkey). *Paléorient* 27(2):105–139.

Greguss, P. 1955. *Identification of living gymnosperms on the basis of xylotomy*. Budapest: Akademiai Kiado.

Hongo, H. and M. Masikini. 2003. Impact of immigrant pastoral herds to fringing wet lands of Lake Victoria in Magu district, Mwanza region, Tanzania. *Physics and Chemistry of the Earth* 28:1001–1007.

Hopkins, L. 2000. *The ethnoarchaeology of northeastern Turkey — Sos Höyük and Yiğittaşı village*. Unpublished PhD thesis. Melbourne: School of Art History, Cinema, Classics and Archaeology, The University of Melbourne.

Horne, L. 1982. Fuel for the metal worker. *Expedition* 25:6–13.

Howell-Meurs, S. 2001. *Early Bronze and Iron Age animal exploitation in northeastern Anatolia: The faunal remains from Sos Höyük and Büyüktepe Höyük*. Oxford: Archaeopress.

Jennings, R. 1976. Urban population in Anatolia in the sixteenth century: A study of Kayseri, Karaman, Amasya, Trabzon and Erzurum. *International Journal of Middle Eastern Studies* 7:21–57.

Kvavadze, E. V. and S. E. Connor. 2005. *Zelkova carpinifolia* (Pallas) K. Koch in Holocene sediments of Georgia — an indicator of climatic optima. *Review of Palaeobotany and Palynology* 133:69– 89.

Kuzucuoğlu, C. 2003. Environmental changes in southern (Antalya) and south-eastern (Euphrates Valley) Turkey, at the end of the 2nd Millenium BC and beginning of the 1st Millennium BC. In B. Fischer, H. Genz, E. Jean and K. Koroglu (eds), *Identifying changes: The transition from Bronze Age to Iron Ages in Anatolia and its neighbouring regions*, pp 271–281. Istanbul: Turk Eskicağ Bilimleri Enstitusu.

Lemcke, G. and M. Sturm. 1997. δ18O and trace element measurements as proxy for the reconstruction of climate changes at Lake Van (Turkey), preliminary results. In H. N. Dalfes, G. Kukla and H. Weiss (eds), *Third Millennium BC climate change and Old World collapse*. NATO ASI Series 1, 49:653–678.

McNeill, J. 1992. *The mountains of the Mediterranean world*. Cambridge: Cambridge University Press.

Miller, N. 1985. Paleoethnobotanical evidence for deforestation in ancient Iran: A case study of urban Malyan. *Journal of Ethnobiology* 5:1–19.

Miller, N. 1986. Vegetation and land use. In G. Algaze, K. Ataman, M. Ingraham, L. Marfoe, M. McDonald, N. Miller, C. Snow, G. Stein, B. Verharen, P. Wattenmaker, T. Wilkinson and A. Yener. The Chicago Euphrates archaeological project 1980–1984: Interim report. *Anatolica* 13:85–89/119–20.

Miller, N. 1990a. Archaeobotanical perspectives on the rural-urban connection. In N. Miller (ed.), *Economy and Settlement in the Near East*, pp 79–83. Philadelphia: MASCA.

Miller, N. 1990b. Clearing land for farmland and fuel: Archaeobotanical studies of the ancient Near East. In N. Miller (ed.), *Economy and Settlement in the Near East*, pp 71–77. Philadelphia: MASCA.

Miller, N. 1991. The Near East. In W. van Zeist, K. Wasylikowa and K.-E. Behre (eds), *Progress in Old World Palaeoethnobotany*, pp 133–160. Rotterdam: Balkema.

Miller, N. 1998. Patterns of agriculture and land use at Medieval Gritille. In S. Redford (ed.), *The Archaeology of the Frontier in the Medieval Near East: Excavations at Gritille, Turkey*, pp 211–252. Philadelphia: University Museum Publications.

Miller, N. 2004. Flotation samples from the 1992 excavation at Tell Jouweif. In T. J. Wilkinson (ed.), *On the Margin of the Euphrates: Settlement and Land use at Tell es-Sweyhat and in the Upper Lake Assad Area, Syria*, pp 157–165. Chicago: Oriental Institute Publication.

Miller, N. and T. Smart. 1984. Intentional burning of dung as fuel: A mechanism for the incorporation of charred seeds into the archaeological record. *Journal of Ethnobiology* 4:15–28.

Newton, J. 2004. *Life in the mountains of Anatolia: A study of a Bronze Age settlement in its environmental context*. Unpublished PhD thesis. Melbourne: School of Anthropology, Geography and Environmental Studies, The University of Melbourne.

NEAAP 1994–2002. North-East Anatolia Archaeological Project Archives, Melbourne.

Parr, E. M., C. Briggs and A. Sagona. 1999. Physical anthropological analysis of skeletons from Sos Höyük. *Ancient Near Eastern Studies* 36:150–68.

Pearsall, D. 2000. *Paleoethnobotany. A handbook of procedures*. San Diego: Academic Press.

Riehl, S. 1999. *Bronze Age environment and economy in the Troad: The archaeobotany of Kumtepe and Troy*. BioArchaeologica 2. Tubingen: Mo Vince Verlag.

Riehl, S. and M. Nesbitt. 2003. Crops and cultivation in Iron Age Near East: Change or continuity? In B. Fischer, H. Genz, E. Jean and K. Koroglu (eds.), Identifying changes: The transition from Bronze Age to Iron Ages in Anatolia and its neighbouring regions, pp 301–312. Istanbul: Turk Eskicağ Bilimleri Enstitusu.

Roberts, N., J. M. Reed, M. J. Leng, C. Kuzucuoğlu, M. Fontugne, J. Bertaux, H. Woldring, S. Bottema, S. Black, E. Hunt and M. Karabıyıkoğlu. 2001. The tempo of Holocene climatic change in the eastern Mediterranean region: New high-resolution crater-lake sediment data from central Turkey. *The Holocene* 11(6):721–736.

Sagona, A. 2000. Sos Höyük and the Erzurum region in late prehistory: A provisional chronology for northeast Anatolia. In C. Marro and H. Hauptmann (eds), *Chronologies des pays du Caucase et de l'Euphrate aux IVè–IIIè millénaires*, pp 329–373. Paris: de Boccard.

Sagona, A., M. Erkmen, C. Sagona and I. Thomas. 1996. Excavations at Sos Höyük, 1995, second preliminary report. *Anatolian Studies* 46:27–52

Sagona, A., M. Erkmen, C. Sagona and S. Howells. 1997. Excavations at Sos Höyük, 1996, third preliminary report. *Anatolica* 23:181–226.

Sagona, A., M. Erkmen, C. Sagona and I. McNiven. 1998. Excavations at Sos Höyük, 1997, fourth preliminary report. *Anatolica* 24:31–64.

Sagona, A. and C. Sagona. 2000. Excavations at Sos Höyük, 1998 to 2000: Fifth preliminary report. *Ancient Near Eastern Studies* 37:56–127.

Sagona, A., C. Sagona and H. Ozkorucuklu. 1995. Excavations at Sos Höyük 1994 — First preliminary report. *Anatolian Studies* 45:193–218.

Sagona, C. 2004. Did Xenophon take the Aras Highroad? Observations on the historical geography of North-east Anatolia. In A. Sagona (ed.), *A view from the Highlands*, pp 299–333. Herent: Peeters.

Schoch, W., I. Heller, F. H. Schweingruber and F. Kienast. 2004. Wood anatomy of Central European species. URL: http://www.woodanatomy.ch

Stern, R. 1983. Human impact on tree borderlines. In W. Holzner, M. Werger and I. Ikusima (eds.), *Man's impact on vegetation*, pp 275–299. The Hague: Dr W. Junk Publishers.

Smart, T. and E. Hoffman. 1988. Environmental interpretation of archaeological charcoal. In C. Hastorf and V. Popper (eds), *Current paleoethnobotany*, pp 167–205. Chicago: University of Chicago Press.

Tournefort, J. P. de. 1718. *A voyage into the Levant: Perform'd by command of the late French King*, Vol 2, (trans. J. Ozell). London: D. Brown and Co.

van Zeist, W. and J. Bakker-Heeres. 1985. Archaeobotanical studies in the Levant. 1. Neolithic sites in the Damascus Basin: Aswad, Ghoraifé, Ramad. *Palaeohistoria* 24:165–256.

van Zeist, W. and J. Bakker-Heeres. 1986. Archaeobotanical studies in the Levant. 4. Bronze Age sites on the North Syrian Euphrates. *Palaeohistoria* 27:247–316.

van Zeist, W. and H. Woldring. 1978. A postglacial pollen diagram from Lake Van in east Anatolia. *Review of Palaeobotany and Palynology* 26:249–276.

Vermoere, M., E. Smets, M. Waelkens, H. Vanhaverbeke, I. Librecht, E. Paulissen and L. Vanhecke. 2000. Late Holocene environmental change and the record of human impact at Gravgaz near Sagalossos, Southwest Turkey. *Journal of Archaeological Science* 27:571–595.

Wasylikowa, K., M. Carciumaru, E. Hajnalova, B. Hartyanyi, G. Pashkevich and Z. Yanushevich. 1991. East-Central Europe. In W. van Zeist, K. Wasylikowa and K.-E. Behre (eds), *Progress in Old World palaeoethnobotany*, pp 207–239. Rotterdam: Balkema.

Wick, L., G. Lemcke and M. Sturm. 2003. Evidence of late glacial and Holocene climatic change and human impact in eastern Anatolia: High-resolution pollen, charcoal, isotopic and geochemical records from the laminated sediments of Lake Van, Turkey. *The Holocene* 13(5):665– 675.

Willcox, G. 1974 A history of deforestation as indicated by charcoal analysis of four sites in Eastern Anatolia. *Anatolian Studies* 24:117–133.

Willcox, G. 2002. Evidence for ancient forest cover and deforestation from charcoal analysis of ten archaeological sites on the Euphrates. In S. Thiébault (ed.), *Charcoal analysis: Methodological approaches, palaeoecological results and wood use*, pp 141–145. Oxford: Archaeopress.

Woldring, H. and R. Cappers. 2001. The origin of the 'Wild Orchards' of Central Anatolia. *Turkish Journal of Botany* 25:1–9.

Xenophon. 1972. *The Persian expedition*, (trans. R. Warner). Aylesbury: Penguin Classics.

Zohary D. and M. Hopf. 2000. *Domestication of plants in the Old World*. Oxford: Oxford University Press

Zohary, M. 1973. *The Geobotanical foundations of the Middle East*, Vol 1–2. Amsterdam: Swets and Zeitlinger.

10

A multi-disciplinary method for the investigation of early agriculture: Learning lessons from Kuk

Tim Denham[1]
Simon Haberle[2]
Alain Pierret[3]

[1] School of Geography and Environmental Science
Building 11
Monash University
VIC 3800, Australia
Tim.Denham@arts.monash.edu.au

[2] Department of Archaeology and Natural History
Research School of Pacific and Asian Studies
Australian National University
Canberra ACT 0200, Australia
Simon.Haberle@anu.edu.au

[3] IRD-IWMI-NAFRI
UR 176-Solutions
BP 06, Vientiane
Lao PDR
apierret@gmail.com

Abstract

The multi-disciplinary methods used to investigate early agriculture at Kuk Swamp in the Highlands of New Guinea are outlined. Methods adopted during the original investigations in the 1970s (directed by Jack Golson, with Philip Hughes from 1974), as well as during renewed investigations at the site (directed by Tim Denham from 1997 under Golson's supervision) are considered. Three methodological contributions to the study of early agriculture and plant exploitation are highlighted: X-radiography, biostratigraphic markers, and the integration of macrofossil and microfossil analyses. One outstanding methodological problem is considered, the representativeness of different components of a feature fill for reconstructions of past cultivation practices. Thoughts on an integrated method (macrofossil, microfossil and molecular) for future research on early agriculture and arboriculture in the region are presented.

Keywords: Multi-disciplinary method; early agriculture; X-radiography; biostratigraphic markers; macrofossils and microfossils

Introduction

Multi-disciplinary investigations at Kuk Swamp in the Upper Wahgi Valley have confirmed New Guinea to be a centre of early and independent agricultural development (Denham et al. 2003, 2004a, 2004b; Golson 1977, 1991; Golson and Hughes 1980; Hope and Golson 1995). These investigations have identified successive periods of manipulation of the wetland margin for plant exploitation and swamp drainage for cultivation. Although claims for agriculture dating to 10,000 years ago are contentious (Phase 1; Denham 2004a; Denham et al. 2004a), there is general agreement that mounded cultivation occurred on the wetland margin at c. 7000/6500 cal. BP (Phase 2; Denham et al. 2003, 2004a). From approximately 4000 years ago to the present (from Phase 3 onwards), the wetland was periodically drained using ditches. The agricultural history at Kuk, at least from approximately 6000 years ago, is corroborated by archaeological and palaeoecological findings at other sites in the New Guinea Highlands (Denham 2003a, 2005a; Golson 1982; Haberle 2003; Powell 1982a).

In this paper, the multi-disciplinary methods used to investigate early agriculture at Kuk are summarised. The rationale of the methodological approach adopted is outlined and three significant methodological contributions are sketched. A subsequent methodological problem arising from the research is considered. The paper concludes with some thoughts on developing new integrated research methodologies for future investigations of plant exploitation in the Pacific.

The multi-disciplinary method: Description, rationale and value

Initial claims for early and independent agriculture in New Guinea were based on multi-disciplinary investigations at Kuk directed by Jack Golson from the 1970s onwards (and with Philip Hughes from 1974; Golson 1977, 1991; Golson and Hughes 1980; Hope and Golson 1995). As discussed elsewhere (Denham 2006), these claims were not universally accepted (Bayliss-Smith 1996; Spriggs 1996). The reasons for scepticism centred on limitations with the multi-disciplinary lines of evidence published for the early and mid-Holocene at Kuk. These limitations were:

- a lack of published archaeological evidence;
- uncertainties regarding the mode of formation and function of archaeological features;
- a lack of palaeoecological evidence contemporary with the earliest claimed agricultural remains; and,
- equivocal archaeobotanical evidence for the presence, use and cultivation of plants (cf. Powell 1982b; Wilson 1985).

Denham's multi-disciplinary investigations of early agriculture at Kuk, initiated from 1997 onwards, were designed to address these evidential deficiencies (Tables 1 and 2; Denham 2003b). A team of researchers from a range of disciplines was required because each contributed essential information to the investigation of early agriculture and plant exploitation:

- Archaeology: evidence of the features and artefacts associated with former cultivation and plant exploitation practices.
- Archaeobotany: evidence for the presence, use and cultivation of edible, and otherwise useful, plants.
- Dating: chronological resolution for interpretations of past human activities.
- Palaeoecology: evidence of environmental transformations associated with former practices, and their differentiation from climatic and tectonic-induced transformations.
- Stratigraphy: characterisation of former palaeosols and sedimentation through time, including evidence for past soil preparation and the effects of post-depositional processes.

1. Archival study of previous multi-disciplinary investigations in 1970s and early 1980s

2. New archaeological excavations at Kuk in 1998 and 1999

3. New multi-disciplinary investigations using a suite of dating, palaeoecological and stratigraphic analyses

Integration of archaeological and multi-disciplinary investigations to determine types of plant exploitation associated with Phases 1, 2 and 3 at Kuk

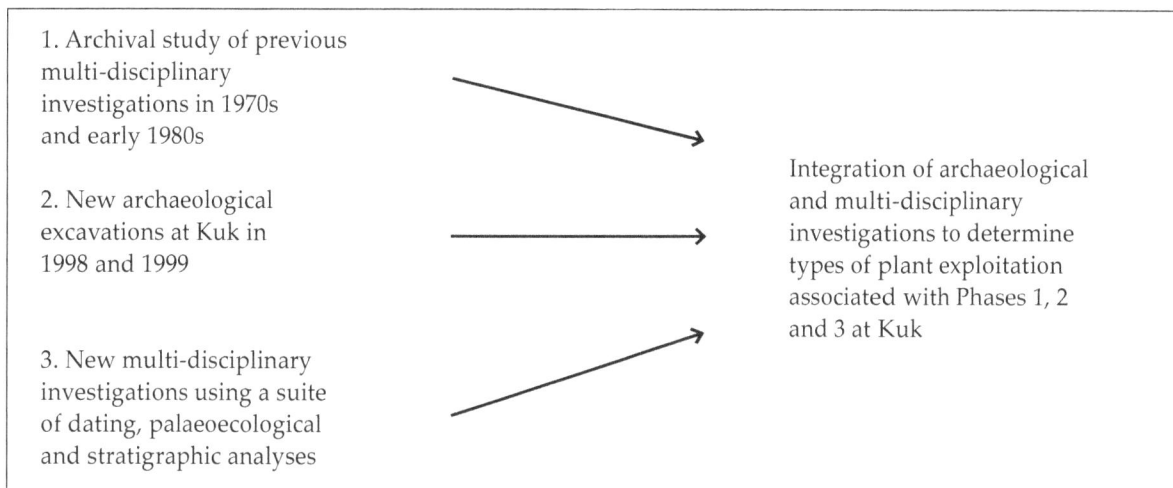

Table 1. Overview of the three main research components for renewed (from 1997) multi-disciplinary investigations of early and mid-Holocene remains at Kuk.

Research Field	Period	Methods
Archaeology	Original	Excavation trenches (n=187) Archaeological and stratigraphic recording in plantation drain walls (n=>15 km)
	Renewed	Excavation trenches (n=19)
Dating[1]	Original	Conventional radiocarbon dating (n=52; RSES, ANU)
	Renewed	Conventional and AMS radiocarbon dating (n=36; RSES and ANSTO)
Palaeoecology	Original	Macrobotany: seeds and wood (n>500; J. Powell/L. Lucking) Phytoliths (n=30; S. Wilson) Pollen (n=31; J. Powell)
	Renewed	Diatoms[2] (n=50; B. Winsborough) Insects (n=10; N. Porch) Phytoliths[2] (n=40; C. Lentfer) Pollen[2] (n=60; S. Haberle) Tool residues (n=12; R. Fullagar, J. Field, C. Lentfer, M. Therin)
Stratigraphy	Original	Deposition rates (P. Hughes) Chemical and physical composition (M. Latham) Ferrimagnetism (R. Thompson and F. Oldfield) Physical composition (J. Powell) X-radiography (R. Blong)
	Renewed	Thin section description (T. Denham) X-radiography (A. Pierret with T. Denham) X-ray diffraction (L. Moore with T. Denham)

Notes:
[1] Conventional and AMS radiocarbon dating undertaken by Research School of Earth Sciences (RSES), Australian National University (ANU) and AMS dating by the Australian Nuclear Science and Technology Organisation (ANSTO).
[2] During the current project, diatom, phytolith and pollen analyses were undertaken by specialists on 40 'paired', or comparable, samples, with additional samples for some techniques. Several of these paired samples were also subject to AMS dating.

Table 2. Summary of archaeological, dating, palaeoecological and stratigraphic work undertaken by individuals and organisations during the original (directed by Jack Golson with Philip Hughes) and renewed (directed by Tim Denham under Golson's supervision) investigations. Note that the renewed investigations solely focussed on early and mid-Holocene remains (Phases 1, 2 and 3).

Of particular concern during the renewed investigations at Kuk were issues concerning the cross-correlation of multi-disciplinary lines of evidence. The results of many previous analyses could not be readily cross-correlated with each other or with precision to stratigraphic and archaeological provenances. A high degree of integration and precision was subsequently sought during renewed investigations.

In the field samples were collected using sections of zinc piping. The location and survey levels of these tins were marked directly on excavation plans and stratigraphic profiles (see Figures 1a). In the laboratory multi-disciplinary sub-sampling of the tins occurred to enable the cross-correlation of results (see Figure 1b). The majority of diatom, phytolith and pollen samples, as well as some AMS samples, were paired, i.e., they were obtained from adjacent, comparable provenances. Three methodological contributions of these multi-disciplinary analyses are considered here.

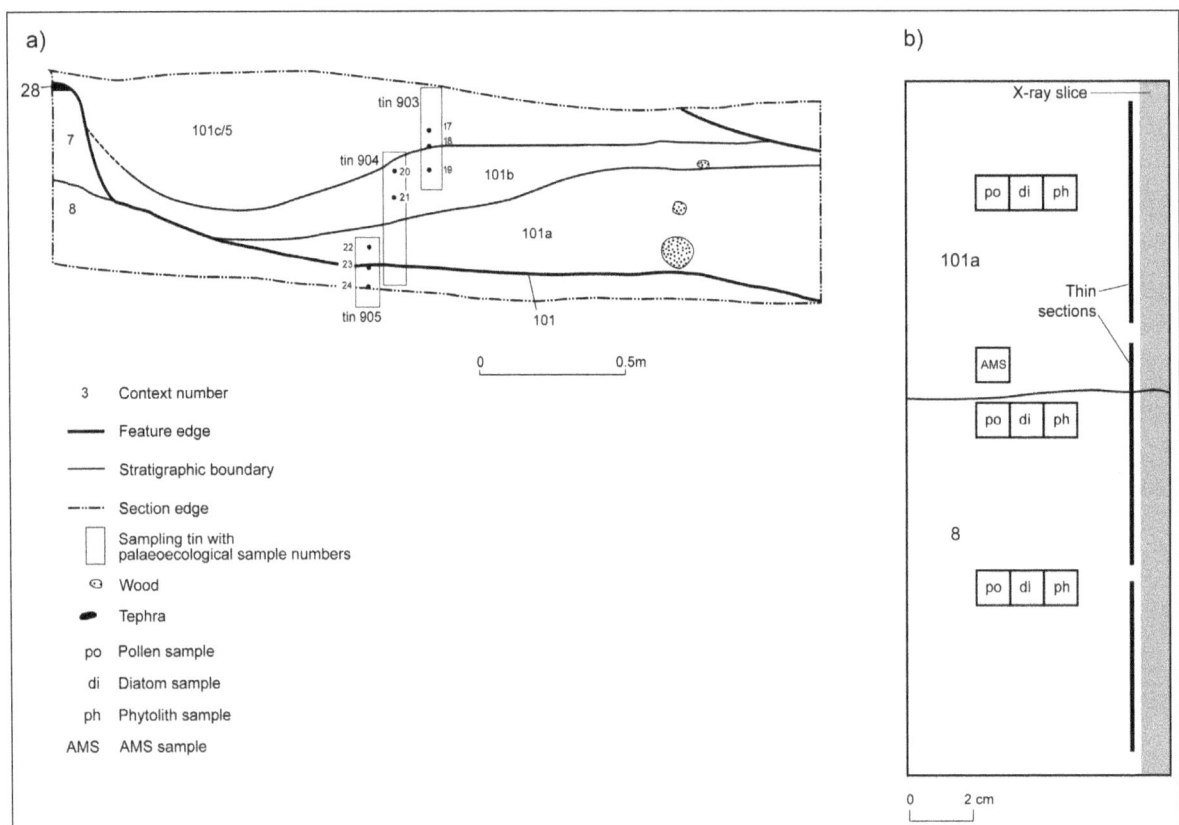

Figure 1. Field and laboratory sampling at Kuk: (a) Section of the stratigraphy indicating the location of monolith samples; and, (b) Idealised representation of multi-disciplinary sub-sampling from a monolith. In practice, thin sections and sometimes X-rays were often derived from a monolith paired with that used for dating, palaeoecological and sedimentological sub-sampling.

1. X-radiography

Comparative X-radiography and photography of undisturbed soil monoliths provide a meso-scale investigation of soil and sediment characteristics often missing in archaeological investigations (Gilbertson 1995). Meso-level analysis links macro-level field descriptions to extremely detailed micromorphological studies, i.e., thin section description. Following Hamblin (1962), Krinitszky has shown that '"thick bedded" or "massive" sedimentary deposits really contain many complex primary structures that are visible with X-rays but are otherwise poorly expressed or invisible' (1970:47). More recently, X-radiography has been used at archaeological sites to detect soil structures and pedoturbation (Butler 1992), to reveal tephra lenses in peats (Dugmore and Newton 1992), and to rapidly assess primary and secondary attributes of deposits prior to subsequent analysis (Barham 1995).

At Kuk, X-radiography has been used to investigate massively structured fills of archaeological features, as well as major stratigraphic units, and to guide further sampling. During recent work, X-radiography has revealed structures associated with former palaeosols (Figure 2a) and the degree of pedogenic homogenisation within a sample. The differentiation of deposits that retain their original stratification (Figure 2b), as opposed to those that have been subject to extensive post-depositional pedogenic modification (Figure 2c), has proven highly significant for the choice of samples for subsequent analysis, for understanding site formation processes, and for the interpretation of analytical results (see below; Denham 2003b).

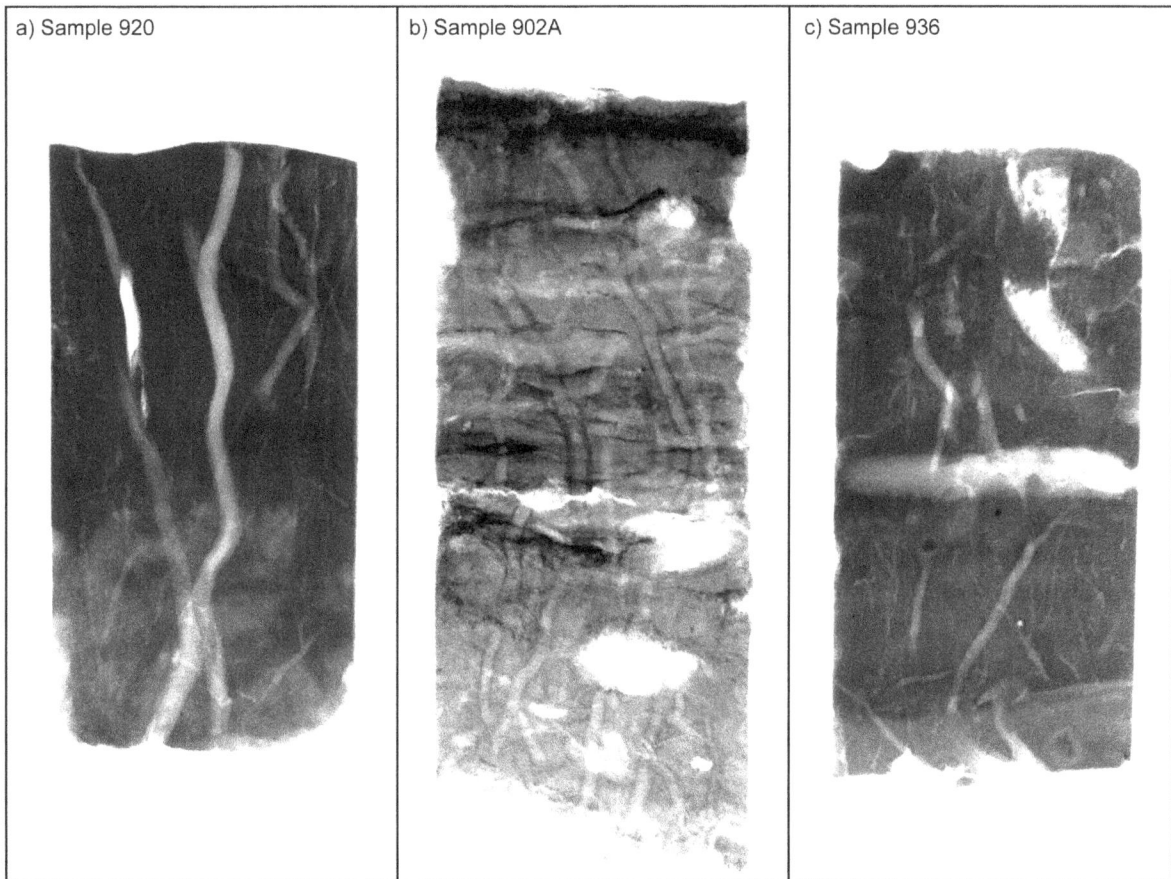

Figure 2. X-ray absorption images for samples from Kuk: (a) X-ray image showing recent vertically-oriented voids (light areas), which are traces of bioturbation by either soil macrofauna or plant roots, and older limited palaeosol development for the fill of a 10,000 year-old palaeosurface feature (Phase 1; sample 920); (b) X-ray image showing high degree of preservation of stratification within a 10,000 year-old palaeochannel fill (Phase 1, channel 101; sample 902A); and, (c) X-ray image showing well-homogenised black clay stratigraphic unit with superimposed vertically and horizontally oriented voids (lighter areas) representing recent root and microfaunal activity (mid-Holocene age; sample 936). Each sample is 8 cm wide.

2. Biostratigraphic markers

Multi-scale and mixed-method analyses (following Canti 1995; Denham 2003b), incorporating the results of the original and renewed investigations, have enabled characterisations of the archaeostratigraphy at Kuk (Denham 2003b). Of concern, and in contrast to previous sedimentary interpretations of the stratigraphy at Kuk (Hughes 1985; Hughes et al. 1991), the majority of the Holocene stratigraphy at the site appears to have been subject to considerable pedogenesis, or soil formation. Soil formation processes generally admix deposits through various forms of biological and mechanical pedoturbation. Consequently, the extent of pedogenesis needs to be taken into account in the interpretation of analytical results on samples used for dating and palaeoecological analysis.

A way of assessing the effects of post-depositional processes, such as pedogenesis, is to compare the archaeostratigraphy of the site against biostratigraphic markers. It can be assumed that if post-depositional processes have predominated, then the characteristics of macrofossil and microfossil assemblages within samples of similar provenance and age would be partially inter-mixed with those from adjacent — most probably higher and lower — contexts. However, if archaeostratigraphic units retain their original palaeoecological characteristics, i.e., those that reflect the ecology of the environment at the time of formation, then samples from similar provenances and age should share similar biostratigraphic traits.

The method for assessing biostratigraphic integrity at Kuk is based on a Principal Components Analysis (PCA) of the pollen analysis (Figure 3; see Haberle et al. n.d. for a full discussion). An examination of groupings clearly shows that samples from the same major stratigraphic units, i.e., black clay, grey clay and Pleistocene organic peat, cluster together. Furthermore, the basal samples from Phase 2 feature fills form a tight cluster, which is significant, because it suggests that the basal fills of Phase 2 features have retained their original biostratigraphic characteristics and are reliable indicators of past environments when the Phase 2 palaeosurface was in use. In contrast, the biostratigraphic signatures for the fills of Phase 1 features are highly variable, which suggests that the Phase 1 palaeosurface is, in part, a palimpsest of features of different ages.

3. A suite of macrofossil and microfossil techniques

A combination of archaeobotanical and palaeoecological techniques (see Table 2) has yielded evidence for the presence, use and cultivation of a range of crop plants at Kuk from the Pleistocene to the recent past (Table 3). Only the earliest occurrences of plants are noted, after which food plants

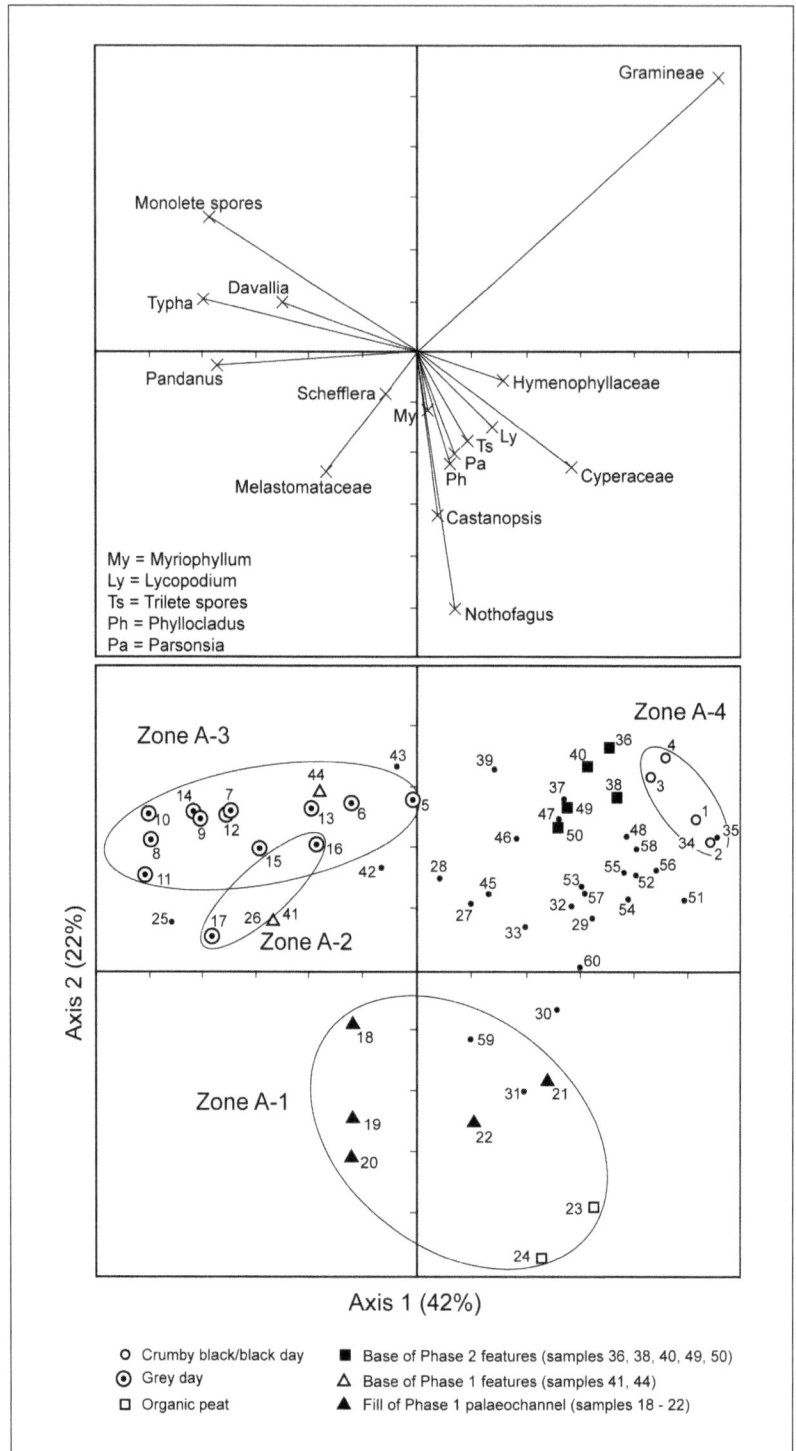

Figure 3. PCA plot of 60 pollen samples showing biostratigraphic groupings including major stratigraphic units and the relatively tight cluster of samples from the base of five Phase 2 features.

are considered to have been continuously present in the vicinity. Previous deficiencies in obtaining evidence of food plants based on macrobotanical and pollen analyses have been overcome by employing phytolith analysis of archaeostratigraphic samples and starch grain evidence of tool residues. These techniques have opened up new avenues for exploring subsistence in the past in New Guinea, as well as other parts of the world (see Piperno 2006 and Torrence and Barton 2005 for recent reviews). Of most significance are phytolith evidence for the cultivation of Musa bananas from 7000–6500 cal. BP (see discussions in Denham et al. 2003, 2004b) and starch grain analysis of residues from stone tools indicating the exploitation of taro (*Colocasia esculenta*) and a yam (*Dioscorea* sp.) from the early Holocene (Fullagar et al. 2006).

Species/Genus[1]	Exploited Form[2]	Edible Part(s)[3]	Evidence[4]	Earliest Record[5]
Abelmoschus sp.[6]	c	l, sh	s	Pleistocene
Acalypha sp.	w, t	l	s, w, p	Pleistocene
Castanopsis sp.	w, t	n	w, p?	Pleistocene
Cerastium sp.	w	p	s	Pleistocene
Coleus sp.	w	l	s	Pleistocene
Ficus cf. *copiosa*[6]	c, w	f, l	s	Pleistocene
Ficus spp.	c, w	l, f	s, w	Pleistocene
Garcinia sp.	w	f, l, b	w	Pleistocene
Hydrocotyle sp.	w	l?	s	Pleistocene
Lycopodium spp.	w	sh	p	Pleistocene
Maesa sp.	w	f	w	Pleistocene
Musaceae	c, w	f, c	ph	Pleistocene
Oenanthe javanica	c, w	l, sh	s, p	Pleistocene
P. antaresensis	w	d	p	Pleistocene
P. brosimos	c, w	d	p	Pleistocene
Pandanus sp.	c, w	d	s, p	Pleistocene
Parsonsia sp.	w	n	p	Pleistocene
Phragmites karka	w	l, r, sh	ph	Pleistocene
Pouzolzia hirta	w	l, st	s	Pleistocene
Rubus moluccanus	w	f	s	Pleistocene
Rubus rosifolius	w	f	s	Pleistocene
cf. *Setaria palmifolia*	c, w	s	ph	Pleistocene
Solanum nigrum	c, w	l, sh	s	Pleistocene
Syzygium sp.	w	f	w	Pleistocene
cf. Zingiberaceae	c, w	r, l, sh	ph	Pleistocene
Colocasia esculenta	c	c, l	st	P1
Dioscorea sp.	c, w	t	st	P1
Elaeocarpaceae	w	n	p	P1
Ipomoea sp.	w	sh	p	P1
Typha sp.	w	st	p	P1
Wahlenbergia sp.	w	p	s, p	P1
Musa section bananas[7]	c, w	f, c	ph	P1?
Ingentimusa section bananas[7]	w	f, c?	ph	P1?
Solanum sp.	c, w	f, l, sh, t	s	pre-P2
Commelina sp.[6]	c, w	l, sh	s	P2

Species/Genus[1]	Exploited Form[2]	Edible Part(s)[3]	Evidence[4]	Earliest Record[5]
Drymaria cordata[6]	w	l?	s	P2
Floscopa sp.[6]	c, w	l, sh	s	P2
Viola arcuata	w	l?	s	P2/P3
Amaranthus sp.[6]	c	l, p	s	P2/P3
Bidens pilosa[6]	w	s	s	P2/P3
Selaginella sp.	w	l	p	P3

Notes:

[1] List of edible species at Kuk based on ethnographically documented use of plants in New Guinea (M. Bourke n.d. and pers. comm. 2002; French 1986; Haberle 1995; G. Hope pers. comm. 2002; Powell 1976:108–12; Powell and Harrison 1982:57–86; Powell et al. 1975:15–39). Edible species have been reported from other early to mid Holocene sites in the Highlands[8, 9, 10].

[2] Exploited form: c = cultivated, w = wild, t = transplanted.

[3] Edible part(s): c = corm, d = drupe, f = fruit, g = gourd, l = leaf, n = nut, p = plant, r = rhizome, s = seed, sh = shoot, st = stem, t = tuber.

[4] Evidence: ph = phytolith (by Carol Lentfer), p = pollen (by Simon Haberle), s = seed (by Jocelyn Powell and Laurie Lucking), st = starch (by Richard Fullagar and Judith Field), w = wood (by Jocelyn Powell and Laurie Lucking). Only those techniques relevant for the earliest recorded occurrence are listed.

[5] Earliest record: Pleistocene (Pleistocene); Phase 1 contexts (P1); possible Phase 1 association or immediately post-dating Phase 1 (P1?); contexts post-dating Phase 1 and pre-dating Phase 2 (pre-P2); Phase 2 contexts (P2); Phase 2 or Phase 3 contexts (P2/P3); and, Phase 3 contexts (P3).

[6] Identification to the genus or species level should be considered provisional because it is based on a single seed sample.

[7] The identification of Australimusa phytoliths from a Phase 2 context at Kuk (Wilson 1985:Table 3) is excluded because it was based on a single sample and no Australimusa phytolith morphotypes were identified by Carol Lentfer from Phase 1–3 contexts at Kuk during renewed investigations (Denham et al. 2003:192).

[8] Gourd from a mid-Holocene context dating to 5665–5325 cal. BP at Warrawau (ANU 288 and 2086; Golson 2002:74) was originally reported as *Lagenaria siceraria* (Powell 1970b:144–5) and *Lagenaria cf. siceraria* (Powell 1970a:199). Golson (2002:73–5) considers it more likely to be *Benincasa hispida* based on the identification of this species from a late Holocene context at Kana (Matthews 2003). Given the lack of macrobotanical investigation conducted on the gourd remains collected in 1966 and 1977, and which could not be located in archival materials during renewed investigations, they are regarded to be 'gourd (species unknown)'.

[9] *Saccharum officinarum* dating to c. 5200 cal. BP at Yuku was reported by Bulmer (1975:31) although the basis for the identification is unknown (Yen 1998:168).

[10] *Benincasa hispida* has been reported from Kana and dated to 2950–2000 cal. BP (ANU 9487; Matthews 2003; Muke and Mandui 2003).

Table 3. List of the earliest occurrence of edible plants documented for all phases at Kuk (augmented and updated version of Denham 2005b: Table 2).

The suite of food plants present at Kuk is similar to those harvested wild and cultivated in gardens across the Upper Wahgi valley today (Powell 1976; Powell et al. 1975). The range of food plants potentially available to former inhabitants includes starchy staples, vegetables, and fruit and nut-bearing trees (see Denham 2005b and Denham and Barton 2006). Although residue analysis provides evidence for the use of taro, a yam, and as yet unidentified plants, most techniques solely indicate the availability of a wide range of edible plants. However, taken together, the suite of plants could have supported broad-spectrum subsistence practices from the beginning of the Holocene, and potentially earlier (Powell 1982a:211).

The findings at Kuk indicate the importance of using a combination of macrofossil and microfossil techniques in the investigation of plant exploitation in the past. Although macrofossil techniques — primarily the macrobotanical investigation of the hard parts of nuts and fruits, as well as seeds and wood — have formerly predominated in New Guinean research, it has long been recognised that these techniques provide only a partial view of plant exploitation in the past (Golson and Ucko 1994; Powell 1970a). Only with recent advances in phytolith (Bowdery 1999; Denham et al. 2003, 2004b; Lentfer and Green 2004), pollen (Haberle 1995), starch grain (Fullagar et al. 2006; cf. Loy 1994) and parenchyma (Hather 2000) research have archaeobotanists been able to investigate people's use of the major starch-rich staples of Pacific agriculture. At Kuk, microfossil analyses have provided evidence from the early Holocene for the presence or exploitation of three of the major starch-rich staples of Pacific agriculture, namely taro, yam and bananas of Musa section (formerly Eumusa). These archaeobotanical finds broadly corroborate phytogeographic hypotheses and genetic interpretations for the domestication of these, as well as other, plants in the New Guinea region (see de Langhe and de Maret 1999; Lebot 1999; Matthews 1995).

An outstanding problem: Sample representativeness

At most sites in the Australasian region, multiple samples from an archaeological context or feature fill are rarely subject to detailed archaeobotanical, dating or palaeoecological analysis. Although there are exceptions, these are usually larger features such as midden deposits, e.g., the Dongan midden (Fairbairn and Swadling 2005; Swadling et al. 1991), and palaeochannels, e.g., at Kuk (Powell 1982b; Wilson 1995). For smaller features, it has been common practice to interpret a subsample of fill — particularly a basal fill — as representative of the human practices accompanying the formation or use of the feature as a whole.

During the renewed investigations at Kuk, multiple samples were taken from larger (i.e., palaeochannels) and smaller (i.e., palaeosurfaces) features. Multiple subsamples of fill were subject to multi-disciplinary and paired analyses. Although the results indicated quite high degrees of variability in the biostratigraphic signatures and age for the fills from several palaeochannels, such results were anticipated based on studies of channel fill composition and chronology in other parts of the world (Brown 1997:48–54). Surprisingly, multiple samples from the fills of shallow palaeosurface features also indicated relatively high degrees of biostratigraphic variability — both within and between features; variability was more marked for Phase 1 features and less so for Phase 2 features. However, even for Phase 2 features, for each of which three subsamples were analysed, there was considerable variability between the upper, intermediate and lower subsamples, even if there was good agreement for the basal fills of all five Phase 2 features analysed (as discussed above). These findings suggest that different environmental conditions and different processes of formation are associated with different parts of a fill. Although it has been assumed that the basal portion of the fill is the most representative of past plant exploitation and cultivation associated with the formation and use of the feature, this assumption requires further investigation.

Different hypothetical interpretations of the processes leading to the formation of, and hence representativeness of, the basal and other parts of the fill can be advanced (see Figure 4). These hypothetical interpretations cast doubt on the reliability of assuming that the basal portion of the fill is representative of — at least in the case of Phase 2 features — cultivation practices associated with formation and use. For example, is the basal portion of the fill formed shortly after initial digging of the feature, thereby comprising largely inwashed residual materials, or does it represent materials that slowly accumulated during cultivation on adjacent mounds (Figures 4a–c)? Additionally, the extent of pedogenesis accompanying and subsequent to the infilling of the feature is unknown (Figure 4d). Depending upon the formation scenario, plant macro- and microfossil assemblages in the basal portions of the feature fill may or may not be representative of cultivation on adjacent mounds (Figure 4e). Although in part these concerns can be addressed through X-radiography and thin section analysis of the relevant fills, some uncertainties remain and are largely a function of trying to understand the rates at which infilling of the features occurred during the year or so of use.

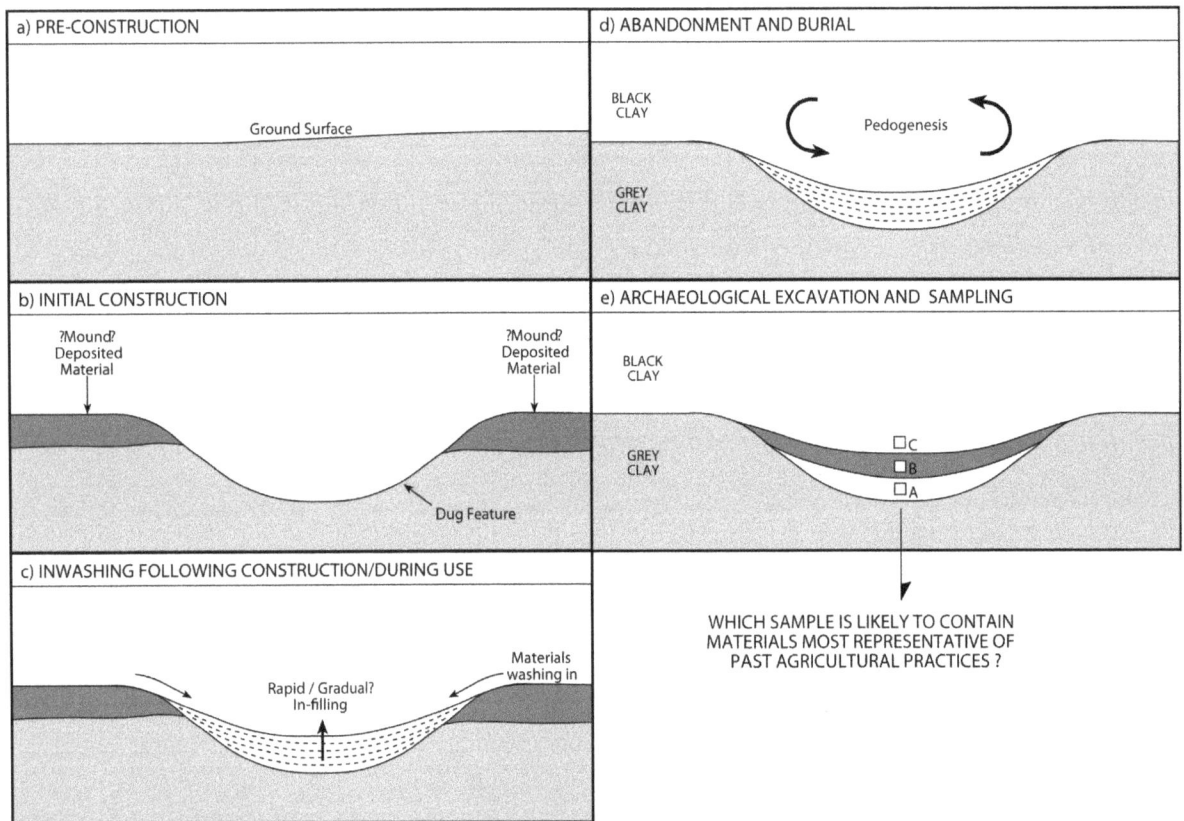

Figure 4. Schematic representation of a former land surface (a), subject to mound construction (b), inwashing following construction and during use (c), abandonment and burial (d), and archaeological excavation and sampling (e).

In an attempt to gain greater interpretative resolution, it has been necessary to adopt continuous, multi-proxy (diatom, phytolith and pollen analyses) and paired sampling strategies through the fills of the feature and underlying deposits (Figure 5). The 1 cm wide slices used for X-radiography of the stratigraphy have been subsampled for Phase 1, Phase 2 and Phase 3 feature fills — with the location of all subsamples marked on the X-ray image. Although the results of these analyses are incomplete, preliminary indications based on diatom and pollen analyses do facilitate greater interpretative resolution and show clear patterns in multi-proxy data that can be readily interpreted with regard to site formation processes and former cultivation practices.

The basal portion of the X-ray image depicts limited aggregate formation at the base of a 10,000 year-old (Phase 1) feature (Figure 5a). The palaeosol was buried beneath massively structured grey clay. The vertically-oriented voids indicate recent pedoturbation. The summary pollen diagram (Figure 5b) shows an increasing and then stable forest signal (samples H–D), which then dramatically declines (samples C–B). The decline in forest cover corresponds to the period when the Phase 1 feature had formed and was in use; manipulation of the wetland was occurring locally in conjunction with a form of plant exploitation. In the uppermost sample (A) the forest signal increases dramatically, although it is

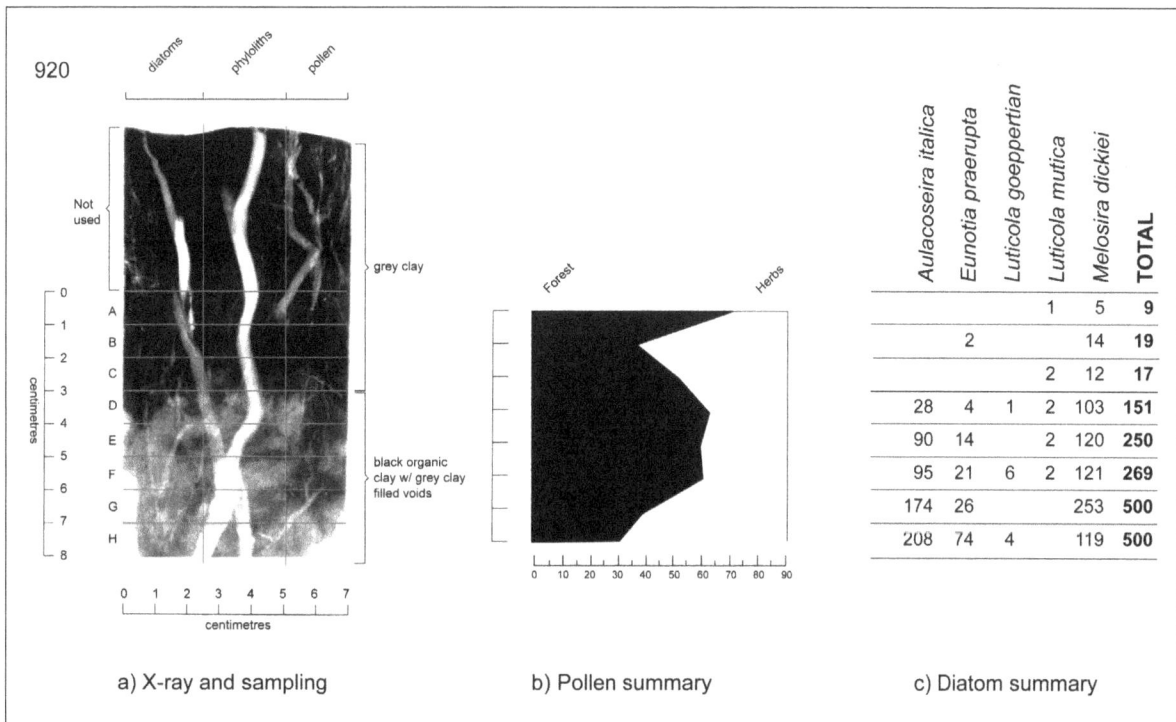

Figure 5. Panel showing the preliminary results of continuous and multi-proxy sampling strategy adopted during ongoing work at Kuk: (a) X-ray absorption image overlain with locations of subsamples for diatom, phytolith and pollen analyses; (b) Summary pollen data (courtesy of Kale Sniderman); and, (c) Summary table of diatom data (courtesy of Barbara Winsborough). Note that the sample had partially desiccated and shrank to c. 7cm wide by the time of continuous sampling.

uncertain whether this is due to forest regeneration following cessation of plant exploitation locally, or inwashing of older sediments from the edge of the feature. The diatom record indicates locally wetter conditions prior to feature formation (samples H–D), but the much lower frequencies of diatoms in samples C–A indicate drier conditions locally when the feature was in use and had infilled.

All three techniques — stratigraphic, pollen and diatom — provide complementary lines of evidence. Together they indicate wetter and forested conditions locally, prior to manipulation of the wetland margin for plant exploitation at 10,000 years ago. The feature was dug and in use during a short-lived period of locally drier conditions that were accompanied by a reduction in forest cover and limited soil formation.

A look to the future

In a series of papers, Bruce Smith has advocated and demonstrated the need for greater rigour in the investigation of the archaeobotanical record (Smith 1998, 2001, 2005; Erickson et al. 2005). Smith has developed a research methodology that utilises archaeobotany (primarily macrobotanical analysis) and the direct AMS dating of identified macrobotanical remains, in conjunction with comprehensive genetic fingerprinting of modern crop plant distributions. By adopting this multi-disciplinary approach, the present-day locations of wild progenitor populations of several domesticated crop plants have been identified using genetics and compared to the earliest archaeological evidence demonstrating the domestication of the same crop plants.

Although a wholesale application of Smith's method to the Pacific context is unlikely in the short term, given the paucity of archaeobotanical data and a reliance on microfossils to identify former starch-rich staple crops, there is certainly a case for partial adoption. Indeed, recent direct AMS dating of macrobotanical remains at the Dongan midden site has already yielded significant results (Fairbairn and Swadling 2005); the original report of mid-Holocene betelnut (*Areca catechu*) at the site (Swadling et al. 1991) has proved incorrect and the betelnut has been shown to be a modern intrusion. There is certainly scope for the greater direct AMS dating of other potentially significant, yet controversial, archaeobotanical finds, such as the putative sugarcane (*Saccharum officinarum*) at Yuku (Bulmer 1975:31; cf. Yen 1998:168).

However, there is an urgent need to improve, perhaps revamp, the methodologies used to investigate plant exploitation in the Pacific. Firstly, the systematic and tandem employment of macrofossil and microfossil techniques is essential to recover evidence for the presence, use and cultivation of a range of edible plants, as discussed above with reference to Kuk. Without a suite of techniques, only a partial view of plant availability and use in the past can be reconstructed.

Secondly, molecular databases for the major Pacific cultivars require to be systematically compiled with accessions from all regions of the Pacific and Southeast Asia. Although this work is ongoing, there is some variability in geographical coverage. For example, Lebot et al.'s (2004) genetic characterisations of taro (*Colocasia esculenta*) include accessions from several regions of Melanesia, Indo-Malaysia and Southeast Asia, whereas a similar study of the genetic relatedness of yam species (*Dioscorea* spp; Malapa et al. 2005) excluded potentially significant centres of yam diversity, namely New Guinea and the Indo-Malaysian archipelago. In the Americas, the resultant phylogenetic data for crop plants have been compared against macrofossil (Smith 2001) and microfossil (Sanjur et al. 2002) records.

Thirdly, ancient DNA research (aDNA) offers numerous potential avenues of research from the identification of archaeobotanical finds to the construction of phylogenetic chronologies that could theoretically track domestication, hybridisation, and the formation of new species and subspecies (e.g., Erickson et al. 2005; Jaenicke-Després et al. 2003). Only once such an integrated methodological approach is adopted in the Pacific will it be possible to fill out recently proposed conceptual frameworks for the emergence of early agriculture (Denham 2004b, 2005b, 2007) and arboriculture (Fairbairn 2005) in the region.

Finally, although the investigation of early agriculture in the Highlands of New Guinea has been ongoing for 40 years (Golson et al. 1967), there are still major research lacunae. Few archaeological excavations have been undertaken in the Highlands over the last 30 years; consequently few wetland agricultural and contemporary occupation sites have been investigated or reported in detail, and most of these studies were undertaken before several microfossil techniques (parenchyma, phytolith and starch grain analyses) were developed and applied. Furthermore, the few readily available specialists in these new technical fields are often hindered by partial reference collections. Given these realities, it is essential that researchers of early plant exploitation practices in Melanesia co-ordinate their activities to maximise outcomes with respect to the limited available resources, and maintain a spirit of co-operation in their pursuit of common goals.

Acknowledgements

We thank Jack Golson, Kale Sniderman and Barbara Winsborough for permission to cite unpublished material. We also thank Mac Kirby (CSIRO – Canberra) for facilitating the X-ray images analysis using an Oxford XTG tube and phase contrast imaging system at CSIRO – Canberra (see Moran et al. 2000). Gary Swinton and Phil Scamp of the School of Geography and Environmental Science, Monash, Melbourne are thanked for production of the graphics. Additional thanks to Wendy Beck, Andy Fairbairn and Jack Golson for comments on an unpublished draft of the manuscript.

References

Barham, A. J. 1995. Methodological approaches to archaeological context recording: X-radiography as an example of a supportive recording, assessment and interpretative technique. In A. J. Barham and R. I. MacPhail (eds), *Archaeological sediments and soils: Analysis, interpretation and management*, pp 145–82. London: Institute of Archaeology, UCL.

Bayliss-Smith, T. P. 1996. People-plant interactions in the New Guinea highlands: Agricultural hearthland or horticultural backwater? In D. R. Harris (ed.) *The origins and spread of agriculture and pastoralism in Eurasia*, pp 499–52. London: University College London Press.

Bourke, R. M. n.d. Altitudinal limits of 220 economic crop species in Papua New Guinea. Unpublished manuscript on file, Research School of Pacific and Asian Studies, Australian National University.

Bowdery, D. 1999. Phytoliths from tropical sediments: Reports from Southeast Asia and Papua New Guinea. *Bulletin of the Indo-Pacific Prehistory Association* 18:159–68.

Brown, A. G. 1997. *Alluvial geoarchaeology: Floodplain archaeology and environmental change*. Cambridge Manuals in Archaeology, Cambridge: Cambridge University Press.

Bulmer, S. 1975. Settlement and economy in prehistoric Papua New Guinea: A review of the archaeological evidence. *Journal de la Société des Océanistes* 31:7–75.

Butler, S. 1992. X-radiography of archaeological soil and sediment profiles. *Journal of Archaeological Science* 19:151–61.

Canti, M. 1995. A mixed-method approach to geoarchaeological analysis. In A. J. Barham and R. I. MacPhail (eds), *Archaeological sediments and soils: Analysis, interpretation and management*, pp 183–90. London: Institute of Archaeology, UCL.

De Langhe, E. and P. de Maret. 1999. Tracking the banana: Its significance in early agriculture. In C. Gosden and J. Hather (eds), *The prehistory of food*, pp 377–96. London: Routledge.

Denham, T. P. 2003a. Archaeological evidence for mid-Holocene agriculture in the interior of Papua New Guinea: A critical review. *Archaeology in Oceania* 38:159–76.

Denham, T. P. 2003b. *The Kuk morass: Multi-disciplinary investigations of early to mid-Holocene plant exploitation at Kuk Swamp, Wahgi Valley, Papua New Guinea*. Unpublished PhD thesis, Australian National University.

Denham, T. P. 2004a. Early agriculture in the Highlands of New Guinea: An assessment of Phase 1 at Kuk Swamp. In V. Attenbrow and R. Fullagar (eds), *A Pacific odyssey: Archaeology and anthropology in the Western Pacific. Papers in honour of Jim Specht*. Records of the Australian Museum, Supplement 29:47–57.

Denham, T. P. 2004b. The roots of agriculture and arboriculture in New Guinea: Looking beyond Austronesian expansion, Neolithic packages and indigenous origins. *World Archaeology* 36:610–20.

Denham, T. P. 2005a. Agricultural origins and the emergence of rectilinear ditch networks in the highlands of New Guinea. In A. Pawley, R. Attenborough, J. Golson and R. Hide (eds), *Papuan pasts: Cultural, linguistic and biological histories of Papuan-speaking peoples*, pp 329–62. Pacific Linguistics 572, Canberra: RSPAS, ANU.

Denham, T. P. 2005b. Envisaging early agriculture in the Highlands of New Guinea: Landscapes, plants and practices. *World Archaeology* 37:290–306.

Denham, T. P. 2006. The origins of agriculture in New Guinea: Evidence, interpretation and reflection. In I. Lilley (ed.), *Blackwell guide to archaeology in Oceania: Australia and the Pacific Islands*, pp 160–88. Oxford: Blackwell.

Denham, T. P. 2007. Early to mid-Holocene plant exploitation in New Guinea: Towards a contingent interpretation of agriculture. In T. P. Denham, J. Iriarte and L. Vrydaghs (eds), *Rethinking agriculture: Archaeological and ethnoarchaeological perspectives*, pp 78-108. Walnut Creek: Left Coast Press.

Denham, T. P. and H. Barton. 2006. The emergence of agriculture in New Guinea: Continuity from pre-existing foraging practices. In D. J. Kennett and B. Winterhalder (eds), *Behavioral ecology and the transition to agriculture*, pp 237–64. Berkeley: University of California Press.

Denham, T. P., J. Golson, and P. J. Hughes. 2004a. Reading early agriculture at Kuk (Phases 1–3), Wahgi Valley, Papua New Guinea: The wetland archaeological features. *Proceedings of the Prehistoric Society* 70:259–98.

Denham, T. P., S. G. Haberle and C. Lentfer. 2004b. New evidence and interpretations for early agriculture in Highland New Guinea. *Antiquity* 78:839–57.

Denham, T. P., S. G. Haberle, C. Lentfer, R. Fullagar, J. Field, M. Therin, N. Porch and B. Winsborough. 2003. Origins of agriculture at Kuk Swamp in the Highlands of New Guinea. *Science* 301:189–193.

Dugmore, A. J. and A. J. Newton. 1992. Thin tephra layers in peat revealed by X-radiography. *Journal of Archaeological Science* 19:163–70.

Erickson, D. L., B. D. Smith, A. C. Clarke, D. H. Sandweiss and N. Tuross. 2005. An Asian origin for a 10,000 year-old domesticated plant in the Americas. *Proceedings of the National Academy of Sciences USA* 102:18315–20.

Fairbairn, A. 2005. An archaeobotanical perspective on plant-use practices in lowland northern New Guinea. *World Archaeology* 37:487–502.

Fairbairn, A. and P. Swadling. 2005. Re-dating mid-Holocene betelnut (Areca catechu L.) and other plant use at Dongan, Papua New Guinea. *Radiocarbon* 47:377–82.

French, B. R. 1986. *Food plants of Papua New Guinea: A compendium*. Privately published book.

Fullagar, R., J. Field, T. P. Denham and C. Lentfer. 2006. Early and mid-Holocene processing of taro (*Colocasia esculenta*) and yam (*Dioscorea sp.*) at Kuk Swamp in the Highlands of Papua New Guinea. *Journal of Archaeological Science* 33:595–614.

Gilbertson, D. 1995. Studies of lithostratigraphy and lithofacies: A selective review of research developments in the last decade and their applications in geoarchaeology. In A. J. Barham and R. I. MacPhail (eds), *Archaeological sediments and soils: Analysis, interpretation and management*, pp 99–144. London: Institute of Archaeology, UCL.

Golson, J. 1977. No room at the top: Agricultural intensification in the New Guinea Highlands. In J. Allen, J. Golson and R. Jones (eds), *Sunda and Sahul: Prehistoric studies in Southeast Asia, Melanesia and Australia*, pp 601–38. London: Academic Press.

Golson, J. 1982. The Ipomoean revolution revisited: Society and sweet potato in the upper Wahgi Valley. In A. Strathern (ed.), *Inequality in New Guinea Highland societies*, pp 109–36. Cambridge: Cambridge University Press.

Golson, J. 1991. Bulmer Phase II: Early agriculture in the New Guinea Highlands. In A. Pawley (ed.), *Man and a half: Essays in Pacific anthropology and ethnobiology in honour of Ralph Bulmer*, pp 484–91. Auckland: The Polynesian Society.

Golson, J. 2002. Gourds in New Guinea, Asia and the Pacific. In S. Bedford, C. Sand and D. Burley (eds), *Fifty years in the field. Essays in honour and celebration of Richard Shutler Jr.'s archaeological career*, pp 69–78. New Zealand Archaeological Journal Monograph 25, Auckland: Auckland Museum.

Golson, J. and P. J. Hughes. 1980. The appearance of plant and animal domestication in New Guinea. *Journal de la Société des Océanistes* 36:294–303.

Golson, J. and P. Ucko. 1994. Foreword. In J. Hather (ed.), *Tropical Archaeobotany*, pp xiv–xix. London: Routledge.

Golson, J., R. J. Lampert, J. M. Wheeler and W. R. Ambrose. 1967. A note on carbon dates for horticulture in the New Guinea Highlands. *Journal of the Polynesian Society* 76:369–71.

Haberle, S. G. 1995. Identification of cultivated Pandanus and Colocasia in pollen records and the implications for the study of early agriculture in New Guinea. *Vegetation History and Archaeobotany* 4:195–210.

Haberle, S. G. 2003. The emergence of an agricultural landscape in the Highlands of New Guinea. *Archaeology in Oceania* 38:149–58.

Haberle, S. G., C. Lentfer and T. P. Denham. n.d. The palaeoenvironments of Kuk Swamp during the formative to late phases of agricultural development (10,220–2250 cal. BP), Papua New Guinea. In prep.

Hamblin, W. M. K. 1962. X-ray radiography in the study of structures in homogenous sediments. *Journal of Sedimentary Petrology* 32:201–10.

Hather, J. G. 2000. *Archaeological parenchyma*. London: Archaeotype Publications.

Hope, G. S. and J. Golson. 1995. Late Quaternary change in the mountains of New Guinea. *Antiquity* 69 (Special Number 265):818–30.

Hughes, P. J. 1985. Prehistoric man-induced soil erosion: Examples from Melanesia. In I. Farrington (ed.), *Prehistoric Intensive Agriculture in the Tropics*, pp 393–408. International Series 232, Part I, Oxford: British Archaeological Reports.

Hughes, P. J., M. E. Sullivan and D. Yok. 1991. Human induced erosion in a Highlands catchment in Papua New Guinea: The prehistoric and contemporary records. *Zeitschrift für Geomorphologie Suppl.* 83:227–39.

Jaenicke-Després, V., E. S. Buckler, B. D. Smith, M. T. P. Gilbert, A. Cooper, J. Doebley and S. Pääbo. 2003. Early allelic selection in maize as revealed by ancient DNA. *Science* 302:1206–8.

Krinitzsky, E. L. 1970. *Radiography in the earth sciences and soil mechanics*. New York: Plenum Press.

Lebot, V. 1999. Biomolecular evidence for plant domestication in Sahul. *Genetic Resources and Crop Evolution* 46:619–28.

Lebot, V., M. S. Prana, N. Kreike, H. van Heck, J. Pardales, T. Okpul, T. Gendua, M. Thongjiem, H. Hue, N. Viet and T. C. Yap. 2004. Characterisation of taro (*Colocasia esculenta* (L.) Schott) genetic resources in Southeast Asia and Oceania. *Genetic Resources and Crop Evolution* 51:381–92.

Lentfer, C. and R. Green. 2004. Phytoliths and the evidence for banana cultivation at the Lapita Reber-Rakival Site on Watom Island. In V. Attenbrow and R. Fullagar (eds), *A Pacific odyssey: Archaeology and anthropology in the western Pacific. Papers in honour of Jim Specht*. Records of the Australian Museum, Supplement 29:75–88.

Loy, T. H. 1994. Methods in the analysis of starch residues on prehistoric stone tools. In J. G. Hather (ed.) *Tropical archaeobotany: Applications and new methods*, pp 86–114. London: Routledge.

Malapa, R., G. Arnau, J. L. Noyer and V. Lebot. 2005. Genetic diversity of the greater yam (*Dioscorea alata* L.) and relatedness to *D. nummularia Lam.* and *D. transversa* Br. as revealed by AFLP markers. *Genetic Resources and Crop Evolution* 52:919–29.

Matthews, P. J. 1995. Aroids and Austronesians. *Tropics* 4:105–26.

Matthews, P. J. 2003. Identification of *Benincasa hispida* (wax gourd) from the Kana archaeological site, Western Highlands province, Papua New Guinea. *Archaeology in Oceania* 38:186–91.

Moran, C. J., A. Pierret and A. W. Stevenson. 2000. X-ray absorption and phase contrast imaging to study the interplay between plant roots and soil structure. *Plant and Soil* 223:90–115.

Muke, J. and H. Mandui. 2003. Shadows of Kuk: Evidence for prehistoric agriculture at Kana, Wahgi Valley, Papua New Guinea. *Archaeology in Oceania* 38:177–85.

Piperno, D. R. 2006. *Phytoliths: A comprehensive guide for archaeologists and paleoecologists*. Lanham: AltaMira.

Powell, J. M. 1970a. The history of agriculture in the New Guinea Highlands. *Search* 1:199–200.

Powell, J. M. 1970b. *The impact of man on the vegetation of the Mount Hagen region, New Guinea*. Unpublished PhD thesis, Australian National University.

Powell, J. M. 1976. Ethnobotany. In K. Paijmans (ed.), *New Guinea vegetation*, pp 106–83. Canberra: CSIRO and ANU Press.

Powell, J. M. 1982a. The history of plant use and man's impact on the vegetation. In J. L. Gressitt (ed.), *Biogeography and ecology of New Guinea*, volume 1, pp 207–27. The Hague: Junk.

Powell, J. M. 1982b. Plant resources and palaeobotanical evidence for plant use in the Papua New Guinea Highlands. *Archaeology in Oceania* 17:28–37.

Powell, J. M. and S. Harrison. 1982. *Haiyapugua: Aspects of Huli subsistence and swamp cultivation*. Occasional Paper No. 1, new series, Port Moresby: Department of Geography, UPNG.

Powell, J. M., A. Kulunga, R. Moge, C. Pono, F. Zimike and J. Golson. 1975. *Agricultural traditions in the Mount Hagen Area*. Occasional Paper No. 12, Port Moresby: Department of Geography, UPNG.

Sanjur, O., D. R. Piperno, T. C. Andres and L. Wessel-Beaver. 2002. Phylogenetic relationships among domesticated and wild species of Cucurbita (Cucurbitaceae) inferred from a mitochondrial gene: Implications for crop plant evolution and areas of origin. *Proceedings of the National Academy of Sciences USA* 99:535–40.

Smith, B. D. 1998. Between foraging and farming. *Science* 279:1651–2.

Smith, B. D. 2001. Documenting plant domestication: The consilience of biological and archaeological approaches. *Proceedings of the National Academy of Sciences USA* 98:1324–6.

Smith, B. D. 2005. Reassessing Coxcatlan Cave and the early history of domesticated plants in Mesoamerica. *Proceedings of the National Academy of Sciences USA* 102:9438–45.

Spriggs, M. 1996. Early agriculture and what went before in Island Melanesia: Continuity or intrusion? In D. R. Harris (ed.), *The origins and spread of agriculture and pastoralism in Eurasia*, pp 524–37. London: University College London Press.

Swadling, P., N. Araho and B. Ivuyo. 1991. Settlements associated with the inland Sepik-Ramu Sea. *Bulletin of the Indo-Pacific Prehistory Association* 11:92–110.

Torrence, R. and H. Barton. (eds) 2005. *Ancient starch research*. Walnut Creek: Left Coast Press.

Wilson, S. M. 1985. Phytolith evidence from Kuk, an early agricultural site in New Guinea. *Archaeology in Oceania* 20:90–7.

Yen, D. E. 1998. Subsistence to commerce in Pacific agriculture: Some four thousand years of plant exchange. In H. D. V. Pendergast, N. L. Etkin, D. R. Harris and P. J. Houghton (eds), *Plants for food and medicine*, pp 161–83. Kew: Royal Botanic Gardens.

11

Dating marine shell in Oceania:
Issues and prospects

Fiona Petchey

Waikato Radiocarbon Dating Laboratory
School of Science and Engineering
University of Waikato
Hamilton
New Zealand

Abstract

Marine shell has several advantages for radiocarbon (^{14}C) dating in the Pacific — it is ubiquitous in archaeological sites, is easy to identify to the species level, and can often be related directly to human activity. Consequently, shells are one of the most commonly dated ^{14}C sample types within this region. The modelled marine calibration curve and associated regional offsets (known as ΔR) originally construct ed by Stuiver et al. (1986), have been widely accepted as the most accurate method for calibrating surface marine ^{14}C dates. The use of published values, however, is not straightforward because the surface ocean ^{14}C reservoir is variable both regionally and over time, and because of additional uncertainties with the reliability of some shell species due to habitat and dietary preferences. This paper presents an overview of ΔR variability in Oceania and highlights areas of caution when using extant ΔR values, and when select ing marine shell for ^{14}C dating. Particular attention is given to the Hawaiian archipelago where numerous ΔR values are available for evaluation and the influence of ocean currents, estuarine environments and geology is apparent.

Keywords: Marine reservoir; ΔR; marine shell; radiocarbon (^{14}C) dating; Oceania.

Introduction

A plant or animal that obtains carbon from a marine source (or reservoir) yields what is termed an 'apparent age'. The surface ocean (down to around 200m depth) has an apparent ^{14}C age that is, on average, 400 years older than the terrestrial (atmospheric) reservoir. This is known as the marine reservoir effect, and is caused by a delay in the ^{14}C exchange between the atmosphere and ocean, and by the mixing of surface waters with upwelled, ^{14}C-depleted deep ocean water (Stuiver et al. 1986:982). This reservoir effect is automatically corrected for when a marine shell conventional radiocarbon age (CRA)[1] is calibrated using the modelled marine ^{14}C calibration curve (e.g. MARINE04: Hughen et al. 2004). The marine calibration curve represents a global average of the surface ocean ^{14}C as it changes over time.

Local and regional deviations from this global average, however, complicate the calibration of marine samples. To account for this deviation a local correction factor, or ΔR — the difference between the model led ^{14}C age of surface water and the actual ^{14}C age of surface water at that locality — needs to be applied to the calibration. This value can be calculated from contemporaneous terrestrial/marine archaeological samples, or from 'historic' marine samples collected prior to 1950, whose age of death is known precisely (i.e. annually banded corals, shells and/or otoliths of surface-dwelling fish) (e.g. Druffel et al. 2001, 2004; Dye 1994; Guilderson et al. 1998, 2000; Higham and Hogg 1995).

Samples for ΔR research must conform to a number of prerequisites regardless of what marine proxy is chosen:

1. The sample must have been collected live, or the date of death independently validated. For pre-1955 shells of known age, this can be by the presence of documentation, or the fleshy remains of an animal, or valves in articulation with the ligament present. For archaeological hells, food remains fulfil this requirement.
2. The location of collection must be known.
3. The sample must be identified to genus level, and the dietary and habitat preferences of that species must closely represent that of the reservoir being investigated (e.g. open ocean, estuarine, etc).
4. The date of collection must be known, and for historic proxies, the date of collection must be before 1955 (i.e. prior to detonation of thermonuclear devices).[2]

Presently, there is only a limited number of published ΔR values for Oceania (see Table 1 for a list of ΔR values from pre-1950 marine proxies), most of which are readily available from the Marine Reservo ir Database (http://radiocarbon.pa.qub.ac.uk/marine/). Unfortunately, for most of these values it is impossible to know if the prerequisites discussed above have been rigorously adhered to. In particular, a number of values have unknown collection dates and are of questionable live collection. There are fewer reliable ΔR values from archaeological terrestrial/shell pairs due to problems of association, and these are not as easily accessible to researchers.

Location	Loc. No.~	Sample Material#	Calen-dar Age	ΔR (yrs)	Lab No.*	Reference	^ΔR Assessment
Kavieng Harbour, New Ireland	1	*Nerita plicata* (AG)	1931	365±50	Wk-8377	Petchey et al. 2004	d
		Barbatia foliata (FF)	1931	305±110	Wk-8379		
		Conus lividus (C)	1919	311±60	OZB-768		a, d
		Conus lividus (C)	1919	511±60	OZB-769		a, d
		Conus lividus (C)	1919	371±50	OZB-770		a, d
Rabaul Harbour, New Britain	2	*Conus sanguinolentus* (C)	1919	411±80	OZB-771	Petchey et al. 2004	a, f
		Conus sanguinolentus (C)	1919	301±50	OZB-772		a, f
		Conus sanguinolentus (C)	1919	401±50	OZB-773		a, f
Duke of York Island	3	*Nassarius camelus* (C)	1905	43±68	Wk-9219	Petchey et al. 2004	a
Bougainville	4	*Conus* sp. (C)	1944	17±40	Wk-8381	Petchey et al. 2004	b
Teop Island, Bougainville	5	*Anadara antiquata* (FF)	1933	104±45	Wk-8380	Petchey et al. 2004	
Vella Lavella	6	*Chicoreus ramosus* (C)	1930	86±40	Wk-7828	Petchey et al. 2004	a
Fauabu, Malaita	7	*Asaphis violascens* (DF)	1932	135±55	Wk-8382	Petchey et al. 2004	c
Ufa Is. Russel Islands	8	*Asaphis violascens* (DF)	1945	216±50	Wk-8383	Petchey et al. 2004	c
Guadalcanal Island	9	Coral (*Porites australiensis*)	1950	-6±27	CAMS series	Guilderson et al. 2004	
Espiritu Santo, Vanuatu	10	Coral (*Diploastrea heliopara*)	1953	25±10	AA series	Burr et al. 1998	
Ambrym Island, Vanuatu	11	*Tellina linguafelis* (DF)	1943	198±80	Wk-8384	Petchey et al. 2004	c, f
New Caledonia	12	*Venus peupera* (FF)	1876	15±45	Wk-8046	Petchey et al. 2004	a, b
		Venus reticulata (FF)	1876	5±45	Wk-8047	Petchey et al. 2004	a, b
Eniwetak Atoll	13	Coral	1946	164±43	L-584A	Broecker and Olson 1961	
Guam	14	Gastropod	1903	19±50	CAMS-8696	Southon et al. 2002	d
		Strombus gibberulus gibbosus (H)	1930	136±50	Beta-14023	Athens 1986	d
Tinian	15	*Cypraea* sp. (H)	1945	320±80	?	Craib 1993	d
Fanning Island	16	Coral (*Porites* sp.)	1952	22±17	LJ series	Druffel 1987	
Nauru Island	17	Coral (*Porites* sp.)	1951	9±5	CAMS series	Guilderson et al. 1998	
Waitangi Beach, Chatham Islands	18	Gastropod	1933	244±40	CAMS-40758	Sikes et al. 2000	a, d
		Gastropod	1933	94±40	CAMS-40759		d
		Gastropod	1933	114±28	CAMS-40760/1		d
Samoa	19	*Turbo petholatus* (H)	1882	79±40	Wk-6383	Phelan 1999	a, b
		Strombus pacificus (H)	1882	29±40	Wk-6384		a, b
		Strombus lentiginosus (H)	1882	89±40	Wk-6385		a, b

Location	Loc. No.~	Sample Material#	Calen-dar Age	ΔR (yrs)	Lab No.*	Reference	^ΔR Assessment
Pangaimotu, Tongatapu	20	*Anadara antiquata* (FF)	1926	-157±68	ANU-6421	Spennemann and Head 1998	
Havelu, Tongatapu (lagoon)		*Gafrarium tumidum* (FF)	1926	87±74	ANU-6420	Spennemann and Head 1998	e
Moorea	21	*Turbo setosus* (H)	1883	82±42	L-576K	Broecker and Olson 1961	a
Tahiti	22	*Turbo setosus* (H)	1957	46±42	L-576E	Broecker and Olson 1961	Post bomb
Easter Island	23	*Porites lobata*	1950	-113±18	AA series	Beck et al. 2003	
Viti Levu, Fiji	24	Coral	1945	43±12	L-series	Toggweiler et al. 1991	
Rarotonga	25	Coral (*Porites lutea*)	1953	-52±27	CAMS series	Guilderson et al. 2000	
Abiang, Tungaru islands	26	Coral (*Porites cylindrica*)	1860	-72±60	Beta-106800	Paulay and Kerr 2001	a, e
		Coral (*Porites cylindrica*)	1860	-62±60	Beta-106801		a, e
Australia - Narooma	27	*Katelysia rhytiphora* (FF)	1950	11±85	SUA-356	Gillespie 1977	
Abraham Reef	28	Coral (*Porites australiensis*)	1850	15±6	WH & AA series	Druffel and Griffin 1993, 1999	
Heron Island	29	Coral (*Porites australiensis*)	1874	8±6	WH & AA series	Druffel and Griffin 1993, 1999	
Elliot Heads, Queensland	30	*Donax deltoides* (FF)	1925	-51±60	Wk-6994	Ulm 2002	a
Port Curtis	31	*Volachlamys singaporina* (FF)	1929	7±60	Wk-8457	Ulm 2002	
		Volachlamys singaporina (FF)	1929	117±60	NZA-12120		
Gladstone	32	*Anadara trapezia* (FF)	1904	30±50	Wk-8456	Ulm 2002	
		Anadara trapezia (FF)	1904	-90±60	NZA-12119		
Torres Strait	33	*Pinctada margaritigera* (FF)	1909	-5±84	SUA-357	Gillespie 1977	a, b
		Pinna bicolour (FF)	1875	61±85	SUA-354/2		a, b
		Mactra obesa (FF)	1875	78±68	SUA-354/1		a, b
O'ahu	34	*Trochus intextus* (H)	1840.5	139±51	L-576J	Dye 1994	b
Waikane, O'ahu	34a	*Tellina palatum* (DF)	1925	229±40	Beta-14024		c, e
Kane'ohe. O'ahu	34b	*Macoma (Scissulina) dispar* (DF)	1818	-479±120	Beta-13805		e
		Conus distans (C)	1947	44±60	Beta-15794		d, e
Waimanalo, O'ahu	34c	*Conus distans* (C)	1936	502±70	Beta-12749		d
Pearl Harbour, O'ahu	34d	*Trochus intextus* (H)	1936	3842±100	L-576D		d
		Tellina palatum (DF)	1927	-22±60	Beta-15793		c

Location	Loc. No.~	Sample Material#	Calendar Age	ΔR (yrs)	Lab No.*	Reference	^ΔR Assessment
Kualakai, Oʻahu	34e	*Nerita picea* (H)	1930	776±80	Beta-54333/ CAMS-3219		c
Barber's Point, Oʻahu	34f	*Cellana exarata* (H)	1914	822±80	Beta-54332/ CAMS-3218		d
		Cypraea caputserpentis (H)	1915	532±80	Beta-54331/ CAMS-3217		d
Pukoʻo, Molokaʻi	35	*Tellina palatum* (DF)	1905	−39±60	Beta-12903	Dye 1994	c, d
Kona, Keauhou Bay, Hawaii Island	36a	Coral rings (*Porites lobata*)	1923	−29±4	UCI 3172-4432	Druffel et al. 2001	
Kaulana, Hawaii Island	36b	*Cellana exarata* (H)	1923	290±100	Beta-54336/ CAMS-3222	Dye 1994	
		Cypraea caputserpentis (H)	1923	280±80	Beta-54334/ CAMS-3220		
Keaʻau, Hawaii Island	36c	*Nerita picea* (H)	1923	159±80	Beta-54335/ CAMS-3221		

~ Individual locations shown in Figures 1 and 2.

Diet preferences (in brackets): FF = filter-feeder; AG = algae grazer; H = herbivore; C = carnivore; D = deposit feeding

* Lab prefixes: Wk = Waikato Radiocarbon Dating Laboratory; OZ = Australian Nuclear Sciences and Technology Organisation; CAMS = Lawrence Livermore National Laboratories; AA = University of Arizona; NZA = Rafter Radiocarbon Laboratory; L = Lamont-Doherty; LJ = Scripps (UCSD) La Jolla; SUA = University of Sydney; WH = National Ocean Sciences AMS Facility; UCI = University of California, Irvine; UCLA = University of California, Los Angeles.

^ΔR Assessment

a. Exact year of death unknown

b. Exact location unknown

c. Unsuitable species (deposit-feeders)

d. Questionable/unidentified species in limestone region

e. Lagoon or estuary — unknown influence

f. Volcano — unknown influence

Table 1. Published ΔR values for Oceania (excluding New Zealand) and assessment of reliability. ΔR values from http://radiocarbon.pa.qub.ac.uk/marine/ except Abiang, Guam and Tinian.

The limited available ΔR values and lack of guidelines governing the selection of an appropriate ΔR value for each island has been a continuing problem for the accurate calibration of dates on marine shell and other animals that subsisted on marine resources. This makes comparison of the chronology between, and sometimes within, archaeological sites difficult (e.g. Specht and Gosden 1997; Summerhayes 2001). Consequently, it is becoming increasingly important that researchers have an understanding of ΔR variability and shellfish suitability for ^{14}C dating. This paper attempts to begin the process of addressing this problem and presents a discussion of ΔR variability in Oceania, here defined as Melanesia, French Polynesia, Polynesia, East Coast of Australia and Micronesia. ΔR values from New Zealand are not included. An assessment of reliability for each ΔR value is given in Table 1 and discussed in more detail below.

Identification and diet

It is widely recognised that shellfish selected for ^{14}C dating and ΔR research must be identified to species level, and the dietary and habitat preferences of that species known. In particular, algae grazers[3] and deposit-feeders may have anomalously high ΔR values caused by the digestion of detritus in areas dominated by limestone geologies (Anderson et al. 2001; Dye 1994). Even in areas where limestone is absent, ^{14}C dating of deposit-feeding species has resulted in anomalous ^{14}C values due to the consumption of carbon from a terrestrial source (e.g. Hogg et al. [1998] recorded unusual values for *Macomona liliana* which can switch between filter-feeding and deposit-feeding modes depending on the environment [Beesley et al. 1998:343]). Despite these findings, deposit-feeding species regularly continue to be dated, potentially to the detriment of reliable ^{14}C chronologies. Limited data are available for carnivorous shellfish, but they are presumed to show an averaging effect depending on the carbon reservoirs of their prey, and could therefore be subject to similar uncertainties.

Petchey et al. (2004) noted anomalous ΔR values for deposit-feeding species from Ufa Island (Table 1, location 8), Malaita (location 7) and Ambrym Island (location 11), all of which are either limestone islands, or located near limestone deposits. The use of unsuitable shell species is also partly responsible for the range of ΔR values available for the Hawaiian Islands (locations 34, 35 and 36). Of the 14 pre-1955 marine shells reported by Dye (1994), 12 are of herbivorous or deposit-feeding shellfish and have a ΔR range of −479±120 yrs to 3842±100 yrs. Of these, Dye identified only three as coming from Pleistocene limestone coastlines (locations 34f and 34e) resulting in anomalously high ΔR values of 822±80 yrs, 532±80 yrs and 776±80 yrs. The value of 3842±100 yrs for one of the two shell samples from Pearl Harbour (location 34d) was considered by Dye (1994:51) to be anomalous and possibly the result of hydrocarbon contamination. The use of a fossil shell from the nearby limestone is also a possibility. Dye did not notice small pockets of limestone on the eastern coast of O'ahu at Waimanalo (Figure 2a, location 34c), which could be responsible for the ΔR of 502±70 yrs for *Conus distans*, a carnivorous gastropod. Similarly, a ΔR value of 229±40 yrs for the deposit-feeding *Tellina palatum* from Waikane at the northern end of Kaneohe Bay (Figure 2a, location 34a) is suspect since Kapapa and Kekepa Islands located within the bay are consolidated dunes consisting of sand from ancient limestone (Stearns 1938).

Suspension-feeders (or filter-feeders) are usually considered the most reliable shells for ^{14}C dating because they consume suspended phytoplankton and dissolved inorganic carbon from seawater, and therefore more closely reflect the ^{14}C content of the surface ocean (Forman and Polyak 1997:888). Some bivalve species will, however, also engage in deposit-feeding activities depending on local circumstances (Snelgrove and Butman 1994). Moreover, suspension-feeding species may give anomalous ΔR values because of a hardwater effect where calcium carbonate from calcareous strata of disparate age becomes incorporated into the shell of animals that inhabit environs nearby (Spennemann and Head 1998). The incorporation of riverine material especially in estuarine environments (Dye 1994; Ingram 1998; Southon et al. 2002), and localized hydrothermal activity are also considerations. The effect of these varying sources of ^{14}C on shellfish will depend upon the degree of water exchange with the open ocean coupled with the specifics of habitat and geology (Hogg et al.1998; Tanaka et al. 1986).

Northern
Mariana
Islands
15
14

NEC

FEDERATED STATES OF MICRONESIA   13   MARSHALL
ISLANDS

NECC

HAWAIIAN 34
ISLANDS 35 36

North Pacific

Ocean

16

26 Tungaru
Islands

SEC

NGCU   Watom   17   NAURU
         Island 1
             33
    54   SOLOMON
          ISLANDS   TUVALU
PAPUA NEW   6 7
GUINEA   8 9
33

KIRIBATI

KIRIBATI

SECC

Tokelau   Marquesas

Santa Cruz
Islands

SAMOA
19

Cook
Islands

LINE ISLANDS

Coral Sea   10 11 VANUATU

SEC (Southern branch)

FIJI 24

TONGA   20   Niue   25

Society
Islands   21 22   Reao Atoll

Tuamotu Archipelago

31 32 28
29   New
30   Caledonia   12

Subtropical Gyre

Tubai Islands

Pitcairn Islands

EAC   Norfolk Island   Kermadec
Islands

South Pacific

Ocean

27   Tasman
Sea

NEW
ZEALAND

Easter Island
113 ± 18

STF

18 Chatham
Islands   ACC

Scale
0   500 Kilometers

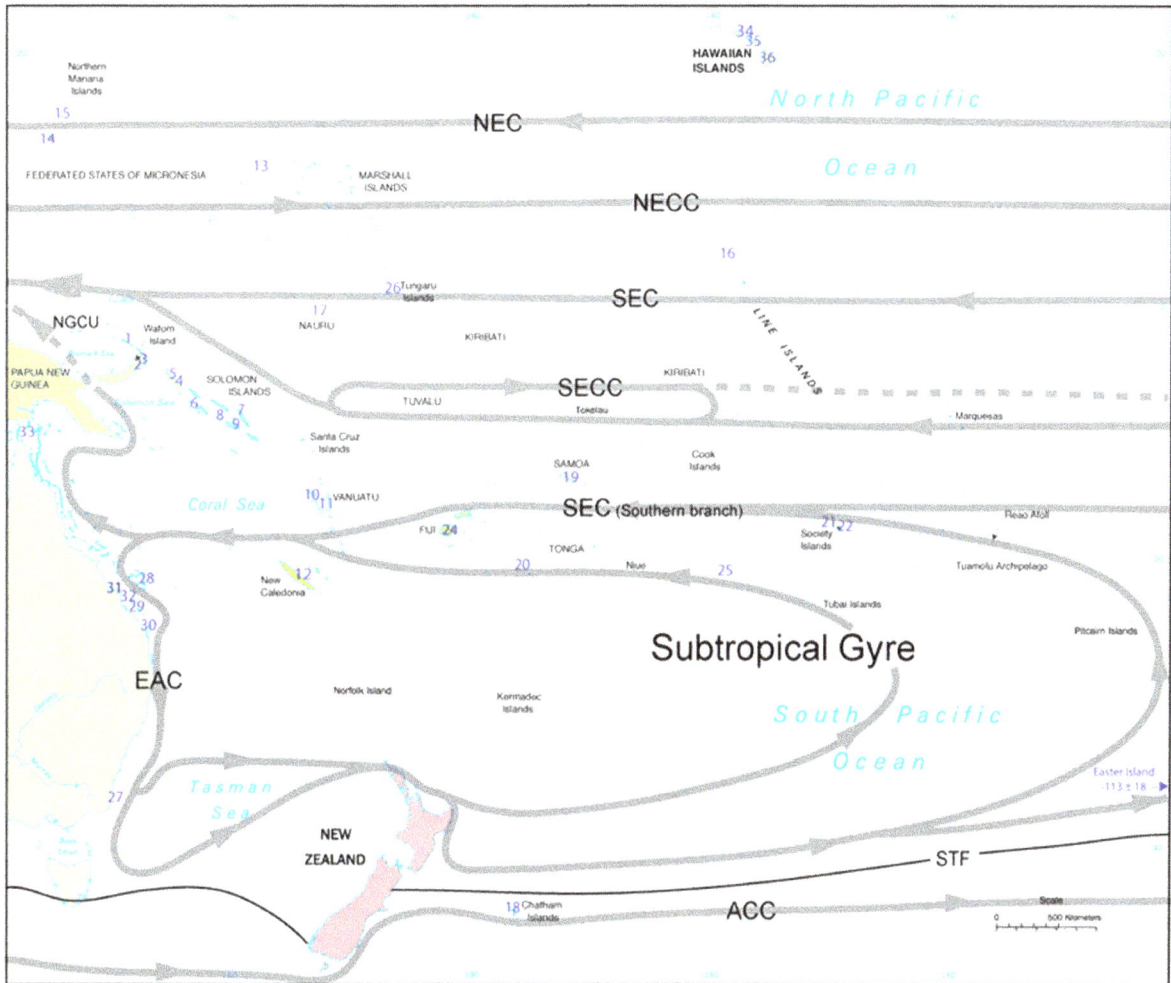

Figure 1. Map showing location of ΔR values given in Table 1 and major oceanic currents within Oceania. Oceanic currents: STF = Subtropical Front; SECC = South Equatorial Counter Current; SEC = South Equatorial Current; NECC = North Equatorial Counter Current; NEC = North Equatorial Current; EAC = East Australian Current; NGUC = New Guinea Coastal Undercurrent; ACC = Antarctic Circumpolar Current (after Tomczak and Godfrey 2001:111).

Geographic location and oceanic variability

The ^{14}C of the surface ocean may also deviate from the modelled marine curve of Hughen et al. (2004) because of variations in upwelling and ocean currents (Stuiver and Braziunas 1993). The South Pacific islands are encircled by the South Pacific Gyre, which is driven by the high latitude eastward-flowing Antarctic Circumpolar Current and the mid latitude westward-flowing South Equatorial Current (SEC) (Figure 1). Coral core records collected over the last 20 years have enabled researchers to build a picture of ocean circulation and ^{14}C variation. Data collected from corals in tropical surface waters, such as the Galapagos Islands tend to have higher ΔR values due to upwelling of old ^{14}C-depleted waters (Druffel 1987). Lower ΔR values have been recorded from southern subtropical waters (e.g. Rarotonga (Table 1, location 25): ΔR = −52±27 yrs) and have been attributed to high air-sea ^{14}C exchange and reduced mixing with older subsurface waters (Guilderson et al. 2000). Mixing of water from tropical and subtropical sources appears to result in midrange ΔR values as seen at Nauru (location 17: ΔR = 9±5 yrs), Guadalcanal Island (location 9: ΔR = −6±27 yrs), Fiji (location 24: ΔR = 43±12 yrs), and the Great Barrier Reef (location 28: ΔR = 15±6 yrs; and location 29: ΔR = 8±6 yrs). There are currently too few ^{14}C values to effectively evaluate the true extent or cause of ΔR variability in the Oceania region, but there is enough data from specific locations within the Pacific to highlight the degree of potential variation possible over relatively small areas.

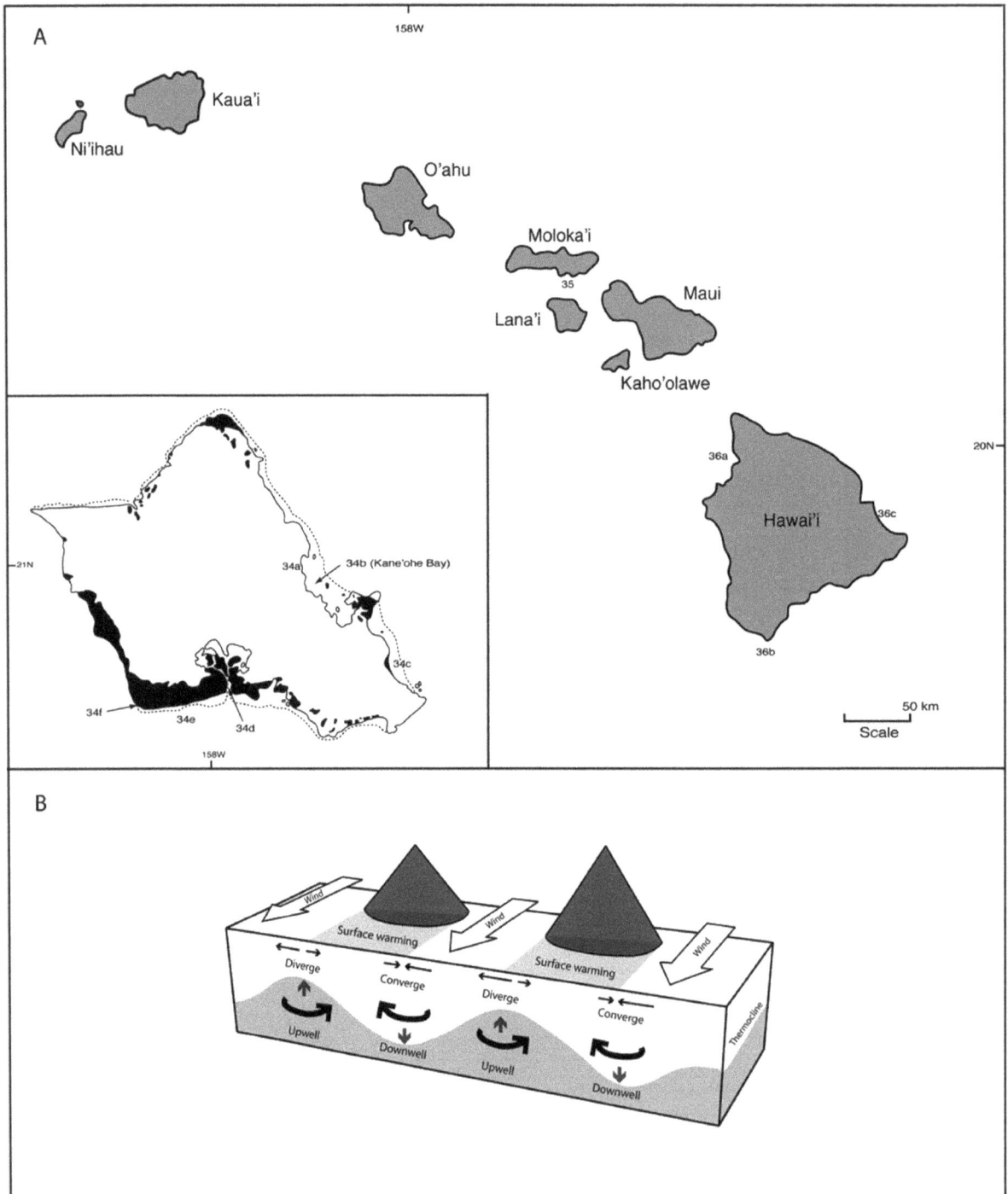

Figure 2. A. Map of the Hawaiian Islands showing places mentioned in the text (Insert: Limestone outcrops around O'ahu [after Stearns 1978]). B. Conceptual diagram of upwelling around the Hawaiian archipelago (from Flament et al. 1996: Plate 10).

To the western edge of the study area, seasonal interaction between the New Guinea Coastal Undercurrent, the SEC, and the North Equatorial Counter Current resulted in high reservoir values for shells from Watom Island (ΔR = 261±101 yrs derived from marine/terrestrial archaeological material) and New Ireland (Table 1, location 1: ΔR = 365±50 yrs and 305±110 yrs) that are indicative of upwelled ^{14}C-depleted water (Figure 1). A ΔR value of 43±68 yrs for the Duke of York Islands (location 3), less than 40 km from Watom Island, is closer to other values recorded for the Coral and Solomon Seas. In this instance, it was hypothesised that the New Guinea Coastal Undercurrent brought higher ^{14}C waters from the Coral and Solomon Seas into the channel between New Britain and New Ireland, shielding the Duke of York Islands from the effects of the upwelling noted above (Petchey et al. 2004, 2005).

The east coast of Australia is likely to be just as complex. On approaching Australia the southern branch of the SEC bifurcates near 18° S with the southern flow feeding the East Australian Current while the northern flow continues northwards along the Great Barrier Reef. This northern flow is suppressed during the summer monsoon season (Tomczak and Godfrey 2001:119) and when combined with ENSO El Niño/Southern Oscillation events (interannual fluctuations in atmospheric and oceanic circulation that occur every two to ten years), shifts in ΔR over relative short periods of time can occur (e.g. coral from the Abraham Reef returned a ΔR of −36 yrs in 1851 compared with 83 yrs in 1865) (Druffel and Griffin 1999: 23, 610). It seems probable that these factors are partly responsible for the range of 'open ocean' values (−90±60 yrs to 30±50 yrs) for the east coast of Australia (Table 1, locations 28, 29, 30 and 32).

Further south, the Chatham Islands (Table 1, location 18) are located within the Subtropical Front (STF). At the STF, there is a transition from mixed Subtropical to ^{14}C-depleted Sub Antarctic Surface Waters (Heath 1985:87; Sparks et al. 1992:729). The high ΔR values recorded for the Chatham Islands (244 ±40 yrs to 94±40 yrs) are typical for regions where ^{14}C depleted water from the deep ocean is upwelled (Stuiver and Braziunas 1993; Toggweiler et al. 1991). Unfortunately, limited information is published about these shells, and additional complications of habitat and diet are likely since they were collected from an area dominated by limestone and peat. A wide fluctuation in ΔR values is possible in areas of up welling which means ^{14}C ages of shell samples from islands in these areas will be difficult to interpret. Data collected from an AD 1760–1771 coral core from Urvina Bay on the west coast of Isabela Island (Galapagos Islands) has presented the most dramatic example of seasonal and yearly ^{14}C variability known in the Pacific (Druffel et al. 2004). Higher ^{14}C values were found during January through March, when upwelling was weak or absent, while low values were obtained mid year during strong upwelling — when the southeast trade winds are the strongest. The extreme range in ΔR values fell at 314±44 yrs in 1765 compared to a value of 26±4 yrs five years later (Druffel et al. 2004:628).

A wide range in ΔR values is also recorded for the Hawaiian archipelago (Table 1). While this variation can be partly explained by the diets of the shellfish selected (see above), unique oceanic conditions that occur around the islands may also be responsible. Northeasterly trade winds flowing through the Hawaiian Islands result in a distinct pattern of upwelling and downwelling in the lee of the islands (Figure 2b) (Flament et al. 1996). Additional complications are caused by the presence of eddies that are generated in the lee of the islands as they impinge on the North Equatorial Current. These conditions create an eastward ocean current (the Hawaiian Lee Counter Current) in the lee of the Big Island that draws warm water from the Asian coast 8000 km away. In response, a cold tongue of upwelled water emanates from the southern tip of the Big Island and extends westward (Xie et al. 2001).

If the available ΔR information for the Hawaii archipelago is reviewed with these observations in mind, a pattern emerges. Coral samples from Kona on the western coast of the Big Island (Figure 2a, location 36a: ΔR = −29±4 yrs) come from a sheltered area in the lee of the island where water is being dow nwelled (Druffel et al. 2001:17). Conversely, samples from Kaulana on the southern tip of the island (location 36b) (ΔR = 290±100 yrs and 280±80 yrs) will have been subject to different oceanic conditions. Given the volcanic nature of the island (i.e. no limestone) these high values appear to be indicative of upwelling of ^{14}C-depleted cold water in keeping with the observations mentioned above. The picture for O'ahu is not as clear because of the presence of Pleistocene-age limestone and the diets of the shell species selected for ΔR (Table 1), but post-1970 coral core data from Kahe Point on the leeward side of

O'ahu Island has also been considered to support downwelling (Druffel 1987:679). It seems likely, therefore, that ΔR values for Hawaii will be highly variable depending on the coast in question. This may also be true for other island locations, though eddies and wakes are not normally as large as those present around Hawaii. One area that may be similarly complex is the southeast coastline of Australia. Here, large scale eddies form at the boundary between warm water of the Coral Sea and the cooler water of the Tasman Sea (Tomczak and Godfrey 2001:126–128). Unfortunately, no data is currently available along this coastline for analysis.

A high level of spatial control is also vital when dating shells from within lagoons. The ΔR value of shells from atoll lagoons will depend on the rate of exchange between the lagoon and the open ocean waters, which is dependant on the number of channels (hoas) as well as their depth and orientation with respect to the prevailing sea swell. The 'residence time' is used to characterize the length of time it takes for water to exchange between the lagoon and the open sea. Throughout the five archipelagos that comprise French Polynesia the residence times of atolls and lagoons vary from five hours to 230 days (Charpy 2002). Consequently, it is possible for the ΔR to vary considerably. Radiocarbon dating of live corals collected from within the lagoon of Reao Atoll (eastern Tuamotu Archipelago) by Pirazzoli et al. (1987:66) suggested a ^{14}C activity that was in equilibrium with the atmosphere, while shells collected from the outer reefs were in equilibrium with sea water (an apparent difference in CRA of 440±70 yrs, equivalent to a near 0 ΔR). Conversely, Paulay and Kerr (2001:1197–8) suggested that the local reservoir effect for the Tarawa Atoll — part of the Tungaru Islands — was 'inconsequential' on the basis of ^{14}C measured from two specimens of *Porites cylindrica* collected in the 1860s from nearby Abaiang Atoll (Table 1, location 26). It is possible, however, that ΔR values for the two atolls may not be comparable since the residence time of Tarawa Atoll has been calculated at around one week (Chen et al. 1995), whereas Abaiang Atoll has more restricted water exchange with the open ocean as well as an influx of freshwater to the south (Smith 1999).

In areas of limited exchange with the open ocean, geology may also play an important role in determining the reservoir value of filter-feeding shellfish, and a hardwater effect may occur in areas where limestone dominates the bedrock. This has been suggested for shellfish collected from Tongatapu (Table 1, location 20). Tongatapu is a raised volcanic atoll with a Pleistocene limestone capping, and an internal lagoon separated from the ocean by a complex system of reefs and channels. A ΔR of 87±74 yrs for *Gafrarium tumidum* a filter feeding bivalve — from within Havelu lagoon was attributed to a hardwater effect caused by evaporation that resulted in ground water discharge into the lagoon, compounded by a long mean residence time of 31 days (Spennemann and Head 1998:1049–50). Conversely, the low ΔR value of –157±68 yrs for the nearby islet of Pangaimotu was considered by Spennemann and Head (1998) to represent ^{14}C values for the open ocean. This value appears, however, to be unusual when compared to other published open ocean values in this area (though a similar value is recorded for *Porites lobata* for Easter Island: ΔR = –113±18 yrs[4]). A possible alternative suggestion is enrichment in ^{14}C caused by wind and wave action, because the reef flat surrounding Pangaimotu is exposed regularly at low tide and water is less than 2m deep (Richmond and Roy 1986). Forman and Polyak (1997:888) have argued that increased wind turbulence may augment transfer of enriched $^{14}CO_2$ from the atmosphere reducing the reservoir effect (and result in a negative ΔR value) by 100 to 200 years. A similar explanation has been given to enriched ^{14}C values for shellfish growing in the open marine inter- tidal zone of Tairua Harbour, New Zealand (Hogg et al. 1998).

Estuarine reservoirs are also potentially complex due to the interaction and incomplete mixing of ^{14}C from both terrestrial and marine reservoirs (Ulm 2002:322). This is especially problematic where circulation is restricted. Kaneohe Bay, O'ahu Island (Figure 2a: location 34b) has a deep lagoon between an outer reef and the shore. To the southeast of the bay low wave energies have enabled the streams entering the bay to form small deltas (Bathen 1968; Moberly 1963:31). The range of ΔR values for Kaneohe Bay includes 44±60 yrs for a sample of *Conus distans*, 229±40 yrs for a sample of deposit-feeding *Tellina palatum* (see discussion above), as well as a low ΔR value of –479±120 yrs for a sample of *Macoma (Scissulina) dispar*. *Macoma dispar* is commonly found in areas of freshwater discharge (Dye 1994:52) and the low ΔR value is likely to be caused by the incorporation of river-borne dissolved and particulate terrestrial organic matter. Ulm (2002:339) also noted that ΔR values calculated from estuarine marine/terrestrial archaeological pairs from central Queensland demonstrated a lack of consistency, which he attributed to estuarine-specific variation in terrestrial carbon input and limited exchange with the open ocean. Two known-age historic samples of *Volachlamys singaporina*, a filter-feeding bivalve from near the mouth of the Boyne River at Port Curtis (Table 1, location 31), gave divergent ΔR results (7±60 yrs and 117 ±60 yrs) which may also reflect estuarine variability, though other seasonal oceanographic factors could be involved (see above).

Change over time

Paleoclimate reconstructions using banded coral core records have indicated that there is long-term marine reservoir variability in some regions of the Pacific (Dunbar and Cole 1996:5; Druffel and Griffin 1993). Archaeological studies (Deo et al. 2004; Ingram 1998; Reimer et al. 2002; Yoneda et al. 2001) have also demonstrated the importance of longer-term ΔR evaluation. For archaeological shells, the age of death is determined by the dating of short-lived charcoal from pene-contemporaneous contexts. Unfortunately, few published archaeological shell/charcoal pairs can demonstrate irrefutable pene-contemporaneity, in part because charcoal is rarely identified to short-lived species increasing the likelihood of inbuilt age (Allen and Wallace n.d.; Kennett et al. 2002), but also because of site disturbance and the misidentification of food shells. Recently, Jones et al. (2007) and Petchey et al. (2005) have used a Bayesian methodology that allows some uncertainty in the dated events to be incorporated in the calculation. This methodology incorporates all chronological data rather than just the paired marine/terrestrial sample approach traditionally used (originally outlined by Stuiver and Braziunas 1993). This has met with initial success for the Santa Cruz Islands (Solomon Islands) and Watom Island, but is not a substitute for well provenanced and identified ^{14}C samples.

Conclusion

There are limited published pre-1950 ΔR values for the islands that make up Oceania, and only a few conform to the prerequisites for ΔR selection listed above. The most problematic values are those for deposit-feeders and other species that may incorporate sediment in their diets. These deposit-feeding shellfish should be avoided for both routine ^{14}C dating and ΔR studies. Where several ΔR values are avail able for a particular region, it becomes apparent that significant variation is possible over short distances. Consequently, it is recommended that the ΔR value used for the calibration of archaeological shell samples belong to the same island. Annually banded corals clearly demonstrate seasonal and longer-term variation, but they are geographically few in number and may not directly relate to the same environmental conditions as the shell species selected from archaeological middens. The use of archaeological marine/terrestrial pairs would go a long way towards alleviating this problem. Unfortunately, there is only a handful of published ΔR values calculated from archaeological marine/terrestrial pairs, (Petchey and Addison in press) and the reliability of these values is currently hindered by problems of association and material suitability.

Notes

1. A conventional radiocarbon age (CRA) is obtained from a radiocarbon measurement following the conventions set out by Stuiver and Polach (1977). A CRA must be calibrated to determine a calendar age. By convention, the symbol BP means 'conventional radiocarbon years before AD 1950', whereas the symbols cal BP or BC/AD are used to express calibrated radiocarbon ages.

2. For ΔR calculated from historic shell, the date of collection should be before 1950 and preferably pre-1850. The 'bomb effect' shows up in coral core records from the north Pacific as early as 1956 (Konishi et al. 1982) and at the very earliest 1957 in the southern Pacific (Druffel and Griffin 1993, 1999; Toggweiler et al. 1991). Post-1850, anthropogenic effects (i.e. Seuss effect), such as dilution in 14C caused by fossil fuel burning has also been noted in both shell and coral records from across the Pacific (Druffel et al. 2001; Druffel and Griffin 1993, 1999; Guilderson et al. 2004; Hideshima et al. 2001) and may affect ΔR values on pre-1955 shells.

3. Algae grazers feeding on a living coral substrate should only incorporate very recent carbon, although this could vary where fossil and/or sub-fossil coral are present. Algal grazers that target species restricted to seaweed surfaces should not have this problem.

4. Of five coral cores collected from around Easter Island, only one exhibited distinct annual growth bands suitable for chronology development (Core Ovahe -97-1). This is attributed to the location of Easter Island at the environmental limits of coral tolerance (Beck et al. 2003; Mucciarone and Dunbar 2003:117, 122) and necessitates caution when using this ΔR value.

Acknowledgements

I would like to acknowledge the help of Gustav Paulay (University of Florida) and Mike Carson (International Archaeological Research Institute, Inc., Honolulu) for drawing my attention to ΔR values not previously listed in the Marine Reservoir Database. Sean Ulm (University of Queensland) and Paula Reimer (Queen's University, Belfast) provided valuable discussion about ΔR.

Bibliography

Allen, M. S. and R. Wallace. n.d. New evidence from the East Polynesian gateway: Substantive and methodological results from Aitutaki, Southern Cook Islands. *Radiocarbon* in press.

Anderson, A., T. F. G. Higham and R. Wallace. 2001. The radiocarbon chronology of the Norfolk Island archaeological sites. *Records of the Australian Museum, Supplement* 27:33–42.

Athens, J. S. 1986. *Archaeological investigations at Tarague Beach, Guam*. Report prepared for Base Civil Engineering, Andersen Air Force Base. International Archaeological Research Institute, Inc., Honolulu.

Bathen, K. H. 1968. A descriptive study of the physical oceanography of Kaneohe Bay, Oahu, Hawaii. *University of Hawaii, Hawaii Institute of Marine Biology Technical* Report 14.

Beck, J. W., L. Hewitt, G. S. Burr, L. Loret and F. T. Hochstetter. 2003. Mata ki Te Rangi: Eyes towards the heavens. Climate and radiocarbon dates. In J. Loret and J. T. Tanacredi (eds), *Easter Island. Scientific Exploration into the World's Environmental Problems in Microcosm*, pp 93–111. New York: Kluwer Academic/Plenum Publishers.

Beesley, P. L., G. J. B. Ross and A. Wells (eds). 1998. *Mollusca: The Southern Synthesis. Fauna of Australia. Vol. 5.* . Melbourne: CSIRO Publishing.

Broecker, W. S. and E. A. Olson. 1961. Lamont radiocarbon measurements VIII. *Radiocarbon* 3:176–204.

Burr, G. S., J. W. Beck, F. W. Taylor, J. Recy, R. L. Edwards, G. Cabioch, T. Correge, D. J. Donahue and J. M. O'Malley. 1998. A high-resolution radiocarbon calibration between 11,700 and 12,400 calendar years BP derived from Th-230 ages of corals from Espiritu Santo Island, Vanuatu. *Radiocarbon* 40:1093–1105.

Charpy, L. 2002. Exchanges between the atolls and the open ocean. The Institute for Research and Development. http://www.com.univ-mrs.fr/IRD/atollpol/fnatoll/ukatocea.htm.

Chen, C. W., D. Leva and W. Kimmerer. 1995. Circulation and the value of passage channels to water quality of Tarawa lagoon. In R. R. Abbott and J. Garcia (eds), *Management plan for Tarawa Lagoon, Republic of Kiribati, Volume 111*. Management Plan. Tiburon CA: Technical Report of BioSystems Analysis Inc.

Craib, J. L. 1993. Early occupation at Unai Chulu, Tinian, Commonwealth of the Northern Mariana Islands *Indo-Pacific Prehistory Association Bulletin* 13:116–134.

Deo, J. N., J. O. Stone and J. K. Stein. 2004. Building confidence in shell: Variations in the marine radiocarbon reservoir correction for the Northwest coast over the past 3,000 years. *American Antiquity* 69(4):771–786.

Druffel, E. R. M. 1987. Bomb radiocarbon in the Pacific: Annual and seasonal timescale variations. *Journal of Marine Chemistry* 45:667–698.

Druffel, E. R. M. 2004. Galapagos coral radiocarbon data. *IGBP Pages/World Data Center for Paleoclimatology Data Contribution Seies #2004–087*. Boulder CO, USA: NOAA/NGDC Paleoclimatology Program. http://www.ncdc.noaa.gov/paleo/coral/galapagos.html

Druffel E. R. M. and S. Griffin. 1993. Large variations of surface ocean radiocarbon: Evidence of circulation changes in the southwestern Pacific. *Journal of Geophysical Research* 98(C11):20,249–59.

Druffel, E. R. M. and S. Griffin. 1999. Variability of surface ocean radiocarbon and stable isotopes in the southwestern Pacific. *Journal of Geophysical Research* 104(C10): 23,607–23,613.

Druffel, E. R. M., S. Griffin, T. P. Guilderson, M. Kashgarian, J. Southon and D. P. Schrag. 2001. Changes of subtropical North Pacific radiocarbon and correlation with climate variability. *Radiocarbon* 43(1):15–25.

Druffel, E. R. M., S. Griffin, J. Hwang, T. Komada, S. R. Beaupre, K. C. Druffel-Rodriguez, G. M. Santos and J. Southon. 2004. Variability of monthly radiocarbon during the 1760's in corals from the Galapagos Islands. *Radiocarbon* 46:627–631.

Dunbar, R. B. and J. E Cole. 1996. Annual records of tropical systems (ARTS). A PAGES/CLIVAR Initiative. Recommendations for Research. *Kauai ARTS Workshop, Sept 1996. Pages Workshop report series 99–1.* http://pangea.standford.edu/Oceans/ARTS/arts_report/arts_report_home.html.

Dye, T. 1994. Apparent ages of marine shells: Implications for archaeological dating in Hawaii. *Radiocarbon* 36:51–57.

Flament, P., S. Kennan, R. Lumpkin, M. Sawyer and E. D. Stroup. 1996. *The Ocean Atlas of Hawaii.* Department of Oceanography, School of Ocean and Earth Science and Technology, University of Hawaii. http://radlab.soest.hawaii.edu/atlas

Forman, S. L. and L. Polyak. 1997. Radiocarbon content of pre-bomb marine mollusks and variations in the ^{14}C reservoir for coastal areas of the Barents and Kara seas, Russia. *Geophysical Research Letters* 24:885–888.

Gillespie, R. 1977. Sydney University natural radiocarbon measurements IV. *Radiocarbon* 19:101–110.

Guilderson, T. P., D. P. Schrag, M. Kashgarian and J. Southon. 1998. Radiocarbon variability in the western equatorial Pacific inferred from a high-resolution coral record from Nauru Island. *Journal of Geophysical Research* 103(C11):24,641–24,650.

Guilderson, T. P., D. P. Schrag, E. Goddard, M. Kashgarian, G. M. Wellington and B. K. Linsley. 2000. Southwest subtropical Pacific surface water radiocarbon in a high-resolution coral record. *Radiocarbon* 42(2):249–256.

Guilderson, T. P., D. P. Schrag and M. A. Cane. 2004. Surface water mixing in the Solomon Sea as documented by a high-resolution coral ^{14}C record. *Journal of Climate* 17:1147–1156.

Heath, R. A. 1985. A review of the physical oceanography of the seas around New Zealand — 1982. *New Zealand Journal of Marine and Freshwater Research* 19:79–124.

Hideshima, S., E. Matsumoto, O. Abe and H. Kitagaawa. 2001. Northwest Pacific marine reservoir correction estimated from annually banded coral from Ishigaki Island, Southern Japan. *Radiocarbon* 43:473–476.

Higham, T. F. G. and A. G. Hogg. 1995. Radiocarbon dating of prehistoric shell from New Zealand and calculation of the ΔR value using fish otoliths. *Radiocarbon* 37(2):409–416.

Hogg, A. G., T. F. G. Higham and J. Dahm. 1998. ^{14}C dating of modern marine and estuarine shellfish. *Radiocarbon* 40(2):975–984.

Hughen, K. A, M. G. L. Baillie, E. Bard, J. W. Beck, C. J. H. Bertrand, P. G. Blackwell, C. E. Buck, G. S. Burr, K. B. Cutler, P. E. Damon, R. L. Edwards, R. G. Fairbanks, M. Friedrich, T. P. Guilderson, B. Kromer, G. McCormac, S. Manning, C. Bronk Ramsey, P. J. Reimer, R. W. Reimer, S. Remmele, J. R. Southon, M. Stuiver, S. Talamo, F. W. Taylor, J. van der Plicht, and C. E. Weyhenmeyer. 2004. Marine04 Marine Radiocarbon Age Calibration, 0–26 Cal Kyr BP. *Radiocarbon* 46:1059–1086.

Ingram, B. L. 1998. Differences in radiocarbon age between shell and charcoal from a Holocene shell mound in Northern California. *Quaternary* Research 49:102–110.

Jones, M., F. Petchey, R. Green, P. Sheppard and M. Phelan. 2007. The marine ΔR for Nenumbo: A case study in calculating reservoir offsets from paired sample data. *Radiocarbon* 49:95–102.

Kennett, D. J., B. L. Ingram, J. R. Southon and K. Wise. 2002. Differences in ^{14}C age between stratigraphically associated charcoal and marine shell from the archaic period site of Kilometre 4, Southern Peru: Old wood or old water? *Radiocarbon* 44:53–58.

Konishi, K., T. Tanaka and M. Sakanoue. 1982. Secular variation of radiocarbon concentration in sea water: Sclerochronological approach. In E. D. Gomez (ed.), *Proceedings of the Fourth International Coral Reef Symposium* 1, pp 181–185, Marine Science Center. Manila: University of the Philippines.

Moberly, R. Jnr. 1963. *Coastal Geology of Hawaii,* Final report prepared for Dept. of Planning and Economic Development, State of Hawaii. Hawaii Institute of Geophysics, Report No. 41. Honolulu: University of Hawaii.

Mucciarone, D. A. and R. B. Dunbar. 2003. Stable isotope record of El Niño-Southern Oscillation events from Easter Island. In J. Loret, and J. T. Tanacredi (eds), *Easter Island. Scientific Exploration into the World's Environmental Problems in Microcosm*, pp 113–132. , New York: Kluwer Academic/ Plenum Publishers.

Paulay, G., and A. Kerr. 2001. Patterns of coral reef development on Tarawa Atoll (Kiribati). *Bulletin of Marine Science* 69(3):1191–1207.

Petchey, F., M. Phelan and P. White. 2004. New ΔR values for the southwest Pacific Ocean. *Radiocarbon* 46:1005–1014.

Petchey, FJ, Addison DJ, in press. Radiocabon dating marine shell in Samoa: A new delta-R from known-age speciment. in: Addison DJ, Asaua TS, Sand C (eds), *Recent Archaeology in the Fiji/West-Polynesia Region: Papers from the Archaeology of the Polynesian Homeland Conference*. Dunedin: Otago University.

Petchey, F., R. Green, M. Jones and M. Phelan. 2005. A local marine reservoir correction value (ΔR) for Watom Island, Papua New Guinea. *New Zealand Journal of Archaeology* 26 (2004):29–40.

Phelan, M. B. 1999. A ΔR correction value for Samoa from known-age marine shells. *Radiocarbon* 41:99–101.

Pirazzoli, P. A, G. Delibrias, L. F. Montaggioni, J. F. Saliège, and C. Vergnaud-grazzini. 1987. Vitesse de croissance latérale des platiers et évolution morphologique récente de l'atoll de Reao, Iles Tuamotu, Polynésie Française. *Annales de l'Institut océanographique* 63(1):57–68.

Reimer, P. J., F. G. McCormac, J. Moore, F. McCormick and E. V. Murray. 2002. Marine radiocarbon reservoir corrections for the mid- to late Holocene in the eastern subpolar North Atlantic. *The Holocene* 12(2):129–135.

Richmond, B. M. and P. S. Roy. 1986. Near shore sediment distribution and sand and gravel deposits in lagoon areas, northern Tongatapu, Tonga. *CCOP/SOPAC* Technical Report 63, June 1986.

Sikes, E. L., C. R. Samso, T. P. Guilderson and W.R. Howard. 2000. Old radiocarbon ages in the southwest Pacific Ocean during the last glacial period and deglaciation. *Nature* 405: 555–559.

Smith, R. 1999. Study for numerical circulation model of Abaiang Lagoon, Kiribati. *SOPAC Preliminary Report* 112.

Snelgrove, P. V. R. and C. A. Butman. 1994. Animal-sediment relationships revisited: Cause vs effect. *Oceanography and Marine Biology: An Annual Review* 32:111–177.

Southon, J, M. Kashgarian, M. Fontugne, B. Metivier and W. W-S. Yim. 2002. Marine reservoir corrections for the Indian Ocean and Southeast Asia. *Radiocarbon* 44:167–180.

Sparks, R. J., G. W. Drummond, G. W. Brailsford, D. C. Lowe, K. R. Lassey, M. R. Manning, C. B. Taylor and G. Wallace. 1992. Radiocarbon measurements in South Pacific ocean waters in the vicinity of the Subtropical Convergence zone. *Radiocarbon* 34:727–736.

Specht, J and C. Gosden. 1997. Dating Lapita pottery in the Bismarck Archipelago. *Asian Perspectives* 36(2):175–199.

Spennemann, D. H. R. and M. J. Head. 1998. Tongan pottery chronology, ^{14}C dates and the hardwater effect. *Quaternary Geochronology* 17:1047–1056.

Stearns, H. T. 1938. Geologic and topographic map of the Island of Oahu, Hawaii. Hawaii Commission on Water Resource Management 1:62,500. http://ngmdb.usgs.gov/ngm-bin/

Stearns, H. T. 1978. Quaternary shorelines in the Hawaiian Islands: *Bernice P. Bishop Museum Bulletin* 237:1–57.

Stuiver, M., G. W. Pearson and T. F. Braziunas.1986. Radiocarbon age calibration of marine samples back to 9000 cal yr BP. *Radiocarbon* 28:980–1021.

Stuiver, M., and T. F. Braziunas. 1993. Modelling atmospheric ^{14}C influences and ^{14}C ages of marine samples to 10,000 BC. *Radiocarbon* 35:137–189.

Stuiver, M. and H. A. Polach. 1977. Discussion: Reporting ^{14}C data. *Radiocarbon* 19:355–63.

Summerhayes, G. R. 2001. Defining the chronology of Lapita in the Bismarck Archipelago. In G. R. Clark, A. J. Anderson and T. Vunidilo (eds), *The Archaeology of Lapita Dispersal in Oceania*, pp 25–38. Papers from the Fourth Lapita Conference, June 2000, Terra Australis 17. Canberra: Pandanus Books.

Tanaka, N., M. C. Monaghan and D. M. Rye. 1986. Contribution of metabolic carbon to mollusc and barnacle shell carbonate. *Nature* 320:520–523.

Toggweiler, J. R., K. Dixon and W. S. Broecker. 1991. The Peru upwelling and the ventilation of the South Pacific thermocline. *Journal of Geophysical Research* 96 (C11): 20,467–20,497.

Tomczak, M. and J. S. Godfrey. 2001. *Regional Oceanography: An introduction. Version* 1.0. Oxford: Elsevier Science Ltd. http://www.es.flinders.edu.au/~mattom/regoc/pdfversion.html

Ulm, S. 2002. Marine and estuarine reservoir effects in central Queensland, Australia: Determination of ΔR values. *Geoarchaeology: An International Journal* V17(4):3119–348.

Xie, S. P., W. T. Liu, Q. Liu and M. Nonaka. 2001. Far-reaching effects of the Hawaiian Islands on the Pacific ocean/atmosphere. *Science* 292:2057–2060.

Yoneda, M., M. Hirota, M. Uchida, K. Uzawa, A. Tanaka, Y. Shibata and M. Morita. 2001. Marine radiocarbon reservoir effect in the western north Pacific observed in archaeological fauna. *Radiocarbon* 43:465–471.

12

Examining Late Holocene marine reservoir effect in archaeological fauna at Hope Inlet, Beagle Gulf, north Australia

Patricia Bourke[1]
Quan Hua[2]

[1] School of Australian Indigenous Knowledge Systems
Charles Darwin University
Darwin, NT 0909, Australia

[2] Australian Nuclear Science and Technology Organisation
PMB1, Menai
NSW 2234, Australia

Abstract

This study examines the marine reservoir effect during the Late Holocene evolution of a small estuary in the Beagle Gulf (12°S, 131°E). The paper aims at refining the local marine reservoir ages (R) and correction values (ΔR), by ^{14}C analysis of stratigraphically associated archaeological fauna (marine shell, charcoal and fish otoliths) from five proximate middens of different chronologies. The results suggest that a marine reservoir age of 340 ± 70 yrs is applicable to the Beagle Gulf for the Late Holocene, which is not significantly different from that determined for nearby Van Dieman Gulf and the north Australian coast.

Keywords: Marine reservoir effect; Holocene; archaeological fauna; dating; middens; northern Australia.

Although there are known uncertainties identified in radiocarbon dating marine samples, particularly shell, related to system complexities (e.g. estuarine fluctuations, oceanic upwelling and organism physiology) of carbonate incorporation, shell offers important advantages to archaeologists, as Higham and Hogg (1995) have outlined:

1. it has the potential to date an event closely. Shellfish are mostly processed close to where they are collected
2. shell remains are ubiquitous in Australian coastal contexts
3. the marine calibration curve is smoother for marine than terrestrial samples, with fewer multiple intercepts and narrower derived calibrated ranges.

By reducing the uncertainties and refining local marine reservoir correction factors for Australia's coasts, we may improve our confidence in the accuracy of ^{14}C based age determinations on this useful sample type. Researchers such as Reimer and Reimer (2001) and Ulm (2002) have indicated a need for localized estuary-specific data in order to accurately calibrate marine radiocarbon dates, given that local variations in marine reservoir corrections (ΔR) may be of the order of several hundred years.

This study examines the marine reservoir effect during the Late Holocene evolution of a small estuary – Hope Inlet, Shoal Bay in the Beagle Gulf (12°S, 131°E). The project addresses questions on the applicability of the Marine Reservoir Age of 384 ± 54 yrs for north Australia (see Reimer and Reimer 2001). These questions have been raised by the close correspondence between some dates obtained on shell and charcoal from this region and for the neighbouring Van Dieman Gulf (Bourke 2000:162; Woodroffe and Mulrennan 1993:40–1; Woodroffe et al. 1988:98). This paper describes the results of investigations aimed at refining the local ΔR value for the Beagle Gulf, by ^{14}C analysis of stratigraphically associated archaeological fauna (marine shell, charcoal and fish otoliths) from five proximate middens of different chronologies.

Projects such as this are becoming increasing important with further refinement of cultural chronologies in Australian archaeology (Ulm 2002:343 and see Ingram 1998; Spenneman and Head 1996). For example, radiocarbon dating of *Anadara* shell mounds on the Beagle Gulf coast suggest rapid formation (a few hundred years or less) and broad contemporaneity of some mounds in terms of human lifespans (Bourke 2004, 2005), that must be interpreted within the limits of error inherent in conventional radiocarbon dating (Head 1991; Ward 1994). Such issues have implications for palaeo-environmental models as well as models of past human coastal economies, often derived in large part from data from shell middens. These models require relatively precise dating if apparent correlations between environmental change and human adaptations and cultural contemporaneity between sites and regions are to be established with any confidence (e.g. Field 2004; Lourandos 1997; Nunn 2000; Veitch 1996; and see Spenneman and Head 1996). The deposits dated in this study comprise a representative sample of some two hundred shell and earth mounds on the Beagle Gulf coast. Many hundreds of these mounds accumulated around 2500 to 500 years BP right across the north Australian coast, from Princess Charlotte Bay to the Pilbara, (e.g. see Bailey 1999; Hiscock 1999; Figure 1) representing a period during the Late Holocene when Aboriginal people followed a tradition of mound building.

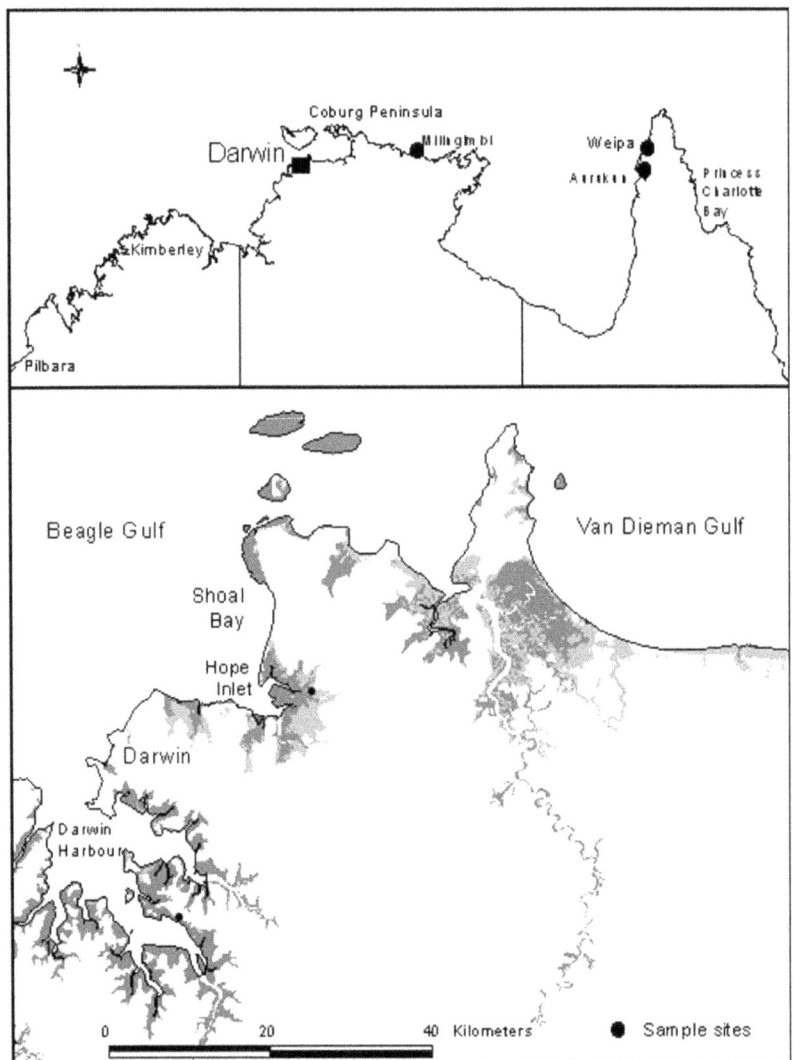

Figure 1. Location of sampled sites on the Beagle Gulf coast and some places mentioned in the text.

Site Description

The data for this study comes from 25 radiocarbon ages on paired samples from five archaeological sites on the Beagle Gulf mainland (Table 1). Most of the dates are on a series of five shell/charcoal pairs and three shell/otolith/charcoal sets, taken from three Aboriginal midden sites at Hope Inlet (HI81, HI83, HI80), which are large stratified mounded shell deposits. These mounds are located within a few hundred metres of each other in adjacent ecological zones, on a hinterland headland and associated saltflats area partitioned by tidal channels, mangroves and swamps (Figure 2). Radiocarbon dates taken from these three deposits show that each accumulated relatively rapidly (within or much less than a few hundred years) and provide an opportunity to examine local marine reservoir factors over a relatively short time span between 2500 to 500 years BP of human occupation.

Site	Lab Code	Sample	Av. Depth (cm)	$\delta^{13}C$ (‰)	^{14}C Age (yr BP)
Hope Inlet					
HI80	*OZC956	A. granosa	3	−3.0	960±80
HI80	*OZC957	Charcoal	3	−25#	590±110
HI80	*OZC958	A. granosa	40	−3.2	1190±90
HI80	*OZC959	Charcoal	40	−24.5	860±80
HI80	*OZC960	A. granosa	48	−3.5	1060±90
HI80	*OZC961	Charcoal	48	−24.7	1020±90
HI80	*OZH889	Otoliths	48–52	−5.3	1165±35
HI83	Wk8252	A. granosa	16	−3.2	2020±90
HI83	*OZH893	Charcoal	16–20	−25.6	1705±40
HI83	*OZI287	Otolith	19	−3.9	1995±40
HI83	Wk6526	A. granosa	67	−2.3	1910±70
HI83	Wk6527	Charcoal	67	−25.3	1850±70
HI81	Wk6524	A. granosa	5	−1.6	1900±70
HI81	*OZH891	Charcoal	5–9	−25.4	1570±35
HI81	*OZH892	Otoliths	5–9	−7.4	1820±40
HI81	*Wk16609	A. granosa	103	−2.3	2005±33
HI81	*Wk16610	Charcoal	103	−25.5	1635±38
HI81	Wk6523	A. granosa	140–42	−2.4	2220±70
HI81	*OZH890	Charcoal	140–42	−24.6	1835±35
HI97	*OZI286	A. granosa	14–16	−2.6	1800±40
HI97	*OZH896	Charcoal	14–16	−26.4	1345±45
Darwin Harbour					
MA7	Beta-95257	A. granosa	5–15	0#	1870±70
MA7	Beta-95256	Charcoal	5–15	−25#	850±80
MA7	Beta-87872	A. granosa	30–40	0#	1220±60
MA7	Beta-87873	Charcoal	30–40	−25#	1070±80

Notes:
*AMS analysis
Assumed $\delta^{13}C$ values
The stable isotope values for the *Anadara* shell indicate that they came from an estuarine environment. More negative $\delta^{13}C$ values indicate less saline environment (Head 1991).

Table 1. 14C ages obtained on shell, otolith and charcoal paired samples.

Excavations revealed that under a compact surface layer of fragmented shell, soil and vegetation, these north Australian shell mounds are unconsolidated deposits, of large irregularly shaped objects (shells) packed loosely together, with interstitial spaces filled with a fine silty matrix. Nonetheless, these stratified middens exhibited a relatively high level of integrity, with little evident post-depositional disturbance, as seen in a stratigraphic profile of HI81 exposed after excavation (Figure 3). Defined layers were observed, of whole, densely packed shell alternating with ashy humic layers of more fragmented *Anadara* shell, other faunal remains and stone artefacts, which was borne out in the laboratory by analysis revealing a low level of shell breakage consistent with rapid deposition.

Site HI80 in particular, an elongated shell mound (5.5 m x 10 m), located on the supra-tidal salt-flats of Hope Inlet, is a buried, salt-encrusted, solidly compacted shell deposit which, due to the cohesive nature of the matrix, suggests a stratigraphic integrity not usually found in north Australian shell middens (Figure 4).

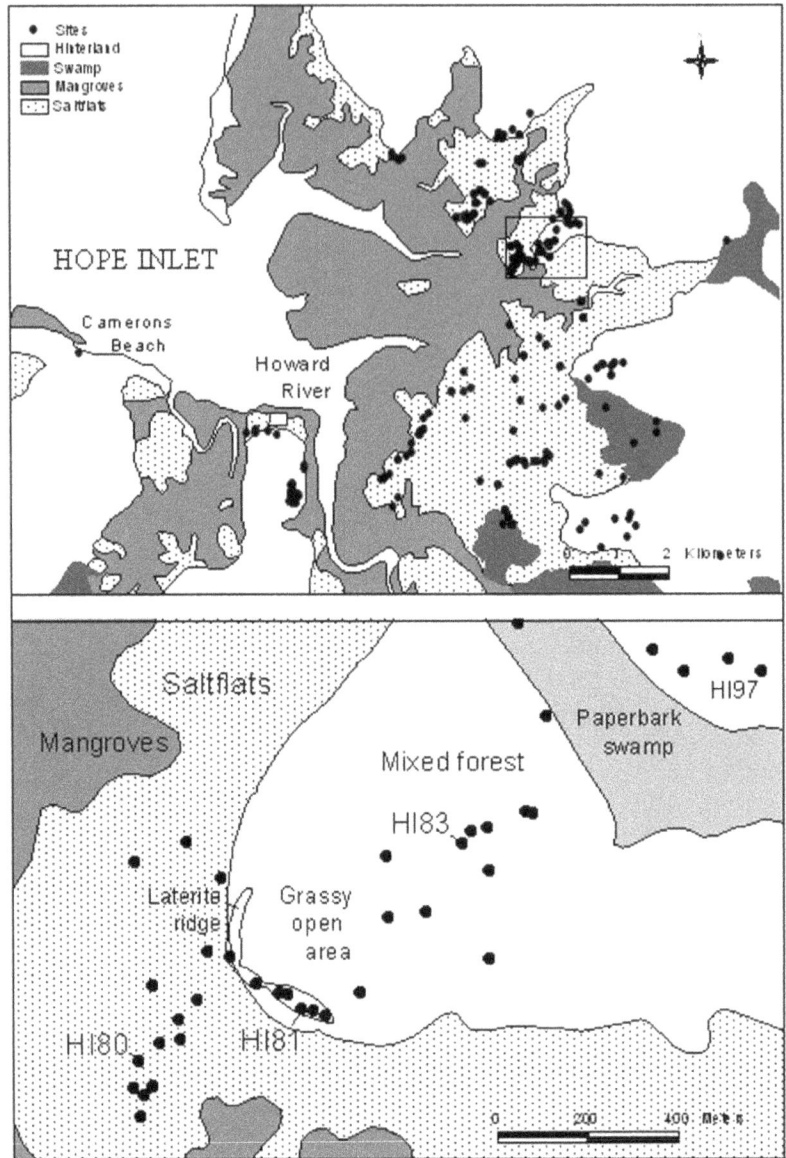

Figure 2. Locality plan of sites HI81, HI83, HI80 and HI97 at Hope Inlet.

Figure 3. Stratigraphic profile of mound HI81, showing defined layers.

Materials and Methods

The paired samples obtained from these deposits were mostly whole shells and charcoal chunks collected during excavations (by increments of average 3 cm spits), where they were observed to be in close stratigraphic proximity, often in situ inside shell valves (except the otoliths and the samples from HI97 which were selected in the lab).

The dates were obtained with limited funds over a number of years (1996–2005). Radiometric radiocarbon mea-

Figure 4. Stratigraphic profile of compacted shell deposit HI80.

surements and AMS analysis were undertaken on the Hope Inlet sites by the Waikato Radio-carbon Dating Laboratory and the Australian Nuclear Science and Technology Organisation (ANSTO) AMS Facility. The estimate on the Darwin Harbour site was obtained in 1996 from Beta Analytic via The NWG Macintosh Centre for Quaternary Dating, the University of Sydney.

Initially, prior to the most recent 2005 grant from AINSE, the samples analysed were limited to four paired samples. Higham and Hogg (1995) argue that for a more robust analysis to refine the local ΔR value, it is necessary to minimise uncertainties associated with small sample size and sample type, by increasing the number of samples and range of faunal species used. We had hoped to obtain AMS analysis of other fauna, such as fish otoliths and terrestrial (macropod) bone/teeth samples also found in close stratigraphic association in the three shell mounds, to cross check against the dates on marine shell and on the charcoal, which was unidentified. Unfortunately the terrestrial samples were too fragmented and degraded, and not suitable for radiocarbon dating, so only the marine otolith samples were analysed. The possibility of a high inbuilt age in charcoal from an unknown species, identified as a problem in places with long-lived trees, such as New Zealand or the wet tropics of northeast Australia, (cf. Higham and Hogg 1995; Ulm 2002), is considered here to be less likely in the semi-arid tropics of this part of north Australia. This region is not known for old trees, due to the annual cycle of cyclones/storms, floods, fires and drought, in combination with the ravages of termites (e.g. Bowman and Panton 1994; Wilson and Bowman 1987).

The samples range from 0.1–1.5 g for the charcoal component and 12–40 g for the shell component. *Anadara granosa*, the dominant species that makes up the shell middens, was used in all cases, reducing possible variation due to difference in the relationship of specific species to the carbon cycle (Ulm 2002:331).

The samples were all within the average size of the *Anadara granosa* shells in these middens, which is between 25–35 mm in length. It takes about 15 months for *Anadara* cockles to reach a size of 18–32 mm and specimens larger than 53 mm are a rare occurrence (Broom 1985), so the samples are likely to represent relatively short life spans of around two years.

Anadara granosa is a marine bivalve cockle with limited mobility, which appears to be both a suspension and surface deposit-feeder, gaining nutrition from a mixture of microorganisms attached to detritus and benthic microalgae that are easily re-suspended from the substrate (Broom 1985). Studies suggest that the use for examining local reservoir conditions, of species that ingest detritus can be problematic, as ingested organic carbon from a mixture of sources (including terrestrial organic carbon introduced into estuaries by rivers), can be metabolised and incorporated into shell (Hogg et al. 1998, Tanaka et al. 1986; cited in Ulm 2002:330).

In this current analysis an additional marine sample type was employed to minimise uncertainties associated with the single species sample of *Anadara granosa* shell. We have used samples of the otoliths of forktailed catfish (*Arius* sp.), to allow a cross check against possible systematic effects of carbon incorporation in the *Anadara* species and thus check the validity of ages obtained on this shell (Spenneman and Head 1996). Helen Larson of the Museum and Art Galleries of the Northern Territory (MAGNT) estimated that forktailed catfish species, which live in the shallow, inshore waters of the estuarine system, have a lifespan of some 10–15 years.

Most of radiocarbon dates in this study were analysed at the ANSTO AMS facility (Table 1). The chemistry procedures for AMS ^{14}C sample preparation at this facility were as follows:

> For charcoal samples, the samples were washed with 2M HCl at 60°C for two hours to remove any possible carbonate contamination. The samples were then treated with 1–2% NaOH at 60°C until all humic acid contamination was removed, followed by another 2M HCl treatment at room temperature for two hours. Before combustion, the pre-treated samples were oven-dried at 60°C for two days.

> For shell and otolith samples, the samples were washed several times with deionised water in an ultrasonic bath, then were leached for 10–15 minutes in dilute HCl to remove any surface contamination and possible secondary carbonate. They were then dried in an oven at 60°C for two days.

> The cleaned samples were combusted (for charcoal) or hydrolysed (for shell and otolith) to CO_2, then converted to graphite using the Fe/H_2 method (Hua et al. 2001). AMS ^{14}C measurements were performed using the ANTARES facility at ANSTO (Fink et al. 2004).

Results and Discussion

Radiocarbon results

Our ^{14}C results for shell, otolith and charcoal samples from the study sites, expressed in conventional radiocarbon ages, after correction for isotopic fractionation using δ^{13}C values, are presented in Table 1. There is a good agreement between ^{14}C ages of shells and those of otoliths as they overlap each other within 1σ uncertainty. This gives us confidence in our ^{14}C ages of marine samples. If there were a proble m then the otoliths would systematically date older than the shell, but there was no difference seen with just the three samples dated.

To calibrate ^{14}C ages of charcoal samples, we used the IntCal04 calibration data set (Reimer et al. 2004) and the CALIB program version 5.01 (http://radiocarbon.pa.qub. ac.uk/). It is well known that the atmospheric ^{14}C levels for different parts of the world are not the same (e.g., Stuiver and Braziunas 1998; Hua et al. 2004a). Therefore, a regional atmospheric ^{14}C offset correction has been applied for our study sites in the tropics during the age calibration process, using IntCal04 data set, which represents atmospheric ^{14}C levels for northern temperate regions. We adopted an atmospheric Δ^{14}C offset between our sites at ~12°S and the northern temperate regions at 40–50°N of ~2‰ (atmospheric ^{14}C level at our sites is lower) from the modelling work of Braziunas et al. (1995). This value is equivalent to ~16 ^{14}C years (terrestrial samples growing at our sites are older) and to simplify the issues we assumed that this value is constant through time. Sixteen years was subtracted from all conventional radiocarbon ages of charcoal samples before these were converted to calendar ages using the CALIB program and IntCal04 data set. Calibrated ages (1σ) of charcoal samples are reported in column 7 of Table 2.

Average depth (cm) (1)	Charcoal samples Lab No. (2)	14C Age (BP) (3)	Marine samples Lab No. (4)	14C Age (BP) (5)	R (14C years)a (6)	Cal BP (1σ)b (7)	Modelled marine 14C agec (8)	ΔR (14C years)d (9)
Hope Inlet HI80								
3	OZC957	590±110	OZC956	960±80	370±136	655–520	1010±94	−50±123
40	OZC959	860±80	OZC958	1190±90	330±120	899–685	1245±117	−55±148
48–52	OZC961	1020±90	OZC960	1060±90	40±127	1046–794	1380±116	−320±147
			OZH889	1165±35	145±97			−215±121
Hope Inlet HI83								
16–20	OZH893	1705±40	Wk8252	2020±90	315±98	1688–1541	2045±61	−25±109
			OZI287	1995±40	290±57			−50±73
67	Wk6527	1850±70	Wk6526	1910±70	60±99	1865–1701	2190±68	−280±98
Hope Inlet HI81								
5–9	OZH891	1570±35	Wk6524	1900±70	330±78	1516–1402	1915±47	−15±84
			OZH892	1820±40	250±53			−95±62
103	Wk16610	1635±38	Wk16609	2005±33	370±50	1553–1418	1950±51	55±61
140–142	OZH890	1835±35	Wk6523	2220±70	385±78	1813–1714	2170±37	50±79
Hope Inlet HI97								
14–16	OZH896	1345±45	OZI286	1800±40	455±60	1298–1184	1695±60	105±72
Darwin Harbour MA7								
5–15	Beta-95256	850±80	Beta-95257	1870±70	1020±106	897–679	1240±120	630±139
30–40	Beta-87873	1070±80	Beta-87872	1220±60	150±100	1064–804	1400±119	−180±133

Notes:

a. Measured reservoir age is the difference between the conventional radiocarbon ages of marine and charcoal samples. The associated 1σ uncertainty is defined as $(\sigma_{marine}{}^2 + \sigma_{charcoal}{}^2)^{1/2}$.

b. Calibrated age ranges (1σ) of charcoal samples with the atmospheric ^{14}C offset correction. The correction is applied by subtracting 16 years from the conventional ^{14}C ages of charcoal samples before they were calibrated using CALIB program version 5.01 and IntCal04 data set.

c. Modelled marine age is the hypothetical age and uncertainty estimate that, when calibrated using CALIB program, Marine04 data set and ΔR=0, produces the associated calibrated age ranges (1σ) for charcoal samples, which are seen in column 7.

d. Marine reservoir correction value is the difference between the measured and modelled marine ^{14}C ages. The associated 1σ uncertainty is defined as $(\sigma_{measured}{}^2 + \sigma_{modelled}{}^2)^{1/2}$.

Table 2. Marine reservoir ages (R) and correction values (ΔR) for the Beagle Gulf.

At Hope Inlet, two shell/charcoal paired samples and one shell/charcoal/otolith sample dated from mound HI81 show that this mound formed on a laterite ridge at the uplands edge of mudflats over a few centuries between ca 1800 and 1400 cal BP (Bourke 2004). The date the mound began to form follows a major phase of mudflat build-out dated ca 2300 cal BP at Shoal Bay (Woodroffe and Grime 1999). Dating of one shell/charcoal pair and one shell/charcoal/otolith sample shows rapid formation of mound HI83, located some 300 m inland of HI81, also within this period. One shell/charcoal pair taken from a 30 cm wide midden layer within an earth mound (site HI97) that lies inland of the three sampled shell mounds, suggests that this mound formed soon after this period. Dating of two shell/charcoal pairs and one shell/charcoal/otolith sample shows that mound HI80 was formed on nearby saltflats between ca 900 and 500 cal BP, many centuries after HI83 and HI81 ceased accumulating, and following another phase of mudflat build-out and chenier building around 1000–900 cal BP at Shoal Bay (Woodroffe and Grime 1999).

In the neighbouring Darwin Harbour estuary system, dating of two shell/charcoal pairs from a smaller shell midden site (MA7), suggests that this mound formed around the same period as HI80, around 900–700 cal BP.

Marine reservoir ages and correction values

A marine reservoir age (R) for a particular region at a time t is determined as:

Measured R(t) = Measured marine ^{14}C age(t) – Measured charcoal ^{14}C age(t)

Measured marine and charcoal ^{14}C ages are presented in columns 5 and 3 of Table 2, respectively. Our R values are reported in column 6 of Table 2.

According to Stuiver and Braziunas (1993), a marine reservoir correction value (ΔR) for a particular region at a time t is defined as:

ΔR = measured R(t) – modelled R(t) + ΔRa
or
ΔR = measured marine ^{14}C age(t) – modelled marine ^{14}C age(t) + ΔRa

where ΔRa is a fraction of ΔR due to a regional atmospheric ^{14}C offset.

For this study, ΔRa is different from zero and was incorporated into the calculation of our ΔR values during the process of estimating modelled marine ^{14}C age(t) following the method described in Owen (2002).

The procedure of ΔR determination is as follows:
- A 16-yr offset correction was applied for all radiocarbon ages of charcoal samples before they were calibrated using CALIB program and IntCal04 data set. Calibrated results are presented in column 7 of Table 2.
- The modelled marine age and uncertainty estimate (reported in column 8 of Table 2) are determined by successive approximation as the values that, when calibrated with Marine04 curve (Hughen et al. 2004) and ΔR=0, result in the calibrated ages of charcoal samples reported in column 7 of Table 2. The smooth shape of Marine04 curve ensures that there is almost always a unique solution (Owen 2002).
- ΔR value is the difference between measured and modelled marine ^{14}C ages, which are presented in columns 5 and 8 of Table 2, respectively. ΔR values of this study are reported in column 9 of Table 2.

R and ΔR values for Hope Inlet, Beagle Gulf estimated from our ^{14}C data are also illustrated in Figures 5 and 6. R values range from 40 to 1020 years. Most values are from 250 to 455 yrs, and only one data point shows a very high value of 1020 yrs for MA7 at 5–15 cm, and four data points show low values ranging from 40 to 150 yrs, for HI80 at 48–52 cm, HI83 at 67 cm and MA7 at 30–40 cm. Similarly, ΔR values vary from –320 to 630 yrs. Most data are in the range of –95 to 105 yrs with the exception of four low values from –320 to –180 yrs, and one high value of 630 yrs. Note that the four pairs showing low R values also show low ΔR values. Similarly, the pair showing high R value also shows high ΔR value.

For MA7, a pair sample at 5–15 cm shows a very high R value of 1020 yrs and the other pair at 30–40 cm shows a low R value of 150 yrs. It seems that there may be a stratigraphic problem for this midden, possibly due to upward vertical movement of shell through post-depositional disturbance by goanna activity that is evident in this mound. The data derived from this mound may therefore not be reliable, at least for MA7 at 5–15 cm.

For HI80, two pair samples at 48–52 cm show low R values of 40 yrs for the charcoal/*Anadara granosa* pair, and 145 yrs for the charcoal/ otolith pair. Contami-nation of shell or otolith samples by secondary carbonate as a result of their contact to groundwater or seawater may cause low R values. However, the fish otolith (OZH889) had a smooth surface indicating no or minimal contamination by secondary carbonate. Both marine samples (OZC960 and OZH889) were treated at ANSTO with dilute acids to remove any surface contami-nation before dating and had similar ^{14}C ages. This implies contamination by secondary carbonate is not the cause for these low R values.

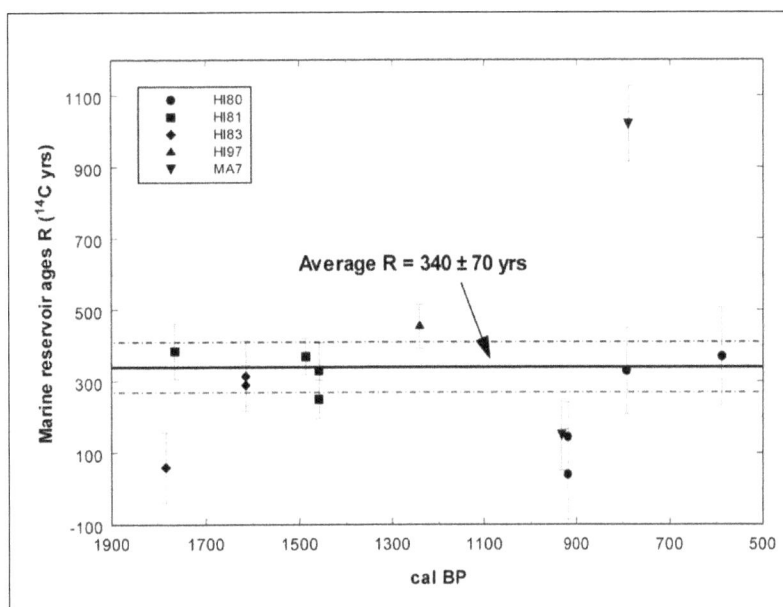

Figure 5. Marine reservoir and average R for the Beagle Gulf coast determined from this study. All symbols are plotted in the middle of calibrated age ranges reported in column 7 of Table 2. Very low (from 40 to 150 years) and high (1020 years) values of R were not included in the estimate of the average value.

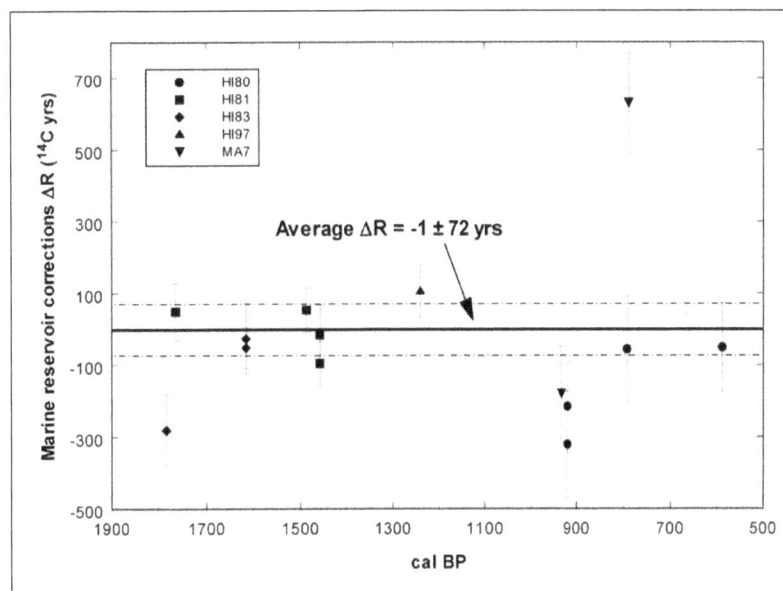

Figure 6. ΔR marine corrections for the Beagle Gulf coast determined from this study. All symbols are plotted in the middle of calibrated age ranges reported in column 7 of Table 2. Very low (from –320 to –180 years) and high (630 years) values of ΔR were not included in the estimate of the average value.

The low R and ΔR values at the base of the HI80 deposit occurs for both the shell and otolith samples. These are taken from the base of a defined layer between 48–52 cm depth, containing a high proportion of the bivalves *Marcia hiantina* and *Anadara inaequivalvis* in the otherwise Anadara dominated deposit (Bourke 2004). These two species thrive in coarser, sandier substrates than the finer, silty muddy substrates preferred by the dominant *Anadara granosa*. In site MA7 too, *Marcia hiantina* makes up 20% more of the total shell taxa than in a nearby mound (MA1) dated to an earlier period (Bourke 2005: Tables 1 and 2). The results suggest that something occurred around 1000–900 cal BP causing low R and ΔR values, as this can be seen from three data points from two different sites (HI80 and MA7). Sites HI80 at Hope Inlet and MA7 on the Darwin Harbour shoreline are approximately 25 km apart in neighbouring estuaries.

Thus these results may reflect short-term environmental change on a regional rather than local micro-scale confined to the Hope Inlet area, and imply high rainfall or storms and changes in relative quantities of sand, mud, river and sea currents and the location of sand banks, as observed on the Arnhem Land coast during the 1970s (Meehan 1982:62,70, 78, 142).

For HI83, a pair sample at 67 cm shows a low R value of 66 yrs. A mismatch between shell and charcoal may cause a low R value, but one author (PB) believes that this is not the case here, as the shell/charcoal pair for site HI83 at 67 cm was recovered in secure stratigraphic association, consisting of a large chunk (5 g) of charcoal cemented together with an *Anadara granosa* valve in a clay base. Contamination of shell by secondary carbonate may also cause a low R value. This contamination can be detected by X-ray diffraction (XRD) analysis. This shell sample (Wk6526) measured at Waikato was analysed by XRD to make sure it had not suffered any secondary carbonate contamination. This indicates secondary carbonate contamination may not be the cause for this low R value. Also, a low R value for HI83 at 67 cm is thought not to be due to environmental changes at ~1800 cal BP, because a pair sample at HI81 site at 140–142 cm shows a reasonable R of 385 ± 75 yrs at that time. A possible cause for a low R value for HI83 at 67 cm is the old wood problem. However, this possibility is unlikely as the semi-arid tropics of north Australia are not known for old trees (see discussions in Materials and Methods) (cf. Higham and Hogg 1995). It is also possible that the charcoal sample is a reworked charcoal that has survived in the surrounding environment for a long period of time before finally ending up in the midden. The low R value for HI83 at 67 cm is not well understood and should be investigated in the future.

If these very high and low values of R are not taken into account, an error-weighted average R for Hope Inlet of 340 ± 70 ^{14}C yrs for the period 1800–600 cal BP can be inferred (Figure 5). Similarly, if these high and low ΔR values are not included, an error-weighted average ΔR for Hope Inlet of −1 ± 72 ^{14}C yrs for the period 1800–600 cal BP can be estimated (Figure 6). The uncertainties associated with error weighted average values of R and ΔR are their standard deviations.

Conclusions

The results of this study, based on a small number of dates obtained on paired samples — shell/charcoal pairs and shell/otolith/charcoal sets from five archaeological deposits — indicate for the Hope Inlet estuary on the Beagle Gulf mainland, values of an average marine reservoir age (R) of 340 ± 70 yrs and ΔR marine correction of −1 ± 72 yrs for the period 1800–600 cal BP. This marine reservoir age for the Beagle Gulf for the Late Holocene is not significantly different from the R value of 384 ± 58 yrs determined for the north Australian coast (see Reimer and Reimer 2001) as the two values overlap each other within 1σ uncertainty. Similarly, the weighted mean ΔR value for Hope Inlet is not significantly different from the regional mean ΔR value of 64 ± 24 yrs quoted in Hua et al. (2004b) for NW Australia and Java, as the two values overlap each other within 1σ uncertainty.

The results also show lower values of R and ΔR ranging from 40 to 150 yrs and from −180 to −320 yrs, respectively, for a short period during 1000–900 cal BP. These low values, based on only 3 paired samples from two different shell middens, may reflect short-term environmental change (e.g. high rainfall or storms) on a regional scale. More data are needed to confirm this observation.

Studies such as this that provide localized estuary-specific data contribute to coastal and estuarine data sets required as a baseline for accurate calibration of marine radiocarbon dates, which are increasingly important in refining archaeological and environmental chronologies. Further studies that may be useful to investigate local variations in marine ^{14}C reservoir ages would be to locate and measure the reservoir ages of historic shells for this region, as well as stable oxygen and carbon isotopic measurements of the historic and archaeological shells.

Acknowledgements

We gratefully acknowledge funding from AINSE for AMS ^{14}C measurements (grants 97/185R and 05/014). We also thank Paula Reimer and Fiona Petchey for their critical comments which improved the manuscript.

References

Bailey, G. N. 1999. Shell mounds and coastal archaeology in northern Queensland. In J. Hall and I. McNiven (eds), *Australian coastal archaeology*, pp 105–112. Canberra: ANH Publications, Department of Archaeology and Natural History, Research School of Pacific and Asian Studies, The Australian National University.

Bourke, P. M. 2000. Late Holocene Indigenous economies of the tropical Australian coast: An archaeological study of the Darwin Region. Unpublished Ph.D. thesis. Darwin: Northern Territory University.

Bourke, P. 2004. Three Aboriginal shell mounds at Hope Inlet: Evidence for coastal, not maritime Late Holocene economies on the Beagle Gulf mainland, northern Australia. *Australian Archaeology* 59:10–22.

Bourke, P. 2005. Archaeology of shell mounds of the Darwin coast: Totems of an ancestral landscape. In P. Bourke, S. Brockwell and C. Fredericksen (eds), *Darwin archaeology: Aboriginal, Asian and European heritage of Australia's Top End*, pp 29–48. Darwin: Charles Darwin University Press.

Bowman, D. M. J. S. and W. J. Panton. 1994. Fire and cyclone damage to woody vegetation on the north coast of the Northern Territory, *Australia. Australian* Geographer 25:32–35.

Braziunas, T. F., I. Y. Fung and M. Stuiver. 1995. The preindustrial atmospheric $^{14}CO_2$ latitudinal gradient as related to exchanges among atmospheric, oceanic, and terrestrial reservoirs. *Global Biogeochemical Cycles* 9:565–584.

Broom, M. J. 1985. *The biology and culture of marine bivalve mollusca of the genus* Anadara. Manila: Philippines: International Centre for Living Aquatic Resources Management.

Field, J. 2004. Environmental and climatic considerations: A hypothesis for conflict and the emergence of social complexity in Fijian prehistory. *Journal of Anthropological Archaeology* 23:79–99.

Fink, D., M. Hotchkis, Q. Hua, G. Jacobsen, A. M. Smith, U. Zoppi, D. Child, C. Mifsud, H. van der Gaast, A. Williams and M. Williams. 2004. The ANTARES AMS Facility at ANSTO. *Nuclear Instruments and Methods in Physics Research B* 223–224:109–15.

Head, J. 1991. The radiocarbon dating of freshwater and marine shells. In R. Gillespie (ed.), *Quaternary dating workshop* 1990, pp 16–18. Canberra: Department of Biogeography and Geomorphology, The Australian National University.

Higham, T. F. G. and A. G. Hogg 1995. Radiocarbon dating of prehistoric shell from New Zealand and calculation of the ΔR value using fish otoliths. *Radiocarbon* 37(2):409–416.

Hiscock, P. 1999. Holocene coastal occupation of western Arnhem Land. In J. Hall and I. McNiven (eds), *Australian coastal archaeology*, pp 91–103. Canberra: ANH Publications, Department of Archaeology and Natural History, Research School of Pacific and Asian Studies, The Australian National University.

Hogg, A. G., T. F. G. Higham and J. Dahm. 1998. ^{14}C dating of modern marine and estuarine shellfish. *Radiocarbon* 40:975–984.

Hua, Q., G. E. Jacobsen, U. Zoppi, E. M. Lawson, A. A. Williams, A. M. Smith and M. J. McGann 2001. Progress in radiocarbon target preparation at the ANTARES AMS Centre. *Radiocarbon* 43:275–82.

Hua, Q., M. Barbetti, U. Zoppi, D. Fink, M. Watanasak and G. E. Jacobsen. 2004a. Radiocarbon in tropical tree rings during the Little Ice Age. *Nuclear Instruments and Methods in Physics Research B* 223–224:489–494.

Hua, Q., C. D. Woodroffe, M. Barbetti, S. G. Smithers, U. Zoppi and D. Fink. 2004b. Marine reservoir corrections for the Cocos (Keeling) Islands, Indian Ocean. *Radiocarbon* 46:603–610.

Hughen, K. A., M. G. L. Baillie, E. Bard, J. W. Beck, C. J. H. Bertrand, P. G. Blackwell, C. E. Buck, G. S. Burr, K. B. Cutler, P. E. Damon, R. L. Edwards, R. G. Fairbanks, M. Friedrich, T. P. Guilderson, B. Kromer, G. McCormac, S. Manning, C. Bronk Ramsey, P. J. Reimer, R. W. Reimer, S. Remmele, J. R. Southon, M. Stuiver, S. Talamo, F. W. Taylor, J. van der Plicht and C. E. Weyhenmeyer. 2004. Marine04 marine radiocarbon age calibration, 0–26 cal kyr BP. *Radiocarbon* 46:1059–1086.

Ingram, B. L. 1998. Differences in radiocarbon age between shell and charcoal from a Holocene shell-mound in northern California. *Quaternary Research* 49:102–110.

Lourandos, H. 1997. *Continent of hunter-gatherers*: New perspectives in Australian prehistory. Cambridge: Cambridge University Press.

Meehan, B. 1982. *Shell bed to shell midden.* Canberra: Australian Institute of Aboriginal Studies.

Nunn, P. 2000. Environmental catastrophe in the Pacific Islands around AD 1300. Geoarchaeology: *An International Journal* 15(7):715–740.

Owen, B. D. 2002. Marine carbon reservoir age estimates for the far south coast of Peru. *Radiocarbon* 44:701–708.

Reimer, P. and R. Reimer 2001. A marine reservoir correction database and on-line interface. *Radiocarbon* 43:461–463. URL:http://calib.org/marine.

Reimer, P. J., M. G. L. Baillie, E. Bard, A. Bayliss, J. W. Beck, C. J. H. Bertrand, P. G. Blackwell, C. E. Buck, G. S. Burr, K. B. Cutler, P. E. Damon, R. L. Edwards, R. G. Fairbanks, M. Friedrich, T. P. Guilderson, A. G. Hogg, K. A. Hughen, B. Kromer, G. McCormac, S. Manning, C. Bronk Ramsey, R. W. Reimer, S. Remmele, J. R. Southon, M. Stuiver, S. Talamo, F. W. Taylor, J. van der Plicht and C. E. Weyhenmeyer. 2004. IntCal04 terrestrial radiocarbon age calibration, 0–26 cal kyr BP. *Radiocarbon* 46:1029–1058.

Spennemann, D. H. R. and M. J. Head. 1996. Reservoir modification of radiocarbon signatures in coastal and near-shore waters of eastern Australia: The state of play. *Quaternary Australasia* 14/1:32–39.

Stuiver, M. and Braziunas, T. F. 1993. Modeling atmospheric ^{14}C influences and ^{14}C ages of marine samples to 10,000 BC. *Radiocarbon* 35:137–189.

Stuiver, M. and Braziunas, T. F. 1998. Anthropogenic and solar components of hemispheric[14]C. *Geophysical Research Letters* 25:329–332.

Tanaka, N., M. C. Monaghan and D. M. Rye. 1986. Contribution of metabolic carbon to mollusc and barnacle shell carbonate. *Nature* 320: 520–523.

Ulm, S. 2002. Marine and estuarine reservoir effects in Central Queensland, Australia: Determination of R values. *Geoarchaeology: An International Journal* 17(4):319–348.

Veitch, B. 1996. Evidence for mid-Holocene change in the Mitchell Plateau, Northwest Kimberley, Western Australia," in P. Veth and P. Hiscock (eds), *Archaeology of Northern Australia. Tempus* 4, pp 66–89. St. Lucia, Queensland: Anthropology Museum, University of Queensland.

Ward, G. K. 1994. On the use of radiometric determinations to 'date' archaeological events. *Australian Aboriginal Studies* 2:106–109.

Wilson, B. A. and D. M. J. S. Bowman 1987. Fire, storm, flood and drought: The vegetation ecology of Howards Peninsula, Northern Territory, Australia. *Australian Journal of Ecology* 12:165–74.

Woodroffe, C. D. and D. Grime. 1999. Storm impact and evolution of a mangrove-fringed chenier plain, Shoal Bay, Darwin, Australia. *Marine Geology* 159:303–321.

Woodroffe, C. D. and M. E. Mulrennan 1993. Geomorphology of the lower Mary River plains, *Northern Territory*. Darwin: North Australia Research Unit, Australian National University.

Woodroffe, C. D., J. M. A. Chappell and B. G. Thom 1988. Shell middens in the context of estuarine development, South Alligator River, Northern Territory. *Archaeology in Oceania* 23:95–103.

13

Archaeological surfaces in western NSW: Stratigraphic contexts and preliminary OSL dating of hearths

Edward J. Rhodes[1]
Patricia Fanning[2]
Simon Holdaway[3]
Cynthja Bolton[4]

[1]Department of Environmental and Geographical Sciences
Manchester Metropolitan University,
Chester Street, Manchester, M5 1GD, UK.
e.rhodes@mmu.ac.uk

[2]Graduate School of the Environment, Macquarie University,
Sydney, NSW 2109, Australia.
pfanning@gse.mq.edu.au

[3]Department of Anthropology, University of Auckland,
PB 92109 Auckland, New Zealand.
sj.holdaway@auckland.ac.nz

[4]Research School of Earth Sciences and Department of Earth
and Marine Sciences, The Australian National University,
Canberra, ACT 0200, Australia.
cynthja.bolton@anu.edu.au

Abstract

A two-phase process for developing a chronology of Aboriginal occupation in arid western NSW, Australia, has been developed over the past ten years by the Western NSW Archaeology Program. Radiocarbon dating of charcoal from the remains of heat-retainer hearths, built by Aboriginal people in the past to cook food, and Optically Stimulated Luminescence (OSL) dating of sediments have been used to construct a chronology of 'archaeological surfaces'. Here we provide preliminary age estimates using OSL dating of stones from heat-retainer hearths which have previously been dated by radiocarbon. Our method is novel in several ways including the rapid preparation method adopted and the approach to estimating the dose rate for surface samples. We discuss the limitations of this virtually non-destructive and efficient OSL dating method, and provide an agenda for future technical development and application.

Keywords : Archaeological surfaces, OSL dating, heat-retainer hearths, Western NSW

Introduction

Surface archaeological deposits are widespread in western NSW (Figure 1), but have received relatively little attention, owing in part to the difficulty of providing chronological constraint. Heat-retainer hearths, or earth ovens, are most often found in association with stone artefact deposits, and radiocarbon dating of charcoal from the hearths is beginning to provide a powerful means of developing a chronology of occupation of the locations in which they are found (e.g. Holdaway et al. 2002, 2005).

Optically-stimulated luminescence (OSL) dating of sediments (Huntley et al. 1985; Rhodes 1988) provides an additional method which may be applied in these contexts. For example, Fanning and Holdaway (2001) used OSL sediment dating combined with ^{14}C dating of hearth charcoal to provide a chronology for archaeological material at Stud Creek in Sturt National Park in far northwest NSW (Figure 1). In later work, Fanning et al. (2008) found that the OSL signals of sediments from Fowlers Gap, NSW, were poorly bleached, but that single grain OSL measurements, combined with the minimum age model of Galbraith et al. (1999), provided reliable age estimates for deposition.

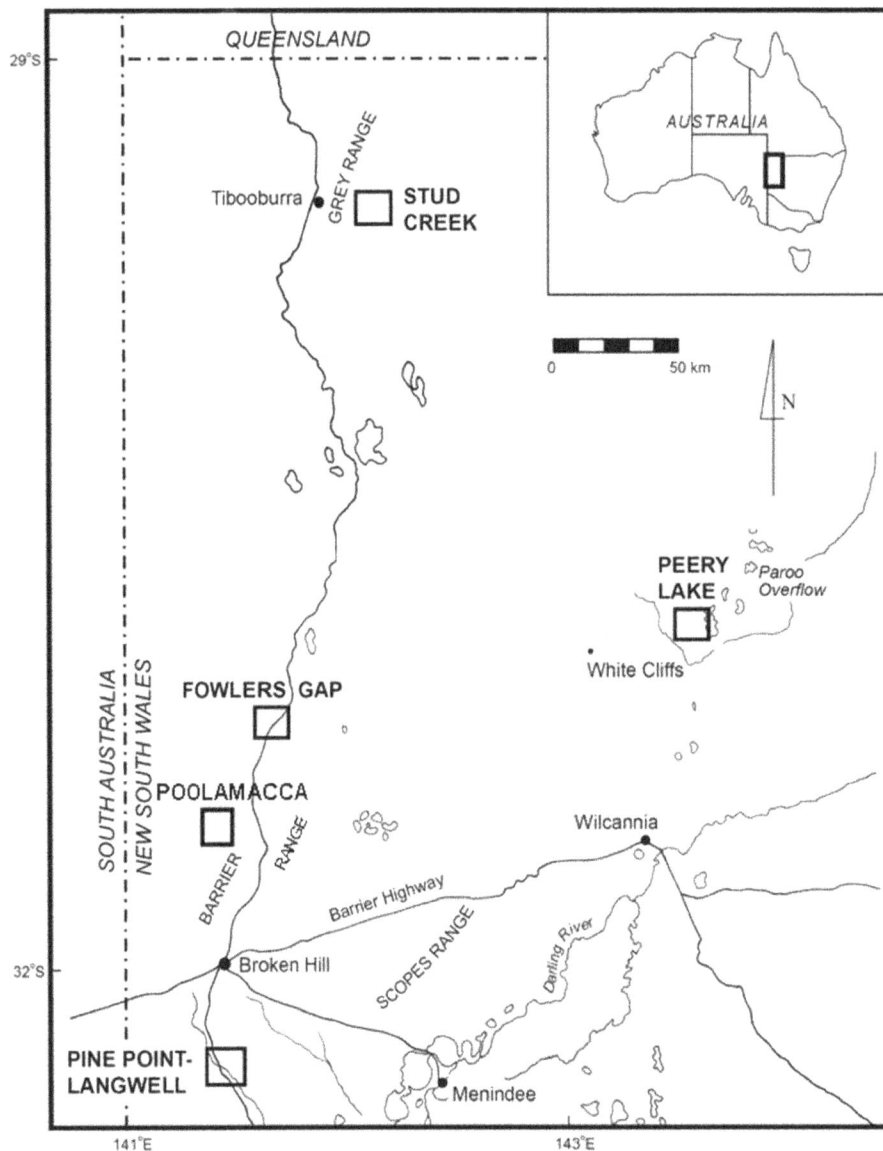

Figure 1. Map of western NSW, Australia, showing the locations of Fowlers Gap Arid Zone Research Station and the pastoral property known as 'Poolamacca'. Other Western NSW Archaeology Program (WNSWAP) research locations are also shown.

In this paper, we present preliminary OSL dating results for heat-retainer hearth stones, providing age estimates of the last heating event. These results are compared with radiocarbon determinations on charcoal from the same hearths, and provide powerful evidence of the potential of OSL for directly dating the hearths. One clear advantage of using OSL dating, in comparison with radiocarbon, is that it is far less destructive: only a small number of stones, between 1 and 4, need to be removed from the hearth for OSL dating, and no excavation is required. In contrast, for reliable radiocarbon dating, a significant proportion (typically one quarter to one half) of a hearth must be fully excavated to retrieve charcoal (see Fanning et al. this volume). Given the conservation concerns of heritage managers and the desire of Aboriginal people for their heritage to be left in place, a reliable alternative method of dating Aboriginal occupation is highly desirable. Moreover, this method, if successful, will allow age determinations of hearths where the charcoal is no longer present, thereby extending the data set from which an occupation chronology is constructed. The methods which we have developed for rapid preparation of the samples and the approach to estimating the past dose rate experienced by grains within hearth stones are both new. With further developmental work and more complete assessment, we consider that this OSL-based approach will provide a new means for archaeologists working in arid environments to determine a chronology of human activity without the requirement to excavate.

Nature of 'archaeological surfaces' in western NSW

The term 'archaeological surfaces' is used here to mean any land surface that contains deposits of stone artefacts and associated heat-retainer hearths. The processes involved in their formation have the potential to vary considerably, depending on the geomorphic setting and human occupation history, but detailed geomorphological and stratigraphic research of numerous surface archaeological contexts in western NSW reveals a similar story (Fanning 1999, 2000; Fanning and Holdaway 2001), as described briefly below.

Typically, a new land surface is constructed from the deposition of alluvial (water-lain) and/or colluvial (slope) sediments within part of a catchment, usually the valley floors, following, for example, a flood event. The 'surface' simply represents the last sedimentary unit deposited at each location. The surface is initially free of archaeology, although the sediments have the potential to contain reworked archaeological material from upstream locations. In western NSW, valley-bottom sediments comprise poorly-sorted silts (including pelletal silts) and sands (Fanning 1999; Fanning and Holdaway 2001), with occasional gravely units. While archaeological material of coarser grade may be redeposited within the gravel units, it is rarely redeposited within sand or silt-dominated contexts.

In the following years, soil formation processes (weathering, bioturbation) will slowly modify the surface, and localised deposition of aeolian material may take place. At the same time, the surface is 'available' for human habitation, with the record of human activities taking place there being preserved in the archaeological material deposited upon it. This includes the construction of heat-retainer hearths, though it is important to note that these structures are constructed by digging down through the surface sedimentary unit, sometimes into underlying strata. Thus, the main part of the hearth may occupy space at a deeper (older) stratigraphic level (Figure 2).

The surface will continue to be available to 'collect' archaeological material until either buried by renewed deposition or eroded away. This may occur in localised patches, for example by limited aeolian erosion and deposition or limited surface fluvial activity, or it may occur over a greater area, for example by floodplain stripping or by widespread deposition of flood deposits caused by another flood event. In such cases, the upper surface of the new sedimentary unit becomes another archaeological surface, in turn retaining archaeological material deposited on it. Under this accumulation-dominated scenario, it is possible that a catchment may contain multiple archaeological surfaces of different ages, possibly stacked up one upon another.

In western NSW, however, the stratigraphic records of upland catchments are dominated by erosion (Fanning and Holdaway 2001; Holdaway et al. 2004). Accelerated erosion caused by overgrazing by cattle and sheep over the past 150 years is the principal determinant of the current high visibility of archaeological material at the surface (Fanning 1999, 2002). Once uncovered, however, these surface archaeological deposits have a very limited lifespan. Much of the archaeological material currently exposed in these contexts is in highly unstable locations, and is unlikely to survive more than two or three decades (Fanning et al. this volume).

Figure 2 represents the typical sedimentary and archaeological relationships that exist in much of western NSW today. The original archaeological surface is represented by the boundary between layers B and C. Archaeological material (lithic artefacts and debitage, bone and other materials) is found at this level as indicated by the symbol. Note that some movement of material within the sediment by bioturbation and other processes such as swelling and contraction during wetting and drying cycles, and very localised erosion and redeposition, means that archaeological material can extend over a depth range of several cm, rather than lying on exactly one surface.

Typical hearth-sediment relationships are also shown in this figure. Hearths are constructed down from the archaeological surface (layer B–C interface) into layer B. Field observation suggests that hearths extend to a range of depths, but rarely penetrate the more consolidated, fine-grained underlying sediments represented by layer A in Figure 2.

Building a geo-archaeological chronology

The Western NSW Archaeology Program (WNSWAP) has, over the past ten years, developed a two-phase dating framework for developing a chronology of both landform development and human occupation at our field sites. The first phase consists of stratigraphic analysis and dating of the valley fill sequences described above, primarily using Optically Stimulated Luminescence (OSL) to date the sedimentary units above and below the 'archaeological surfaces'. This is described in detail in Fanning et al. (2008). The second phase consists of age determinations on the heat-retainer hearths excavated by Aboriginal people into the 'archaeological surfaces' in the past. Up to now, we have relied on partial excavation of these hearth remains to extract a sample of charcoal

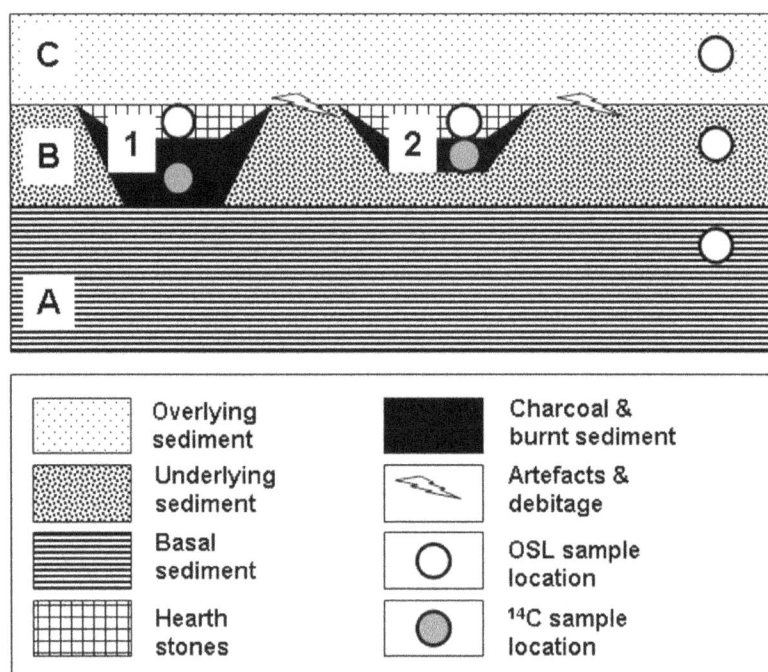

Figure 2. Typical sedimentary and archaeological relationships that exist in much of western NSW today. Three sedimentary units are indicated with different hatching and by the letters A, B and C. The archaeological surface is represented by the boundary between layers B and C, and was available to retain archaeological material after the deposition of unit B, but before the deposition of unit C. Archaeological materials (lithic artefacts and debitage, bone and other materials) are found at this level as indicated by the symbol. Hearths are constructed down from the archaeological surface (B–C interface) into layer B. Field observation suggests that hearths extend to a range of depths, but rarely penetrate the more consolidated, fine-grained underlying sediments represented by layer A. Also shown is our simplified dating strategy, dating the charcoal of the hearths using radiocarbon, and dating the heating of the hearth stones and the deposition of the underlying and overlying sediment units using OSL.

for radiocarbon dating. Here, we describe a new alternative method, using OSL techniques, which may in the future eliminate the need for excavation. We explore the potential of this approach by comparing OSL and ^{14}C age estimates from the same hearths from two separate locations in western NSW (Fowlers Gap Arid Zone Research Station and the pastoral property known as 'Poolamacca' — Figure 1).

1. Dating heat-retainer hearths using [14]C

The approach to dating heat-retainer hearths using ^{14}C is described by Holdaway et al. (2002, 2005). Not all hearth remains contain charcoal; only one third of all hearths excavated in western NSW over the past ten years contained identifiable charcoal suitable for dating. This limits the detail with which a chronology can be constructed, and demands excavation of a large number of hearths. The excavation process can be relatively slow, and the charcoal requires significant laboratory processing to isolate an uncontaminated sample. Excavation is at least in part destructive, and is understandably discouraged by heritage managers and Aboriginal people. Hearths with significant quantities of charcoal can be dated using conventional methods, but more expensive AMS dating is required for small samples (i.e. less than a few grams).

The chronological data provided by ^{14}C dating are of high accuracy and reliability, though plateau effects and the anthropogenic release of dead carbon reduce resolution at certain periods, especially for material from the last 250 years. It is possible that the preservation of charcoal is significantly better for younger hearths, biasing the chronological dataset towards younger age estimates. However, preservation is controlled by the degree of erosion at individual hearth locations (Fanning et al. this volume), which has been demonstrated to be independent of age within the context of a single archaeological surface (Holdaway et al. 2002).

2. OSL dating of hearth stones — a potential non-destructive alternative

OSL signals are reset by heating as well as by light exposure (Smith et al. 1986). This provides the possibility of using OSL to date the last heating of hearth stones. For a five minute heat treatment, virtually all the OSL signal of quartz is removed at a temperature of 300°C (Smith et al. 1986; 1990). For longer heating durations, lower temperatures have a similar zeroing effect. These temperatures are well within the range of those experienced within typical wood-burning fires, and penetration of temperatures in excess of several hundred degrees Celsius to the interiors of cobble-sized and smaller hearth stones is considered likely. If insufficient heating were experienced, age over-estimation would result.

This method has the potential to be significantly less invasive than ^{14}C, as stones currently exposed at the ground surface within the hearth structure can be collected for dating, and no excavation is required. Possible limitations of the technique are encapsulated in the following questions:

1. Was the OSL signal zeroed?
2. Are the OSL signals of light-coloured and translucent rocks bleached (i.e. reduced) by light exposure at the surface following erosive exposure of the hearth before sample collection?
3. Is it possible to deal adequately with the complex geometries of hearth and surrounding sediment and missing overburden when estimating the environmental dose rate?

2.1 Sample selection and collection procedure

For this pilot project, primarily designed to assess the feasibility of using OSL for dating heat-retainer hearth structures, we deliberately selected stones from a range of available lithologies, including some highly translucent rock types such as vein quartz. For some hearths, only one stone was selected, while for others up to four stones were collected. It is important that we have a high degree of confidence that the selected stones were part of the identified hearth structure. For this reason, stones showing obvious signs of burning, including colour change, fire-cracking or conjoining, and pot-lid or other thermal fracturing structures were favoured, and stones from outside the hearth cluster were generally avoided. For this assessment project, both large (cobbles of up to 15 cm diameter) and small stones (pebbles down to 2 cm diameter) were selected.

Using a portable gamma spectrometer with a 3″ NaI crystal, a 900 s (15 minute) measurement was made at the surface of the hearth. This provides an environmental gamma radiation determination with a 2π geometry. Gamma rays enter the crystal from the underlying sediment and hearth, travelling approximately 30 to 40 cm from a half sphere beneath the crystal. No contribution can be measured from sediment that previously covered the hearth, as this has already been eroded and redistributed elsewhere in the landscape. Thus, for the age estimates presented below, this 2π geometry measurement has simply been doubled to provide a proxy 4π geometry estimate. Note that this procedure implicitly assumes that the samples were previously buried by at least 30 cm of overburden with a similar composition to the surface material on which the hearth lies. We consider this one of the main significant unknown factors in this approach, and the comparison with radiocarbon dating is intended to assess the validity of the methods which we have adopted. No correction for airborne radionuclide contributions has been made, though it is possible that ^{222}Rn and daughters, from the ^{238}U decay series, provide a small contribution to the measured ^{214}Bi and total gamma dose rate measurements.

2.2 Sample preparation

Our aim in preparing samples for OSL measurement is to achieve a quartz-dominated separate, and to employ a single-aliquot regenerative-dose (SAR) protocol (Murray and Wintle 2000). The degree of quartz purity required for reliable OSL age determination depends on the luminescence characteristics of the quartz and other minerals present within the available lithologies. Some rocks selected are pure quartz (e.g. vein quartz), others are quartz-dominated (e.g. quartzite), while others contain significant quantities of other minerals (e.g. metamorphic rocks, sandstone). Quartz OSL signals have been demonstrated to undergo the 'predose effect' (Aitken 1985; Smith et al. 1986; Smith and Rhodes 1994), which leads to a significant increase in luminescence sensitivity as a result of heating for many samples. As all of the samples selected have been heated in the past, we expected to encounter high sensitivity OSL signals.

With these considerations in mind, and with the desire to develop a procedure which could be applied to a large number of samples, but where accuracy is limited by environmental dose rate estimation, we have opted not to undertake a full 'quartz inclusion' preparation protocol (Fleming 1970). Instead, we simply isolated rock from at least 3 mm inside each stone, and ground this to sand-size particles under acetone using an agate mortar and pestle. After grinding, samples were treated with dilute HCl and subsequently dried. Resultant grains were mounted on stainless steel discs using silicone oil for OSL measurement.

2.3 OSL measurement

All measurements were made in a Risø TL-DA-15 mini-sys automated luminescence reader. OSL measurements were performed at 125°C, and were preceded by IRSL measurements at 60°C. Natural and regenerative-dose measurements were preceded by a preheat of 220°C for 10 s, while sensitivity measurements were preceded by a heat of 200°C for 10 s, as used by Rhodes et al. (2003). Three discs of each sample were measured, using a post-IR OSL protocol, similar to that described by Banerjee et al. (2001).

For many samples, intense, rapidly decaying natural and regenerated OSL signals were observed, typical of quartz. Samples showinging bright OSL signals displayed a linear dose response, and good agreement between the three measured aliquots was observed. Several samples had only very low OSL signal sensitivity (Tables 1 and 2), and some samples displayed significant IRSL signals. This latter observation may suggest that a more rigorous preparation including HF treatment may be advantageous for those lithologies.

2.4 Age calculation

Age estimates were calculated using the weighted mean of the three measured equivalent dose (De) determinations for each sample. For the dose rate calculation, the signal was assumed to come dominantly from sand-sized grains with negligible internal dose rate, and insignificant external alpha particle contributions. These assumptions may be worthy of further investigation in the future. The beta dose rate for each rock was calculated using ICPMS measurement of U, Th and K content, and the gamma dose rate used was simply twice the measured 2π geometry measurement at the surface of each hearth. An overburden thickness of 50 cm was assumed for every rock, in order to calculate the cosmic dose rate contribution using the formulae of Prescott and Hutton (1994). This assumption is difficult to support in every case: for most samples there is no information available regarding previous burial depth. However, for higher dose rate samples, the cosmic ray contribution represents a rather small component, and uncertainty in this parameter does not lead to a significant uncertainty in the age estimate. Where the burial depth was less than around 30 cm, the doubling of the measured surface gamma dose rate would not be valid, and the assumption of >30 cm burial may require further examination in the future. It is possible that the measurement of both low and high internal dose rate rocks from the same hearth might be used to overcome uncertainty in burial depth, by taking an isochron approach to subtract the more poorly known external (gamma and cosmic) dose rate.

Preliminary OSL dating results
The results shown in Tables 1 and 2 represent all those so far measured, and include samples from hearths without radiocarbon age control. These results are useful in assessing the technique, as they provide information regarding sample purity, OSL signal brightness and D_e consistency, and for hearths with more than one OSL age estimate, the degree of external consistency can be assessed. Note that OSL dating results are presented as years before AD 1950, so that they are directly comparable with calibrated ^{14}C age estimates. Where ^{14}C calibration leads to multiple peaks, the full range from maximum to minimum 2 sigma range is quoted.

Some of the OSL results listed in Tables 1 and 2 agree well with the corresponding calibrated ^{14}C age estimates (e.g. K0118, K0141, K0519, K0524), while others do not agree (e.g. K0115, K0120, K0526, K0547). Some OSL-^{14}C sample pairs do not have overlapping 2 sigma age ranges, but do not disagree substantially beyond these (e.g. K0131, K0143, K0146, K0523, K0535, K0548).

At Fowlers Gap (Table 1), 19 OSL samples have corresponding ^{14}C age estimates. Of these 37% disagree strongly, 16% disagree mildly, while for 47% OSL and ^{14}C have overlapping 2 sigma uncertainty ranges. This sample suite contains a relatively high proportion of quartz and quartzite samples, and both over- and under-estimates in OSL age are observed. At Poolamacca (Table 2), 34 OSL samples have corresponding ^{14}C age estimates. Of these 24% disagree strongly, 9% disagree mildly, while for 67% OSL and ^{14}C have overlapping 2 sigma uncertainty ranges. This sample suite contains a lower proportion of quartz and quartzite samples, and under-estimates in OSL age are observed more frequently than over-estimates, of which there is just one (K0542).

In total, 53 OSL samples have corresponding ^{14}C age estimates. Of these 28% disagree strongly, 11% disagree mildly, while for 60% OSL and ^{14}C have overlapping 2 sigma uncertainty ranges.

Discussion

These preliminary OSL age estimates are very encouraging, as 67% of paired OSL and ^{14}C have overlapping 2 sigma uncertainty ranges at Poolamacca, and 60% for the two datasets combined. However, around one quarter of the OSL age estimates do not agree with their corresponding ^{14}C control, and it is clear that the method still requires a little refinement.

Within both datasets, several OSL age under-estimates are observed for quartz and quartzite samples. It is possible that these relate to the reduction of the OSL signal by daylight (bleaching) which penetrated the rock while it was exposed at the surface after erosion of the overburden and before collection. For other lithologies, this is considered very unlikely, owing to the presence of dark minerals, in particular iron oxides, which are strongly light-absorbent. It is clear, however, that signal bleaching has not adversely affected other quartz and quartzite samples, such as K0117 and K0118. Some rocks classified as quartzite are relatively dark, and therefore light-absorbent, though this is not usually the

Field code	Lab. code	Hearth code	Lithology	OSL initial intensity (cts/mgGy)	Internal dose rate (mGy/a)	External dose rate (mGy/a)	Total dose rate (mGy/a)	1 sigma uncertainty (mGy/a)	Equivalent dose (Gy)	1 sigma uncertainty (Gy)	OSL age (years before AD1950)	1 sigma uncertainty (years)	Radiocarbon lab. code	Uncalibrated C-14 age (BP)	1 sigma uncertainty	Calibrated 2 sigma minimum (cal BP)	Calibrated 2 sigma maximum (cal BP)
FG03-03	K0113	SC69	quartz	20	0.01	1.38	1.39	0.04	9.30	0.72	6640	560	Wk-10279	588	59	490	650
FG03-04	K0114	SC69	quartz	20	0.24	1.38	1.62	0.04	2.58	0.84	1540	520	Wk-10279	588	59	490	650
FG03-05	K0115	SC69	quartzite	4240	0.90	1.38	2.27	0.08	0.81	0.02	300	15	Wk-10279	588	59	490	650
FG03-06	K0116	SC69	quartzite	520	0.84	1.38	2.22	0.08	0.00	0.04	-50	20	Wk-10279	588	59	490	650
FG03-07	K0117	SC13	quartz	20	0.01	1.30	1.31	0.04	1.35	0.19	980	150	Wk-10272	904	47	670	910
FG03-08	K0118	SC13	quartzite	3570	1.60	1.30	2.90	0.13	2.63	0.07	850	50	Wk-10272	904	47	670	910
FG03-10	K0120	SC24	quartz	100	0.20	1.30	1.50	0.05	1.18	0.17	730	120	Wk-10274	1349	69	1050	1330
FG03-11	K0121	SC24	quartzite	2870	1.46	1.30	2.76	0.12	4.45	0.09	1560	75	Wk-10274	1349	69	1050	1330
FG03-15	K0125	ND20	quartz	60	0.02	0.81	0.83	0.04	1.93	0.49	2260	600					
FG03-16	K0126	ND20	quartz	1540	0.01	0.81	0.83	0.04	0.17	0.02	140	20					
FG03-17	K0127	ND20	quartzite	8610	0.23	0.81	1.04	0.04	0.29	0.01	220	15					
FG03-20	K0130	MD158	sandstone	7780	0.24	0.70	0.94	0.04	0.98	0.02	990	55	Wk-10264	990	81	680	980
FG03-21	K0131	MD183	sandstone	31920	0.36	0.76	1.13	0.05	1.62	0.04	1380	65	Wk-10265	1144	45	920	1130
FG03-22	K0132	MD153	quartzite	190	0.19	0.74	0.94	0.04	1.38	0.03	1420	75	Wk-10261	960	47	730	930
FG03-23	K0133	MD153	sandstone	17370	0.12	0.74	0.87	0.04	0.97	0.03	1070	70	Wk-10261	960	47	730	930
FG03-24	K0134	NH1	quartzite	79650	0.19	0.94	1.13	0.04	0.34	0.01	240	15					
FG03-28	K0138	NH2	sandstone	8420	0.46	0.74	1.20	0.05	0.31	0.01	200	15					
FG03-31	K0141	NN05	quartz	970	0.16	1.30	1.45	0.04	0.69	0.08	420	55	Wk-10257	468	50	320	550
FG03-32	K0142	NN17	quartz	40	0.02	1.11	1.13	0.04	1.65	0.44	1410	390	Wk-10259	957	44	730	920
FG03-33	K0143	NN17	conglomerate	3360	0.22	1.11	1.32	0.04	1.45	0.03	1040	45	Wk-10259	957	44	730	920
FG03-34	K0144	NN20	quartz	40	0.04	1.22	1.26	0.04	0.66	0.11	460	85	Wk-10260	514	65	320	640
FG03-35	K0145	FC5	quartz	10	0.01	1.20	1.20	0.04	4.28	0.30	3500	280	Wk-9353	3392	49	3440	3700
FG03-36	K0146	FC22	quartz	40	0.00	1.26	1.27	0.04	3.46	1.60	2680	1270	Wk-9348	5243	164	5600	6300
FG03-38	K0148	FC460	quartz	540	0.03	1.25	1.28	0.04	0.46	0.05	300	40	Wk-10255	1963	55	1710	1990

Table 1. Preliminary OSL (bold) and ^{14}C age comparison from Fowlers Gap heat-retainer hearths, in years before AD 1950. Also shown are lithology, OSL test dose sensitivity; internal, external and total dose rates; equivalent dose values; radiocarbon codes, uncalibrated ^{14}C age, maximum and minimum calibrated ^{14}C age estimates. Note that multiple peaks in ^{14}C calibrations are not shown.

Field code	Lab. code	Hearth code	Lithology	OSL initial intensity (cts/mGy)	Internal dose rate (mGy/a)	External dose rate (mGy/a)	Total dose rate (mGy/a)	1 sigma uncertainty (mGy/a)	Equivalent dose (Gy)	1 sigma uncertainty (Gy)	OSL age (years before AD1950)	1 sigma uncertainty (years)	Radiocarbon lab. code	Uncalibrated C-14 age (BP)	1 sigma uncertainty	Calibrated 2 sigma minimum (cal BP)	Calibrated 2 sigma maximum (cal BP)
PL05-BS01	K0518	15004	meta-sst	13110	0.30	1.58	1.89	0.05	1.96	0.05	980	40	Wk-17577	5794	34	6430	6650
PL05-BS02	K0519	15011	meta-sst	4380	1.28	1.65	2.93	0.09	1.96	0.05	610	25	Wk-17569	647	48	530	660
PL05-BS03	K0520	15014	quartzite	6990	2.45	1.55	4.00	0.20	2.84	0.11	650	45	Wk-17570	1030	36	800	960
PL05-BS04	K0521	15014	quartzite	5	0.27	1.55	1.83	0.05	1.29	0.48	650	270	Wk-17570	1030	36	800	960
PL05-BS05	K0522	15027	quartzite	7460	0.34	1.57	1.91	0.05	1.18	0.05	560	30	Wk-17571	933	34	730	910
PL05-BS06	K0523	15028	quartzite	6460	0.82	1.48	2.30	0.07	0.76	0.03	270	15	Wk-17572	376	34	310	490
PL05-BS07	K0524	15028	meta-sst	6470	0.38	1.48	1.86	0.05	0.81	0.03	380	20	Wk-17572	376	34	310	490
PL05-BS08	K0525	15033	quartzite	17770	0.64	1.69	2.33	0.06	1.69	0.03	670	25	Wk-17573	912	36	680	910
PL05-BS09	K0526	15033	quartz	50	0.18	1.69	1.87	0.05	0.27	0.04	90	20	Wk-17573	912	36	680	910
PL05-BS10	K0527	15036	quartz	3360	0.32	1.56	1.87	0.05	0.19	0.01	50	5	Wk-17574	206	34	-11	310
PL05-BS11	K0528	15036	meta-sst	3050	1.57	1.56	3.13	0.13	0.62	0.02	140	10	Wk-17574	206	34	-11	310
PL05-BS12	K0529	15044	meta-sst	4760	0.66	1.56	2.22	0.06	0.71	0.05	270	25	Wk-17575	915	37	680	910
PL05-BS13	K0530	15045	meta-sst	72900	0.26	1.66	1.92	0.05	1.33	0.03	640	25					
PL05-BS14	K0531	15050	meta-sst	6780	1.68	1.82	3.50	0.14	5.77	0.12	1590	75	Wk-17578	1627	33	1380	1540
PL05-BS15	K0532	15050	quartzite	22370	0.57	1.82	2.40	0.06	3.47	0.08	1390	50	Wk-17578	1627	33	1380	1540
PL05-BS16	K0533	15059	quartz	5	0.00	1.72	1.72	0.05	0.53	0.30	250	170	Wk-17576	256	36	140	440
PL05-BS17	K0534	15059	meta-sst	2710	0.58	1.72	2.30	0.06	0.69	0.02	240	15	Wk-17576	256	36	140	440
PL05-BS18	K0535	15062	quartzite	6820	0.34	1.86	2.20	0.05	1.18	0.07	480	35	Wk-17579	698	38	550	670
PL05-BS19	K0536	15063	quartzite	130	0.14	1.77	1.91	0.05	0.23	0.03	60	15	Wk-17580	170	35	-11	280
PL05-BS20	K0537	15068	schist	180	3.26	1.64	4.90	0.30	0.98	0.10	150	25	Wk-17581	168	35	-11	280
PL05-BS21	K0538	15071	quartzite	15330	0.43	1.58	2.01	0.05	0.29	0.01	90	5	Wk-17582	115	33	-11	260
PL05-BS22	K0539	15097	quartzite	780	1.85	1.68	3.52	0.12	3.74	0.08	1010	40	Wk-17583	1059	46	790	1050
PL05-BS23	K0540	15116	quartzite	370	0.70	1.73	2.43	0.07	11.57	0.76	4700	340					
PL05-BS24	K0541	15116	quartz	60	0.05	1.73	1.78	0.05	1.33	0.07	690	45					
PL05-BS25	K0542	15126	quartz	15	0.01	1.89	1.90	0.05	4.15	0.72	2130	390	Wk-17584	587	46	500	640
PL05-BS26	K0543	15126	quartzite	17990	1.98	1.89	3.87	0.14	2.74	0.06	650	30	Wk-17584	587	46	500	640
PL05-BS27	K0544	15154	schist	2930	0.20	1.82	2.02	0.05	0.40	0.08	150	40	Wk-17585	175	34	-11	290
PL05-BS28	K0545	15154	quartzite	250	1.64	1.82	3.46	0.12	1.03	0.04	240	15	Wk-17585	175	34	-11	290
PL05-BS29	K0546	15154	schist	2420	1.58	1.82	3.40	0.15	0.68	0.03	140	10	Wk-17585	175	34	-11	290
PL05-BS30	K0547	15164	quartz	1840	0.25	1.80	2.05	0.05	0.15	0.00	20	5	Wk-17586	395	35	320	500
PL05-BS31	K0548	15164	schist	3530	2.59	1.80	4.39	0.23	1.36	0.05	260	20	Wk-17586	395	35	320	500
PL05-BS32	K0549	15176	meta-sst	950	1.67	1.84	3.51	0.12	2.36	0.07	620	30	Wk-17587	651	34	540	660
PL05-BS33	K0550	15176	quartz	30	0.07	1.84	1.91	0.05	1.83	0.25	900	130	Wk-17587	651	34	540	660
PL05-BS34	K0551	15180	quartz	140	0.03	1.54	1.57	0.05	0.94	0.19	540	120	Wk-17588	883	33	670	800
PL05-BS35	K0552	15180	quartz	610	0.21	1.54	1.75	0.05	0.44	0.02	200	15	Wk-17588	883	33	670	800
PL05-BS36	K0553	15180	meta-sst	1840	0.68	1.54	2.22	0.06	1.44	0.04	600	25	Wk-17588	883	33	670	800
PL05-BS37	K0554	15180	quartz	1	0.01	1.54	1.54	0.05	4.93	23.50	3140	15250	Wk-17588	883	33	670	800

Table 2. Preliminary OSL and ^{14}C comparison from Poolamacca heat-retainer hearths, in years before AD 1950. Also shown are lithology (meta-sst = meta-sandstone), OSL test dose sensitivity; internal, external and total dose rates; equivalent dose values; radiocarbon codes, uncalibrated ^{14}C age, maximum and minimum calibrated ^{14}C age estimates. Note that multiple peaks in ^{14}C calibrations are not shown.

case for samples classified as quartz. No attempt to relate sample colour or size to the degree of OSL age under-estimation has been undertaken, though careful re-examination of each sample may reveal patterns in this trend. In the future, it may be useful to explore sub-samples or single grains from different depths within individual rocks; surface OSL ages may be expected to be lower than those from central parts for samples which have experienced partial zeroing by light exposure before collection.

Some samples displayed significant IRSL signals (e.g. K0528), although no obvious relationship of IRSL magnitude to OSL age was observed. A brief HF treatment might reduce the IRSL signal, and possibly also improve OSL age agreement.

A wide variation in natural sensitivity of the OSL signal was observed, spanning around 5 orders of magnitude; quartzite sample K0134 had a sensitivity of 79,650 $c.s^{-1}.mg^{-1}.Gy^{-1}$ (Table 1), while quartz sample K554 had one less than 1 $c.s^{-1}.mg^{-1}.Gy^{-1}$ (Table 2). In general, quartz samples had the lowest OSL sensitivities of any lithology, leading to reduced precision in the equivalent dose determination for several samples.

For the preliminary OSL age estimates shown here, the beta dose rate was assumed to originate outside the measured quartz grains, and no alpha dose rate contribution has been included. In Tables 1 and 2, internal dose rate refers to the rock sample, not the quartz grains. These assumptions may require modification in future. Note that if further alpha and beta dose rate contributions were included, the total dose rate would increase, and the estimated OSL age would decrease.

Our preliminary results are very encouraging, and clearly demonstrate the potential of OSL for dating exposed heat-retainer hearths in western NSW. The method requires some further refinement to improve the degree of agreement between OSL and independent chronological control provided by radiocarbon. However, we can now provide provisional answers to the three questions we initially posed:

4. We observe no clear indication of incomplete signal zeroing, though some samples do provide OSL age over-estimates.

5. Some translucent or pale-coloured lithologies provide OSL age under-estimates, suggesting that daylight bleaching occurred while samples lay at the surface.

6. The good agreement between OSL and ^{14}C for many of the samples suggests that complex geometry does not represent a significant obstacle for OSL dating. Most of the samples which provide OSL age estimates not in agreement with ^{14}C are significantly in error, rather than just a little outside the estimated uncertainty limits, which suggests other causes for these errors.

We plan to re-measure these samples after a brief HF treatment to reduce possible feldspar contamination, and to examine the dependence of age on burial depth, internal dose rate, sediment moisture content, and other parameters. We will also examine the role of sample size and colour in controlling surface bleaching. From the large dataset of paired OSL and ^{14}C samples presented here, it will be possible to estimate meaningful values for inter-sample variation beyond the estimated uncertainties, which may be applied to similar OSL age estimates in the future. We will also produce guidance on optimal lithology and sample size to assist efficient collection.

Conclusions

From the data presented here, we are confident that OSL can provide a very useful technique for estimating the age of heat-retainer hearths exposed at the surface in arid Australia. This technique is likely to be popular with Aboriginal groups and heritage managers as it does not require excavation and causes only minimal damage to the hearths. It is also requires relatively little laboratory preparation time and chemical treatment, and the OSL measurement time is short. Thus, there are likely to be considerable cost savings compared with radiocarbon dating of hearth charcoal. This method is suitable for application to large numbers of stones, which in turn will lead to meaningful chronologies of landscape occupation.

Acknowledgements

We would like to thank Maureen O'Donnell, Chairperson of the Broken Hill Local Aboriginal Lands Council, who encouraged us to undertake the project, and provided a very warm welcome and support during our stay at Poolamacca. We also thank Raymond O'Donnell and Christine and Tahnee Tester, and other members of the O'Donnell family who worked with us at Fowlers Gap and at Poolamacca, including Bernie, young Bernie, Maxine, Raelene, Latisha, Courtney and Jade O'Donnell, Malcolm McKellar, Peter and Ty Farnham, and Pauly Menz. Bridget Mosley, Margaret Rika-Heke, Irene Wallis, Phil Deacon and John Pickard are thanked for their contributions to the fieldwork. We also thank the University of NSW for access to Fowlers Gap Arid Zone Research Station, and the NSW National Parks and Wildlife Service for permission to excavate hearths and collect hearth stones. Thanks also to Norman Hill at the ANU for laboratory assistance and the Waikato Radiocarbon Laboratory, particularly Alan Hogg and Fiona Petchey, for providing the radiocarbon determinations, and to Jim Feathers for comments on the manuscript. This research was supported in part by AIATSIS grant G2004/6955, an ARC Large Grant to Holdaway and Fanning, and a Macquarie University Research Grant to Fanning.

References

Aitken, M. J. 1998. Introduction to optical dating. Oxford: Oxford University Press.

Banerjee, D., A. S. Murray, L. Bøtter-Jensen and A. Lang. 2001. Equivalent dose estimation using a single aliquot of polymineral fine grains. Radiation Measurements 33:73–94.

Fanning, P. C. 1999. Recent landscape history in arid western New South Wales, Australia: A model for regional change. Geomorphology 29:191–210.

Fanning, P. C. 2002. Beyond the Divide: A new geoarchaeology of Aboriginal stone artefact scatters in Western New South Wales, Australia. PhD thesis, Macquarie University, Sydney, Australia.

Fanning, P. C. and S. J. Holdaway. 2001. Temporal limits to the archaeological record in arid Western NSW, Australia: Lessons from OSL and radiocarbon dating of hearths and sediments. In M. Jones and P. Sheppard (eds), Australasian connections and new directions: Proceedings of the 7th Australasian Archaeometry Conference, pp. 85–104. Research in Anthropology and Linguistics 5, Auckland: Department of Anthropology, University of Auckland.

Fanning, P. C., Holdaway S.J. and Rhodes, E. J. 2008 A new geoarchaeology of Aboriginal artefact scatters in western NSW, Australia. establishing spatial geomorphic controls on the surface archaeological record. Geomorphology 101, 524-532.

Fanning, P. C., S. J. Holdaway and R. Phillips. Heat-retainer hearth identification as a component of archaeological survey in western NSW, Australia. This volume.

Fleming, S. J. 1970. Thermoluminescence dating: Refinement of the quartz inclusion method. Archaeometry 15:133–147.

Galbraith, R. F., R. G. Roberts, G. M. Laslett, H. Yoshida and J. M. Olley. 1999. Optical dating of single and multiple grains of quartz from Jinmium rock shelter, northern Australia: Part I. Experimental design and statistical models. Archaeometry 41:339–364.

Holdaway, S. J., P. C. Fanning, D. Witter, M. Jones, G. Nicholls and J. Shiner. 2002. Variability in the chronology of Late Holocene Aboriginal occupation on the arid margin of southeastern Australia. Journal of Archaeological Science 29:351–363.

Holdaway, S. J., P. C. Fanning and J. Shiner. 2005. Absence of evidence or evidence of absence? Understanding the chronology of indigenous occupation of western New South Wales, Australia. Archaeology in Oceania 40:33–49.

Holdaway, S. J., J. Shiner and P. C. Fanning. 2004. Hunter-gatherers and the archaeology of the long term: An analysis of surface stone artefact scatters from Sturt National Park, New South Wales, Australia. Asian Perspectives 43(1):34–72.

Huntley, D. J., D. I. Godfrey-Smith and M. L. W. Thewalt. 1985. Optical dating of sediments. Nature 313:105–107.

Murray, A. S. and A. G. Wintle. 2000. Luminescence dating of quartz using an improved single-aliquot regenerative-dose protocol. Radiation Measurements 32:57–73.

Prescott, J. R. and J. T. Hutton. 1994. Cosmic ray contributions to dose rates for luminescence and ESR dating: Large depths and long term time variations. Radiation Measurements 23:497–500.

Rhodes, E. J. 1988. Methodological considerations in the optical dating of quartz. Quaternary Science Reviews 7: 395–400.

Rhodes, E. J., C. Bronk Ramsey, Z. Outram, C. Batt, L. Willis, S. Dockrill and J. Bond. 2003. Bayesian methods applied to the interpretation of multiple OSL dates: High precision sediment age estimates from Old Scatness Broch excavations, Shetland Isles. Quaternary Science Reviews 22:1231–1244.

Smith, B. W., M. J. Aitken, E. J. Rhodes, P. D. Robinson and D. M. Geldard. 1986. Optical dating: Methodological aspects. Radiation Protection Dosimetry 17:229–233.

Smith, B. W. and E. J. Rhodes. 1994. Charge movements in quartz and their relevance to optical dating. Radiation Measurements 23:581–585.

Smith, B. W., E. J. Rhodes, S. Stokes and N. A. Spooner. 1990. The optical dating of sediments using quartz. Radiation Protection Dosimetry 34(1):75–78.

14

HPLC-MS characterisation of adsorbed residues from Early Iron Age ceramics, Gordion, Central Anatolia

Todd Craig[1]
Peter Grave[1]
Stephen Glover[2]

[1]Department of Archaeology and Palaeoanthropology
University of New England
Armidale NSW 2351, Australia

[2]School of Biological, Biomedical and Molecular Sciences
University of New England
Armidale, NSW, 2351, Australia

Abstract

High performance liquid chromatography with mass spectrometry (HPLC-MS) is a powerful method for characterising modern fats and oils at the molecular level (triacylglycerols). It is a technique that should have great potential for the analysis of archaeological fatty residues. However, its archaeological potential has yet to be systematically evaluated. This paper presents the results of an HPLC-MS study of food residues adsorbed into archaeological ceramics. Experimentally three classes of materials were studied: raw plant and animal foods likely to be available to the prehistoric population of Anatolia (Turkey), the same foods after cooking and archaeological residues from Late Bronze and Early Iron Age (ca 1500–900 BCE) contexts from ceramics excavated at the site of Gordion, Central Turkey. Modern foods were used as a baseline; modern cooked residues provided a measure of the effects of cooking on the same fatty residues. These datasets were then compared to the HPLC-MS results for the archaeological residues. The study found that HPLC-MS does offer important new information, but does not provide a 'magic bullet'. Like established techniques using gas chromatography and mass spectrometry, it is best used in tandem with other techniques in a multi-stranded approach to archaeological residue characterisation and identification.

Keywords: HPLC-MS; residue identification; lipids; cooking residues; fatty acids; triacylglycerols; Gordion

Introduction

Fatty residues recovered from within the porous fabrics of archaeological ceramics have been studied for almost 30 years. Condamin and coworkers (1976) first demonstrated that the porous fabric of earthenware ceramics provides an excellent microenvironment for the preservation of fatty residues. Initially, chemical exploration of archaeological residues focused on fatty acid derivatives due to their archaeological ubiquity and analytic simplicity (Condamin et al. 1976, Morgan et al. 1984, Patrick et al. 1985). In simple terms fatty acids are hydrocarbon chains with a terminal carboxyl group (Markley 1960). These hydrocarbon chains occur in a range of configurations. However, straight-chain saturated fatty acids and carbon-carbon double bond unsaturated fatty acids are most common (Figure 1b, 1c) (Christie 2003). The relative ubiquity of most fatty acid species in modern plant and animal products limits their diagnostic utility (Hilditch and Williams 1964). Less than 25 fatty acid species can account for more than 90% of fatty acids in nature. Of these, some are highly susceptible to degradation and therefore unlikely to survive over archaeological time (Evershed et al. 1992, Evershed et al. 2001).

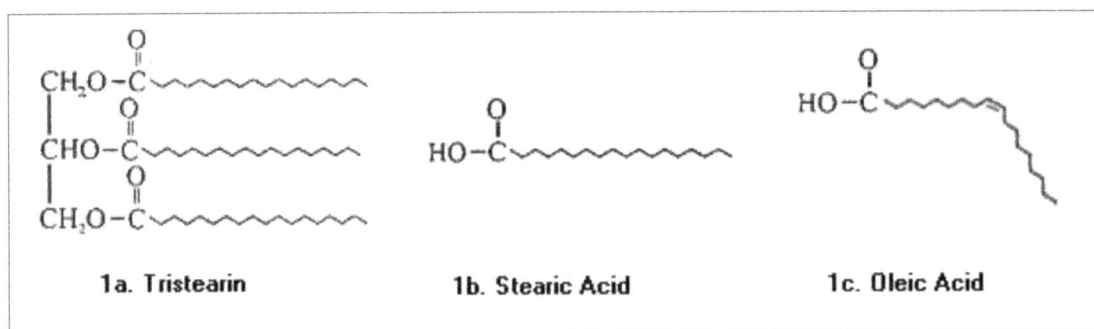

Figure 1. 1a. tristearin; 1b. stearic acid; 1c. oleic acid. Adapted from Evershed et al. (2001).

Improvements in column chromatography and the increasingly common technique of gas chromatography with mass spectrometry (GC-MS) in laboratories in the 1990s led to the rapid expansion of types of archaeological molecules analysed, including fatty acids (Copley et al. 2005, Craig et al. 2005, Eerkens 2005, Evershed et al. 2002, Kimpe et al. 2004, Malainey et al. 1999), waxes (Charters and Evershed 1997, Evershed et al. 1997, Gariner et al. 2002), alcohols (Reber and Evershed 2004), terpenes (Eerkens 2002, Fox et al. 1995) and sterols (Evershed, et al. 1992). These analyses of new classes of molecules provided important diagnostic tools, such as the presence of cholesterol and campesterol to identify the presence of animal and plant residues respectively.

Gas chromatography relies on molecular volatility, primarily separating molecules according to their evaporative point (Abian 1999). This limits the capacity for GC to separate low-volatility molecules, such as triacylglycerols. Triacylglycerols are the most common component of modern fats and oils forming a diverse and diagnostic class of lipid (Christie 2003, Holcapek et al. 2003, Mottram et al. 1997, Mottram et al. 2001). Triacylglycerols are composed of a glycerol backbone with three fatty acids attached via ester bonds (Figure 1a). Around 150 species of triacylglycerol would be required to account for 90% of natural variation, a much larger and more diverse range than fatty acids. Some researchers have observed the presence of triacylglycerols in archaeological residues (Charters et al. 1995, Copley et al. 2005, Dudd et al. 1999, Evershed et al. 2001, Heron et al. 1991). These triacylglycerols may provide valuable information for elucidating the original substances contributing to fatty archaeological residues and other aspects of ancient technologies. However, the technical limitations of GC-MS have meant their potential has not yet been evaluated.

A recent advance in mass spectrometry has been the coupling of high performance liquid chromatography and mass spectrometry (HPLC-MS). HPLC is ideally suited to the separation of large, non-volatile molecules such as triacylglycerols, separating molecular species by polarity rather than evaporative point (Abian 1999). HPLC-MS can also analyse fatty acid species typically analysed by GC-MS, with the advantage of a simplified preparative process. This study therefore sought to systematically evaluate the utility of triacylglycerol data derived from HPLC-MS analyses for archaeological research.

Methodology

This study utilised three datasets: (1) a set of fats and oils extracted from modern plant and animal products, based on a literature review of Anatolian archaeological finds; (2) a set of residues recovered from the fabric of purpose-built ceramics after experimental cooking of foods; and (3) an archaeological dataset of residues recovered from Late Bronze and Early Iron Age ceramics from Gordion, central Turkey (Figure 2). Fats and oils were extracted from the first dataset by ultrasonicating 500 mg of material in 9 ml of a 1:2 methanol/chloroform mixture for 15 minutes. The supernatant was filtered through coarse filter paper and the process repeated once. These extracts allowed the assessment of HPLC-MS separation and quantification methods, provided a baseline for comparison with experimental cooking data and provided a reference database for the interpretation of archaeological samples.

Experimental residues were prepared by boiling food products in earthenware ceramic pots for five hours. After this, the pots were allowed to cool and contents removed. The cooked-in residue was then extracted by removing a piece of the ceramic from around the 'waterline' of the pot, where the most appreciable volumes of fatty residue accumulate. The ceramic was cleaned with a dental drill and crushed to a fine

Figure 2. Location of Gordion and other key Anatolian sites.

powder with a porcelain mortar and pestle. Eight grams of powdered ceramic was ultrasonicated in 36 ml 1:2 methanol/chloroform for 15 minutes, followed by centrifuging at 3500 rpm. After this, the supernatant was collected and the process repeated once. The experimental database was designed to measure the effects of cooking on residues and the survivability of triacylglycerol species.

Finally, 45 archaeological ceramics were selected for residue extraction, 15 from the Late Bronze Age and 15 from each of two Early Iron Age periods (EIA 7B and EIA 7A). From these 18 were selected for residue analysis based on residue recoveries. Archaeological residues were extracted in the same manner as experimental residues. This dataset allowed the assessment of the primary focus of this study: the utility of HPLC-MS derived triacylglycerol species data for the characterisation of archaeological residues.

HPLC-MS characterisation
Residue and extract molecules were chromatographed on a Varian 150 x 2 mm (i.d.) C18-A (10 μm packing) column fitted with a 20 x 2 mm C18-A (5 μm packing) pre-column guard. Mass spectrometry was conducted on a Varian 1200L Quadrupole Mass Spectrometer.

Residues and extracts were first analysed for triacylglycerols using atmospheric pressure chemical ionisation (APCI). APCI produces predictable, diagnostic parent [M+H]+ and fragment ions [M+H-RCOOH]$^+$ from which triacylglycerol species could be definitively identified (Holcapek et. al. 2003, Mottram and Evershed 1996). A gradient of acetonitrile (MeCN) and dichloromethane (DCM) were used as a mobile phase on the following program: 0 min — 100% MeCN, 80 min — 100% MeCN, 200 min 80% MeCN 20% DCM, 210 min 80% MeCN 20% DCM, with a flow rate of 0.8 ml/minute.

After triacylglycerol characterisation, the remaining sample was converted into free fatty acids and analysed by treating the molecules in 0.5 mol methanolic sodium hydroxide at 70°C for 3 hours. Following this, the mixture was acidified and free fatty acids extracted by washing with hexane (Christie 1987). Fatty acids were analysed using electron spray ionisation, producing detectable free fatty acid ions [M-H]$^-$ with minimal fragmentation, allowing identification of molecular species by their mass and elution order. A gradient of methanol (MeOH) and water (H_2O) was used as a mobile phase on the following program: 0 min — 95% MeOH 5% H_2O, 26min — 100% MeOH, 40 min — 100% MeOH, with a flow rate of 0.05 ml/minute.

Results were calculated by integrated areas of relevant molecules expressed as percentages of all lipids identified. For triacylglycerols, the integrated area of the molecular ion [M+H]$^+$ and relevant fragment ions [M+H-RCOOH]$^+$ were summed. Free fatty acids were quantified by calculating the integrated area of the molecular ion [M-H]$^-$.

Results

Food extracts

HPLC-MS analysis of modern plant and animal products revealed complex mixtures of triacylglycerol species. From the 13 foods analysed, a total of 108 triacylglycerol species were identified. Samples contained between 21 (wheat) and 41 (pea) identifiable species of triacylglycerol, with a mean of 29.5. A total of 13 fatty acid species were identified, ranging from 6 (linseed) to 13 (chickpea, pea and wheat) fatty acid species per sample, with a mean of 10.8.

The relative ubiquity of most fatty acid species is immediately recognisable from the dataset. Only two of the samples possessed less than 10 of the 13 free fatty acid species identified. By contrast, triacylglycerols presented a large and divergent dataset. No sample possessed more than 40% of all identified triacylglycerol species. Further, some triacylglycerol species occurred in only one sample, or a group of related samples. For example, MaSS (margaric-stearic-stearic) was found only in ruminant samples (beef and lamb), while APS (arachidic-palmitic-stearic) was only observed in pork samples. Further, MyPP (myristic-palmitic-palmitic) and PSS (palmitic-stearic-stearic) were found in all animal products studied, while plant products could generally be characterised by high levels of highly unsaturated linoleic- and linolenic-bearing triacylglycerols. Considering that the vast bulk of free fatty acids studied were derived from triacylglycerols, it is clear that reducing triacylglycerols to their component fatty acids results in a significant loss of characteristic information.

Despite the apparent simplicity of the fatty acid dataset, some simple statistics still appear to provide characteristic information. The proportion of unsaturated to saturated fatty acid species appears to differentiate modern plant and animal products (Figure 3). Animal products contained between 55 and 70% unsaturated fatty acids. This range can be reduced to 55 to 61% if chicken, which appears unusually high in unsaturated fatty acid species, is removed. Plant products possess generally higher levels of unsaturates, containing between 69 and 86% unsaturated fatty acids. Therefore, in unmodified foods the proportion of unsaturated to saturated fatty acid species seems an appropriate general indicator of plant or animal origin.

Experimental residues

The experimentally produced residues revealed structural changes in lipid composition relative to their original plant and animal sources (Figure 4). Analysis of triacylglycerols reveals a simplification of the dataset, with only 59 of the 108 original triacylglycerols recovered from experimental ceramics. This relates to preferential destruction of the most unsaturated triacylglycerol species through the hydrolysis of the ester bond between the component fatty acids and the glycerol backbone.

Fatty acid species were similarly altered, with a general reduction in the proportion of highly unsaturated fatty acids. While volumes varied considerably, the complete removal of lipid species was uncommon. Overall the fatty acid species present remain similar to their original food products. Experimental residues contained between 9 (olive) and 13 (chickpea, wheat) fatty acid species, with a mean of 11.5. However, the relative proportions of these species are sometimes radically altered.

Comparing triacylglycerol and free fatty acid results, triacylglycerols provide more detailed characteristic markers, especially for animal products. Generally, triacylglycerols resembled fatty acids, with the preferential degradation of highly unsaturated molecules. Plant products were preferentially degraded due to the higher unsaturation of their triacylglycerol components. No sample possessed more than 55% of all identified triacylglycerol

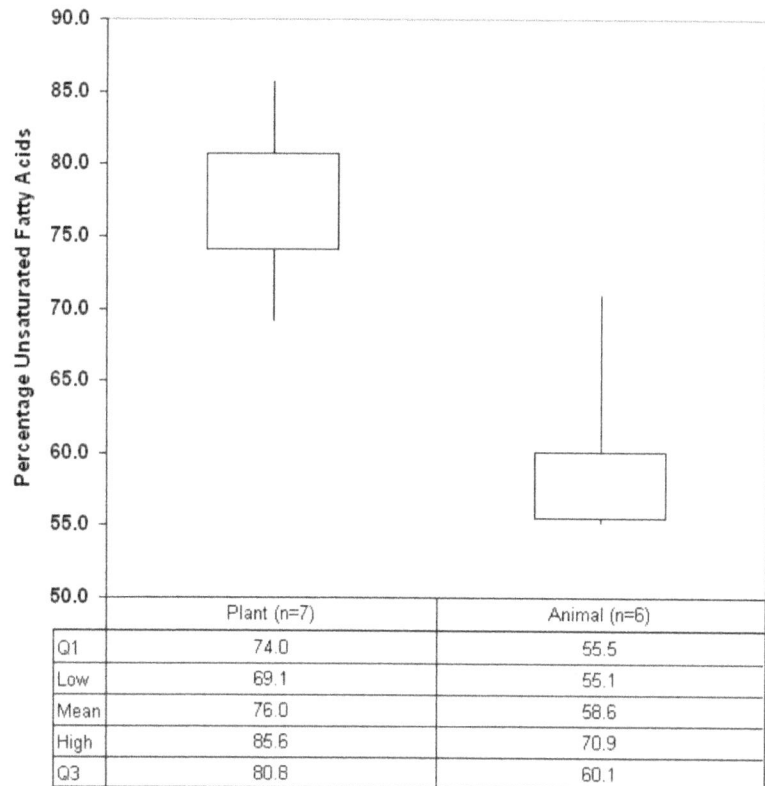

	Plant (n=7)	Animal (n=6)
Q1	74.0	55.5
Low	69.1	55.1
Mean	76.0	58.6
High	85.6	70.9
Q3	80.8	60.1

Figure 3. Boxplots of percentage unsaturated fatty acids in unprocessed plant and animal products.

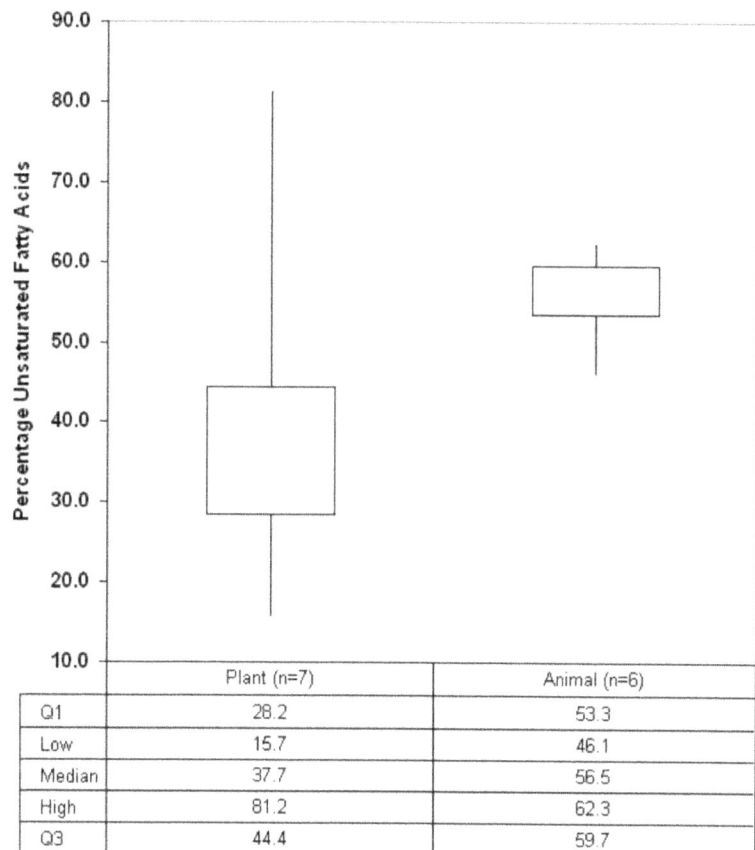

	Plant (n=7)	Animal (n=6)
Q1	28.2	53.3
Low	15.7	46.1
Median	37.7	56.5
High	81.2	62.3
Q3	44.4	59.7

Figure 4. Boxplots of percentage unsaturated fatty acids in experimentally cooked plant and animal products.

species, with the retention of some characteristic molecules, especially more saturated forms. In contrast, no experimental residue contained less than 9 of the 13 fatty acid species identified and only two contained less than 11.

The alteration of lipids by experimental cooking also had an interesting effect on the proportions of unsaturated to saturated fatty acids. After cooking, plant residues contained between 18 and 81% unsaturated species, compared with 69 to 86% in unprocessed extracts. Animal products contained between 46 and 62% unsaturated fatty acids. Experimental cooking seems to have removed the logical structure of the dataset noted in unprocessed extracts. While losing some unsaturated fatty acids, animal products roughly approximate their unprocessed sources. On the other hand, plant products vary across a wide range, most of which occur below the range for animal product residues. The relative instability in plant proportions most likely relates to the higher unsaturation of plant fatty acids. While most unsaturated fatty acids in animal products contain only one unsaturated double bond, plant products have much higher levels of polyunsaturated fatty acids. The susceptibility of fatty acids to degradation increases exponentially with increasing unsaturation (Eerkens 2005), rendering plant lipids more liable to degradation in the prolonged, high-energy environment of the cooking experiment.

In conclusion, the experimental dataset reveals a simplification, with the loss of characteristic triacylglycerols and free fatty acid data. However, triacylglycerols remain a diverse dataset, retaining important characteristic information. Fatty acids undergo dramatic changes in composition, rendering the proportions of unsaturates to saturates meaningless and potentially misleading. From this, it would be expected that archaeological lipid datasets would have undergone further simplification due to the effects of ancient processing and degradation over archaeological time.

Archaeological residues

Of the 18 residues analysed, only 6 contained analytically viable volumes of triacylglycerols. These residues contained between 5 and 16 identifiable triacylglycerol species, with a mean of 8. Triacylglycerols recovered tended towards the saturated end of the spectrum, containing only saturated and monounsaturated component fatty acids. The dataset does not display the variability of the unprocessed or experimentally cooked dataset, with one residue sample containing 76% of the triacylglycerol species identified. However, by reference to the first two datasets, some of the triacylglycerols identified appear characteristic. Triacylglycerol species only observed in beef and lamb (MaSS) were present in 5 of the 6 residues, while a marker identified only in pork (APS) was observed in one sample. The sample that contained a pork marker also contained the beef/lamb marker. Markers that appeared generally indicative of animal products (MyPP, PSS) were observed in 5 of the 6 residue samples, with the one lacking these markers containing the beef/lamb marker. While this evidence is not definitive due to the limits of the reference dataset, archaeological triacylglycerols appear to provide a specific, narrow range of potential candidates for a given archaeological residue.

Fatty acids are again less informative. All 18 residues analysed contained analytically viable volumes of free fatty acids, containing between 9 and 13 fatty acid species. The potential effects of ancient reuse, mixing and differential preservation complicate the picture. However, the proportion of unsaturated fatty acids seems to reveal a logical picture. There appears to be a range separation similar to that observed in the food baseline, with an apparent separation around 25–30% unsaturated fatty acid species (Figure 5). However, the experimental dataset demonstrates that this may not be meaningful. The apparent pattern could result from a relatively predictable degradative shift of plant and animal residues, or as a false positive caused either by differential processing and degradation of a single class of product or a more unpredictable degradative process similar to the experimental dataset. Without additional data, this problem appears insoluble.

Comparison with the triacylglycerol data suggests that this difference is not simply the result of differential preservation or processing technology. If degradation differences were the cause of this spread, these degradative changes would also extend to triacylglycerols. A correlation between high levels of unsaturated fatty acids and the presence of triacylglycerols should be observed. The opposite is, in fact, the case. Residues that contain analytically viable volumes of triacylglycerols have generally lower proportions of unsaturated fatty acids than those without (Figure 5). This data is consistent with two distinct groups of residue undergoing degradative processes dissimilar to the experimental dataset; one predominated by animal products and another by plant products. Archaeological residues therefore underwent a different process of alteration and degradation to that simulated in our experiments. While a detailed explanation of this difference is beyond the scope of this paper, it is likely that archaeological residues were not subjected to the prolonged

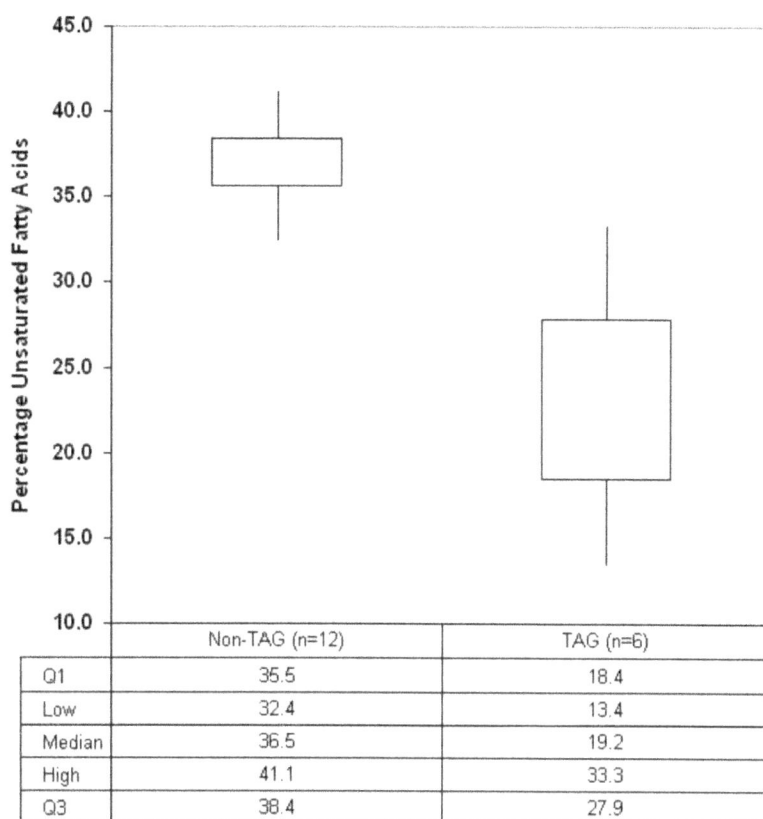

	Non-TAG (n=12)	TAG (n=6)
Q1	35.5	18.4
Low	32.4	13.4
Median	36.5	19.2
High	41.1	33.3
Q3	38.4	27.9

Figure 5. Boxplots of percentage unsaturated fatty acids in triacylglycerol-bearing and non-triacylglycerol bearing archaeological residues.

exposure conditions utilised in the experimental dataset. Without the benefit of triacylglycerol data, it would not have been possible to characterise archaeological residues with confidence. Triacylglycerol data therefore gives us an important new insight not only into residue sources, but also ancient processing technologies.

Conclusion

This study aimed to assess the utility of HPLC-MS analysis and characterisation of triacylglycerol species for archaeological residue research. We have demonstrated that archaeological triacylglycerols can provide significant diagnostic information, reducing the potential sources of some residues to a narrow band of related species.

Archaeological triacylglycerol data also provided important information for a puzzle presented by fatty acid data. Resolving this puzzle allows a new understanding of ancient processing technology and supporting information for the characterisation of non-triacylglycerol bearing residues. Inconsistencies between unprocessed and experimentally cooked fatty acids indicated that even a broad identification of residue contributors would be impossible for residues exposed to similar conditions to the experimental dataset. However, archaeological triacylglycerol data demonstrates a real difference between residues and a real difference in processing technology and degradation relative to the experimental dataset. The incomprehensibility of the experimental dataset calls into question previous archaeological applications of ratios used in the absence of triacylglycerol or other supporting data.

However, HPLC-MS of archaeological triacylglycerols does not provide an exhaustive method of residue characterisation. For the archaeological residues, two-thirds of the dataset would have been meaningless if triacylglycerol characterisation were relied on in isolation. Further, compared with GC-MS techniques where fatty acids, waxes, hydrocarbons and other classes of molecules can be analysed in a single chromatographic run, the HPLC-MS technique for fatty acid identification was less than ideal. The technique was unable to identify any classes of molecule other than fatty acids, potentially resulting in the loss of important diagnostic information.

In conclusion, HPLC-MS of archaeological fatty residues does provide significant information that could not be attained with current GC-MS techniques. Further, triacylglycerol data may reveal important details of ancient processing technologies. These advantages lead us to conclude that HPLC-MS should be considered as an analytic technique for archaeological fatty residues wherever practical. However, HPLC-MS cannot be considered a replacement for established methodologies and should be used in tandem with GC-MS.

Acknowledgements

We would like to thank Dr Robert C. Henrickson and Dr Mary M. Voigt for their permission to use the ceramics in this study. The Australian Research Council supported this research with discovery grant #DP0558992. This study was also supported by a University of New England Internal Research Grant.

References

Abian, J. 1999. The coupling of gas and liquid chromatography with mass spectrometry. *Journal of Mass Spectrometry* 34:157–168.

Charters, S., R. P. Evershed, P. W. Blinkhorn and V. Denham. 1995. Evidence for the mixing of fats and waxes in archaeological ceramics. *Archaeometry* 37:113–127.

Charters, S. and R. P. Evershed. 1997. Simulation experiments for determining the use of ancient pottery vessels: The behaviour of epicuticular leaf wax during boiling of leafy vegetable. *Journal of Archaeological Science* 24:1–7.

Christie, W. W. 2003. *Lipid analysis.* Bridgwater: The Oily Press.

Christie, W. W. 1987. *High-performance liquid chromatography and lipids.* Oxford: Pergamon Press.

Condamin, F., M. O. Formenti, M. Michel and P. Blond. 1976. The application of gas chromatography to the tracing of oil in ancient amphorae. *Archaeometry* 18:195–201.

Copley, M. S., R. Berstan, S. N. Dudd, V. Straker, S. Payne and R. P. Evershed. 2005. Dairying in antiquity. I. Evidence from absorbed lipid residues dating to the British Iron Age. *Journal of Archaeological Science* 32:485–503.

Craig, O. E., G. Taylor, J. Mulville, M. J. Collins and M. Parker Pearson. 2005. The identification of prehistoric dairying activities in the Western Isles of Scotland: An integrated biomolecular approach. *Journal of Archaeological Science* 32:91–103.

Dudd, S. N., R. P. Evershed and A. M. Gibson. 1999. Evidence for varying patterns of exploitation of animal products in different prehistoric pottery traditions based on lipids preserved in surface and absorbed residues. *Journal of Archaeological Science* 26:1473–1482.

Eerkens, J. 2002. The preservation and identification of Pinon resins by GC-MS in pottery from the Western Great Basin. *Archaeometry* 44:95–105.

Eerkens, J. W. 2005. GC-MS analysis and fatty acid ratios of archaeological potsherds from the Western Great Basin of North America. Archaeometry 47:83–102.

Evershed, R. P., C. Heron, S. Charters and L. J. Goad. 1992. The survival of food residues: New methods of analysis, interpretation and application. *New developments in archaeological science*, pp 187–208. Oxford: Oxford University Press.

Evershed, R. P., S. J. Vaughan, S. N. Dudd and J. S. Soles. 1997. Fuel for thought? Beeswax in lamps and conical cups from Late Minoan Crete. *Antiquity* 71:979–985.

Evershed, R. P., S. N. Dudd, M. J. Lockheart and S. Jim. 2001. Lipids in archaeology. In D. R. Brothwell and A. M. Pollard (eds), *Handbook of archaeological sciences*, pp 332–349. Chichester: John Wiley and Sons.

Evershed, R. P., S. N. Dudd, M. S. Copley, R. Berstan, A. W. Stott, H. Mottram, S. A. Buckley and Z. Crossman. 2002. Chemistry of archaeological animal fats. *Accounts of Chemical Research* 35: 660–668.

Fox, A., C. Heron and M. Q. Sutton. 1995. Characterization of natural products on Native American archaeological and ethnographic materials from the Great Basin region, U.S.A.: A preliminary study. *Archaeometry* 37:363–375.

Gariner, N. C., C. Cren-Olivé, C. Rolando and M. Regert. 2002. Characterization of archaeological beeswax by electron ionization and electrospray ionization mass spectrometry. *Analytical Chemistry* 74:4868–4877.

Heron, C., R. P. Evershed and L. J. Goad. 1991. Effects of migration of soil lipids on organic residues associated with buried potsherds. *Journal of Archaeological Science* 18:641–659.

Hilditch, T. P. and P. N. Williams. 1964. *The chemical constitution of natural fats.* London: Chapman and Hall.

Holcapek, M., P. Jandera, P. Zderadicka and L. Hrubá. 2003. Characterization of triacylglycerols and diacy lglycerol composition of plant oils using high-performance liquid chromatography-atmospheric pressure chemical ionization mass spectrometry. *Journal of Chromatography* A 1010:195–215.

Kimpe, K., C. Drybooms, E. Schrevens, P. A. Jacobs, R. Degeest and M. Waelkens. 2004. Assessing the relationship between form and use of different kinds of pottery from the archaeological site Sagalassos (southwest Turkey) with L lipid analysis. *Journal of Archaeological Science* 31:1503–1510.

Malainey, M. E., R. Przybylski and B. L. Sherriff. 1999. Identifying the former contents of Late Precontact period pottery vessels from Western Canada using gas chromatography. *Journal of Archaeological Science* 26:425–438.

Markley, K. S. 1960. Historical and general. In K. S. Markley (eds), *Fatty acids*, pp 1–22. New York: Interscience Publishers.

Morgan, E. D., L. Titus, R. J. Small and C. Edwards. 1984. Gas chromatographic analysis of fatty material from a Thule midden. *Archaeometry* 26:43–48.

Mottram, H. R. and R. P. Evershed. 1996. Structure analysis of triacylglycerol positional isomers using atmospheric pressure chemical ionisation mass spectrometry. *Tetrahedron Letters* 37:8593–8596.

Mottram, H. R., S. E. Woodbury and R. P. Evershed. 1997. Identification of triacylglycerol positional isomers present in vegetable oils by high performace liquid chromatography/atmospheric pressure chemical ionization mass spectrometry. *Rapid Communications in Mass Spectrometry* 11:1240–1252.

Mottram, H. R., Z. M. Crossman and R. P. Evershed. 2001. Regiospecific characterisation of the triacylglycerols in animal fats using high performance liquid chromatography-atmospheric pressure chemical ionisation mass spectrometry. *The Analyst* 126:1018–1024.

Patrick, M., A. J. de Koning and A. B. Smith. 1985. Gas liquid chromatographic analysis of fatty acids in food residues from ceramics found in the Southwestern Cape, South Africa. *Archaeometry* 27:231–236.

Reber, E. A. and R. P. Evershed. 2004. How did the Mississippians prepare maize? The application of compound specific carbon isotopic measurement to absorbed pottery residues from several Mississippi Valley sites. *Archaeometry* 46:19–33.

15

Melting Moments: Modelling archaeological high temperature ceramic data

Peter Grave

Archaeology and Palaeoanthropology
University of New England
Armidale, NSW 2351, Australia
pgrave@une.edu.au

Abstract

A fundamental concept in compositional studies in archaeology is that an elemental fingerprint most broadly reflects region- or area-specific petrogenetic conditions (the provenance postulate). In this study, a sample population from a single shipwreck assemblage of trade ceramics (stoneware jars from East and Southeast Asia), can be separated into clear compositional groups. While the multivariate elemental definition of these groups is not controversial, their interpretation is. The study uses a method of elemental data optimisation where mathematical and statistical techniques are combined to explicitly evaluate the significance of these groups. Contrary to expectations, group elemental signatures emerge as hybrids of provenance and temperature. The results highlight the value of explicit modelling approaches to archaeological ceramic elemental datasets, as well as underscore the general influence of temperature on stoneware compositional profiles.

Keywords: Principal components analysis, polynomial regression, data optimisation, ICP-OES, provenance postulate, East and Southeast Asia

Introduction

In the elemental analysis of archaeological ceramics the 'provenance postulate' remains a major conceptual link between geochemical pattern and archaeological interpretation (Olin et al. 1978; Kolb 1982; Maggetti et al. 1984; Jones 1986; Vitali and Franklin 1986; Ferring and Perttula 1987; Mason and Keall 1988; Middleton et al. 1992; Mommsen et al. 1992; Mallory-Greenough et al. 1998; Neff 2000; Wilson and Pollard 2001; Gomez et al. 2002; Hein et al. 2002; Grave 2004). Experimental studies of earthenwares have shown that interpretation can be complicated by additional factors such as technological treatment or post-depositional alteration (Matson 1971; Kilikoglou et al. 1988; Gosselain 1992; Cogswell et al. 1996; Buxeda i Garrigos 1999; Buxeda i Garrigos et al. 2001; Matsunaga and Nikai 2004; Schwedt et al. 2004). These qualifications of the provenance postulate have largely dismissed firing temperature as a variable with any substantial effect on elemental composition. As a result archaeological ceramic elemental profiles are generally considered unalterable by the varying temperature regimes required for the successful production of a ceramic vessel.

A major difference between clay compositions used for earthenware and stoneware relates to the relative concentration of elements (fluxes) that lower or raise the melting (eutectic) point. Mismatch between firing temperature and flux-rich clays can result in vessel failure with development of excessive glass phases and gas formation ('bloating') causing partial or complete vessel collapse ('slumping'). For stonewares — vitreous ceramics fired between ~1100-1400°C — the relatively low concentrations of fluxin g elements allow firing in the higher temperature ranges required without premature melting and slumping of the ceramic vessel (Kerr and Wood 2004).

Most archaeological characterisation work on high-temperature ceramics from East and Southeast Asia has adopted the provenance postulate (and associated caveats) more or less uncritically (Impey and Tregear 1983; Guo and Li 1986; Liu 1986; Pollard and Hall 1986; Pollard and Hatcher 1986; Yang and Wang 1986; Stenger 1992; Yap and Hua 1992; Pollard and Hatcher 1994; Hughes et al. 1999; Chen et al. 1999). However elsewhere, I have shown that the high firing temperatures required for stoneware production can have a measurable effect on composition (Grave et al. 2000). The observations underpinning that argument were limited to a small population of decorated vessels and the kaolinitic white clays they were produced from, from a single production centre in central northern Thailand, using a sensitive but relatively low-dimensional technique (PIXE/PIGME). Here I evaluate a more robust elemental dataset for a shipwreck assemblage of stoneware jars from several East and Southeast Asian production centres in order to demonstrate the importance of firing temperature as a variable that structures the elemental composition data derived from high-temperature ceramics (Grave et al. 2005a). An analytic advantage for elemental analysis of stoneware storage jars is that these fabrics are typically derived from clay sources that tend to be more elementally complex than the clays used for white bodied 'fine' ceramics. Greater compositional complexity increases the likelihood of detecting the range of factors that modulate elemental signatures.

The objectives of this paper are twofold and interrelated. The first is to evaluate the significance of compositional groups in this assemblage. The second more methodological objective is to provide a means to explicitly test this multivariate dataset against a number of working hypotheses.

Limitations
This study is an empirical presentation and interpretation of relative data patterns and trajectories rather than an effort to reconstruct 'original' ingredients or technological parameters. The complex relationship between composition and eutectic point, for example is beyond its scope. Therefore I have dispensed with the convention of reporting elements as oxides and make no attempt to relate empirical observations of elemental behaviour to absolute estimates of initial clay composition or firing temperature.

Background

East and Southeast Asian archaeological assemblages of the last thousand years or so typically contain stonewares from a wide range of production centres of varying technical proficiency. Stoneware technology, initiated in China prior to the first millennium AD, was widely adopted in East Asia and mainland Southeast Asia from the early second millennium AD (Hayashiya and Hasebe 1966; Impey 1972; Mikami 1972; Medley 1976; Li 1990; Stevenson and Guy 1997). This technology was responsible for goods that defined the character of one of the most extensive trade and exchange systems of the pre-modern period (Bronson 1990), a system fed by numerous small production centres, that gave rise to some of the largest industrial-scale specialised production complexes of the pre-modern world (Harrisson 1986; Long

1992; Valdes et al. 1992; Dupoizat 1996; Descantes et al. 2002). The ready spread of stoneware technology can be at least partly attributed to the relative ease with which the temperatures required can be achieved. Stoneware kilns ranged from simple in-ground bank kilns to larger more elaborate hill-side brick structures ('Dragon' kilns) (Vallibhotama 1974; Hughes-Stanton and Kerr 1980; Garnsey and Alley 1983; Shaw 1985; Liu 1986; Hein and Barbetti 1988; Ho 1990; Impey 1996; Grave et al. 2000).

The current work is part of a wider research project to understand the impact of the transition from the pre- to early-modern economy in East and Southeast Asia. Stoneware compositional data from high resolution chronological contexts (shipwrecks) can be used as an archaeological proxy for changes in craft organisation for East and Southeast Asia and is detailed elsewhere (Grave et al. 2005b). Here I focus on achieving an appropriate level of analytic detail to articulate these production dynamics.

Methodological considerations

The provenance approach to ceramic elemental analysis is best applied where an extended system of trade/exchange is thought to have operated over a region, and would seem ideally suited to East and Southeast Asian archaeological stoneware assemblages. The advantage of establishing an unambiguous geochemical signature is that it provides a means independent of more formal criteria such as typology to distinguish between different production centres, even where the actual source area or centre remains unknown. This characteristic is especially useful where popular forms were widely emulated, as is the case for the numerous competing centres of Asian stoneware production (Brown and Sjostrand 2002).

Analytical requirements normally involve a) a relatively large and representative sample population, b) instrumental technique(s) that can provide accurate and precise (reproducible) measurement of elements; and c) a large number of elements for a high dimensional dataset to maximise identification of elements or elemental suites that are distinctive of different production centres. Substantial technical advances in measurement sensitivity as well as greater flexibility and availability of analytic facilities have no doubt contributed to the growth in popularity of this approach. Criticism of the adoption of novel analytical techniques may be justified where there is no proven advantage in larger numbers of elements or higher sensitivity measurements.

Recently, Baxter and Jackson (2001) suggested that technical developments, rather than archaeological requirements, have driven the move to use increasingly large numbers of elements that may needlessly complicate interpretation. They introduce the idea of data optimisation and adopt a multivariate modelling solution that seeks the most parsimonious explanation of data structure by reducing the number of elements required to maintain sample relationships (Krzanowski 1987). While most techniques of multivariate analysis involve data reduction, a drawback of this method is that its criteria are focussed on the sample or object side of a dataset. A major shortcoming is that it discounts the interpretative potential of elemental/variable behaviour for understanding overall data structure.

Schematic Models

Out of the extensive empirical and experimental work on the factors contributing to ceramic compositional signatures, three major themes emerge: provenance; technological modification (usually focussed on additive or subtractive technologies); and post-depositional alteration. Assessing these major compositional models (provenance, technological and environmental alteration) requires both internal and independent evaluation criteria:

- The type of structure(s) (shape) of the sample;
- The behaviour of elements that describe this structure;
- A measure of statistical goodness-of-fit for elemental behaviour;
- Expectations for an observation independent of elemental behaviour that would unambiguously support a specific hypothesis such as the provenance postulate.

For the typical multidimensional interpretative environment used for high-dimensional datasets — principal components analysis (PCA) — the schematic appearance of these themes can be broadly defined, based on expectations of data structure and behaviour (Figure 1; Grave et al. 2005a). In this technique, contiguity is essential for group identification. In a normally distributed sample group, shape tends to be ellipsoidal, with deviations from normality reflected in more or less distorted group shapes. In a multi-group sample the orientation of group ellipsoids to each other reflects the degree of similarity or distinctiveness between the underlying geochemistry of each group.

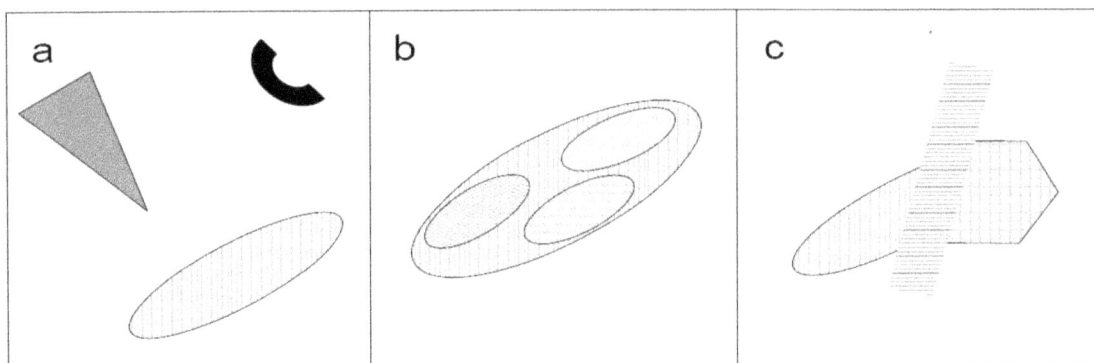

Figure 1. Schematic representation of models discussed in the text: a: provenance — groups defined by distinct elemental suites operating independent of each other and reflecting different parent catchments; b: technological alteration — varied technological treatments producing a number of groups that are subsets of elemental range and orientation of parent catchment; c: post-depositional alteration — elemental signature of parent catchment altered by interaction with mobilised elements of a different catchment to produce a new hybrid signature

For provenance-determined data structure, group range and shape is primarily influenced by the distinctive geochemistry of different catchments. This difference is reflected in independent group orientation as well as distribution (Figure 1a). Where systematic differences in technological modification is the dominant influence, such as removal of or addition of inclusions, separation between groups will be constrained by the orientation and elemental range of the parent geochemical signature (Figure 1b). Finally, post-depositional alteration, where a distinctive provenance-determined signature is altered by a subsequent geochemically distinct depositional environment, will produce a hybrid elemental signature (Figure 1c). Without good control over these subsequent processes, post-depositional alteration may obscure the relationship to unaltered groups from the same parent catchment.

Material and Methods

Sample Population

Typical archaeological contexts involve anything from decadal to centuries-old accumulations of ceramic debris that may reflect a wide range of production centres, each of which can vary greatly in technical proficiency over time. In order to circumvent these issues, this study uses a representative sample population of stoneware storage jars, from an absolutely dated historical context, the Spanish Manila galleon, 'Nuestra Señora de la Concepción', wrecked August 1638 in the northern Marianas Islands whil e on the return leg to Acapulco from the Philippines (Mathers et al. 1990). In line with the 'time capsule' nature of the wreck, the jar assemblage is of limited typological range but appears to represent a number of production centres (Rinaldi 1990). Prior to sampling, scaled digital imaging recorded jar type. A portable circular diamond saw was used to remove samples from an inconspicuous area (typically from the base) of 157 jars.

ICP-OES Analysis

The applicability of inductively coupled plasma-optical emission spectrometry (ICP-OES) to compositional analysis of stonewares has been reported elsewhere (Grave et al. 2005a). The technique relies on superheated plasma (8000–10000° centigrade) to reduce compounds to atomic constituents. The current generation of optical emission spectrometers uses a charge-coupled detector (CCD) to measure the optical properties of the plasma. After spectral splitting, the intensities of light wavelengths that are element-specific are simultaneously measured as an analogue of elemental concentrations. Typically ICP-OES enables both a high sample throughput (100–200 per day) and long periods between routine servicing. Theoretically, ICP-OES is capable of simultaneously measuring up to 70 elements of the periodic table but in practice is normally used for only 20–30 elements.

The method requires careful sample preparation (removal of surfaces with a tungsten carbide high speed burr and ultrasonically cleaned) followed by drying and powdering (grain size <10 microns) in a tungsten carbide ball mill (SPEX 5000) for 10 minutes. Samples are then reduced to an aqueous solution through digestion of a small amount of sample powder (200 mg) with a high pressure microwave system (Milestone Ethos Plus) and a combination of hydrofluoric and nitric acids based on a modified EPA method. Following digestion the aqueous samples are loaded into an auto sampler for nebulisation and ICP-OES analysis (a radial Varian Vista MPX ICP-OES system).

SEM Analysis

The physical and mineralogical characteristics of a representative subsample were examined using scanning electron microscopy (SEM). This involved carbon coating fresh fracture surfaces of samples under vacuum and examination for SEM and X-ray element-specific mapping using a Joel 35C SEM equipped with an energy dispersive spectrometer.

Accuracy and Precision

For ICP-OES measurement performance is gauged by reference to external standards to determine precision and accuracy experimentally. Ideally these standards are as closely matched to the general matrices of the actual samples as possible. Because no internationally accepted standards exist for stonewares, replicates of two standard reference materials (SRM) from the U.S. National Institute of Standards and Technology (NIST) (SRM679 Brick Clay and SRM2711 Montana Contaminated Soil) were used. These clay and soil standards were selected because they roughly approximate the matrices of the archaeological samples and cover a wide range of elements. Table 1 gives experimental recoveries for the NIST standards against their certified/published values. Of a total of 41 measured elements we retain 33 elements that have very good measurement precision (≤10%). Twenty four of these compare well with NIST certified/published SRM values (≤25%), indicating a good level of measurement accuracy.

Multivariate Analysis

Multidimensional data reduction with principal components analysis (PCA) enables standardi-zation of widely divergent elemental ranges (z-scoring) such that the majority of data variation is accounted for on the first few components ((Pollard 1986; Tangri and Wright 1993; Neff 1994). Each successive component describes variation not otherwise accounted for. Typically, with highly structured ceramic compositional data, the first 4 components describe at least 70% of total variation (Grave et al. 2005a). As PCA highlights the most significant structures as well as the contribution of elements in a reduced number of dimensions, it facilitates graphic (scatterplot) interpretation (Tufte 1983; Grave et al. 2000).

A strength of PCA is its intrinsic ability to mathematically isolate distinct features of a dataset by component (i.e. it does not require a priori assumptions about data structure). In anthropometric studies of cranial data, for example, PCA is used to distinguish a general size feature (typically accounted for on component 1) from other diagnostic (and non-size related) data structures on subsequent components (Hennessy and Moss 2001; Adriaens and Verraes 2002; Bruner and Manzi 2004). For archaeological ceramics this capacity — the relationship between elemental pattern and a general feature — has not

Element	ICP-OES Emmission line (nm)	SRM 679 (n=40) Average (ppm)		% Stdev	NIST Cert/Pub		% Stdev	% Recovery	Element	SRM2711 (n=30) Average (ppm)		% Stdev	NIST Cert/Pub		% Stdev	% Recovery
Ba	493.408	462.55	±	6.06	432.2	±	2.27	107.02	Ba	537.3	±	3.98	726	±	5.23	74.01
Bi	222.821	73.3	±	7.05					Bi	18.96	±	7.6				
Ca	317.933	1559.94	±	5.44	1628	±	0.8	95.82	Ca	18781.21	±	4.67	28800	±	2.78	65.21
Cd	226.502	0.52	±	7.95					Cd	3.94	±	2.45	41.7	±	0.6	9.45
Ce	418.659	102.8	±	7.72	105	±	—	97.9	Ce	62.4	±	4.35	69	±	—	90.43
Co	228.615	27.09	±	6.77	26	±	—	104.19	Co	9.5	±	2.95	10	±	—	95
Cr	276.653	88.63	±	6.35	109.7	±	4.47	80.79	Cr	35.06	±	2.69	47	±	—	74.6
Cu	324.754	33.3	±	6.95					Cu	103.65	±	2.69	114	±	1.75	90.92
Fe	238.204	92552.26	±	5.72	90500	±	2.32	102.27	Fe	25864.25	±	2.47	28900	±	2.07	89.5
Ga	417.204	29.8	±	5.98					Ga	13.96	±	2.88	15	±	—	93.07
Gd	342.246	20.46	±	6.22					Gd	7.29	±	2.75				
Hf	264.141	4.05	±	14.47	4.6	±	—	88.04	Hf	4.85	±	9.15	7.3	±	—	66.44
K	766.491	24445.25	±	6.99	24330	±	1.93	100.47	K	22249.12	±	1.94	24500	±	3.26	90.81
La	379.477	48.79	±	5.54					La	25.99	±	1.98	40	±	—	64.98
Li	670.783	75.89	±	8.47	71.7	±	8.65	105.84	Li	21.09	±	5.11				
Lu	261.541	2.98	±	6					Lu	0.95	±	3.75				
Mg	280.27	7483.58	±	5.95	7552	±	1.16	99.09	Mg	6906.31	±	4.02	10500	±	2.86	65.77
Mn	257.61	1810.85	±	6.01	1730	±	—	104.67	Mn	607.9	±	2.95	638	±	4.39	95.28
Na	588.995	585.33	±	7.78	1304	±	2.91	44.89	Na	4795.66	±	2.05	11400	±	2.63	42.07
Nb	309.417	12.89	±	7.36					Nb	9.67	±	4.01				
Nd	401.224	23.83	±	5.85					Nd	11.5	±	2.04	31	±	—	37.1
Ni	231.604	59.09	±	6.3					Ni	17.42	±	4.37	20.6	±	5.34	84.56
Pb	220.353	16.68	±	8.77					Pb	1106.35	±	2.45	1162	±	2.67	95.21
S	181.972	293.59	±	10.95					S	381.68	±	1.48	420	±	2.38	90.88
Sc	335.372	23.03	±	5.61	22.5	±	—	102.36	Sc	8.19	±	1.67	9	±	—	91
Sr	407.771	72.87	±	5.69	73.4	±	3.54	99.28	Sr	185.14	±	1.88	245.3	±	0.28	75.47
Th	283.73	48.86	±	6.12	14	±	—	349	Th	17.65	±	3.43	14	±	—	126.07
Ti	334.941	5722.7	±	6.1	5770	±	5.72	99.18	Ti	2682.23	±	2.82	3060	±	7.52	87.65
V	311.07	188.4	±	6.33					V	84.96	±	2.59	81.6	±	3.55	104.12
Y	371.029	37.5	±	5.69					Y	18.57	±	1.71	25	±	—	74.28
Yb	328.937	5.15	±	6.08					Yb	2.51	±	1.15	2.7	±	—	92.96
Zn	213.857	115.82	±	5.91	150	±	—	77.21	Zn	298.53	±	2.19	350.4	±	1.37	85.2
Zr	343.823	150.16	±	5.83					Zr	107.71	±	3.09	230	±	—	46.83

Table 1. Statistics for the accuracy and precision of the ICP-OES analysis showing total number of elements used in this study and optimal spectral lines in nanometers (nm). Averages based on number of replicates (n=#) with % standard deviations (calculated by taking the standard deviation for the element and dividing it by the element mean X 100. The purpose of this calculation is to make the standard deviation figures directly comparable between elemental concentrations that can differ by orders of magnitude). Data for the certified standards NIST 679 and 2711 includes % standard deviations and % recoveries (calculated by dividing the mean of the NIST value by 100 then dividing the unknown by this result). Highlighted elements are those that are both precise and accurate. Reported but non-certified data are without errors shown by −.

been explored. The significance of a generalised feature lies in the possibility that group differentiation on the first component would reflect a property that if demonstrated, would apply to all samples, irrespective of origin, technical treatment or environmental effects. In other words more specific effects, such as provenance-specific patterning, should be sought on subsequent components.

Polynomial Regression Analysis and 'goodness of fit'

Determining the probability either that the first principal component isolates a generalised effect, or of the best fit between the data and a particular model, requires statistical (rather than mathematical) evaluation. Regression analysis and its associated R^2 score is an effective way to test both (Fan and Gijbels 1996). The non-linear transformations involved in PCA require a polynomial regression approach. For modelling, a hypothesised governing property of the multidimensional behaviour of elements determines their respective positions along one axis (e.g. valence state, atomic weight, physical behaviour) against the first component order of elements on the other axis (the response). The fit of a line through the observed points of this model provides a measure of the extent to which the model explains the distribution of elements. This is given by an R^2 value, an estimate of the proportion of the variation in the response around the mean that can be attributed to the model rather than random error. An R^2 value of 1 reflects a perfect fit and an R^2 value of 0 means that the fit predicts the model no better than the overall response mean. For evaluating how well a general phenomenon is isolated on the first component, then, this should be reflected in a high R^2 score. Conversely, if the first component reflects compound effects, then this should produce a low R^2 value.

Preliminary Modelling Expectations

Having defined the test population, method and quality of instrumental measurement, and the multivariate and statistical means to evaluate data structure and elemental behaviour, we can now turn to defining provisional expectations for each of the three schematic models presented earlier.

Model A (Figure 1a): Provenance postulate — elemental signature that reflects spatially and geologically discrete production centre:

- Groups defined by distinct elemental suites operating independent of each other and reflecting local petrogenetic conditions (i.e. each group defined by a different elemental trajectory in multidimensional space).
- Independent correlates: typology — each group has distinct typological attributes.

Model B (Figure 1b): Technological modification of elemental signature (e.g. addition or subtraction of inorganic or organic inclusions):

- Group defined by distinct elemental suite that reflects change in compositional range as a result of technological alteration. Different technological treatments can produce a number of groups constrained within the same provenance-specific elemental trajectory.
- Independent correlates: physical evidence for additions; correlations with independent measurements of elemental behaviour (e.g. elemental enrichment associated with purposefully added inclusions).

Model C (Figure 1c): Post-depositional alteration of elemental signature (e.g. environmental enrichment/ depletion):

- Elemental signature(s) altered by interaction with mobilised elements to produce a new hybrid signature. If a many-to-one relationship, then this effect would tend to blur provenance-related distinctions.
- Independent correlates: evidence of accretion of salts in pore spaces, dissolved minerals.

Results

Variation in summed elemental totals for each sample conflates both systematic as well as random effects. Because of the difficulties in separating the two, sample measurements were normalised to Σ 50000 ppm. PCA indicates that the normalised dataset is highly structured. It readily separates into five discrete groups on the first two components, with the eigenvalues for these components accounting for almost 65% of overall variation (Figure 2; Table 2). In the accompanying variable plot for these components two elemental trends are evident. The first ranges from high K samples to high Ca samples and describes groups 1–3 and 5 along component 1. The second distinguishes high Ti from the K-Ca trend and separates group 4 samples along component 2. No further new groups are evident on subsequent components.

Polynomial regression of the accompanying elemental plot produces a low R^2 value suggesting that, while the object plot clearly distinguishes 5 groups, there is little correlated behaviour between variables. Optimisation of these results determines how many elements can be removed while retaining group integrity. This step involves iterative removal of redundant or noisy elements using improvements in the polynomial R^2 value as a guide. Elements that exhibit little systematic behaviour due to poor measurement statistics or because of sample outliers are obvious targets for exclusion by optimisation.

Following R^2 optimization the revised dataset was reanalysed by PCA (Figure 3). The results show a substantial 15% improvement in the variation accounted for by the first two components (from ~65% to ~80%). Little substantive difference is evident in the group object scatterplot of component 1 and 2. The polynomial regression R^2 values for the accompanying element plot shows an improvement from 0.3 to almost 0.9 indicating a high level of descriptive efficiency in the element-reduced dataset. While element end members in the optimised dataset appear to retain a high level of redundancy, removal of any of these reduced the R^2 value.

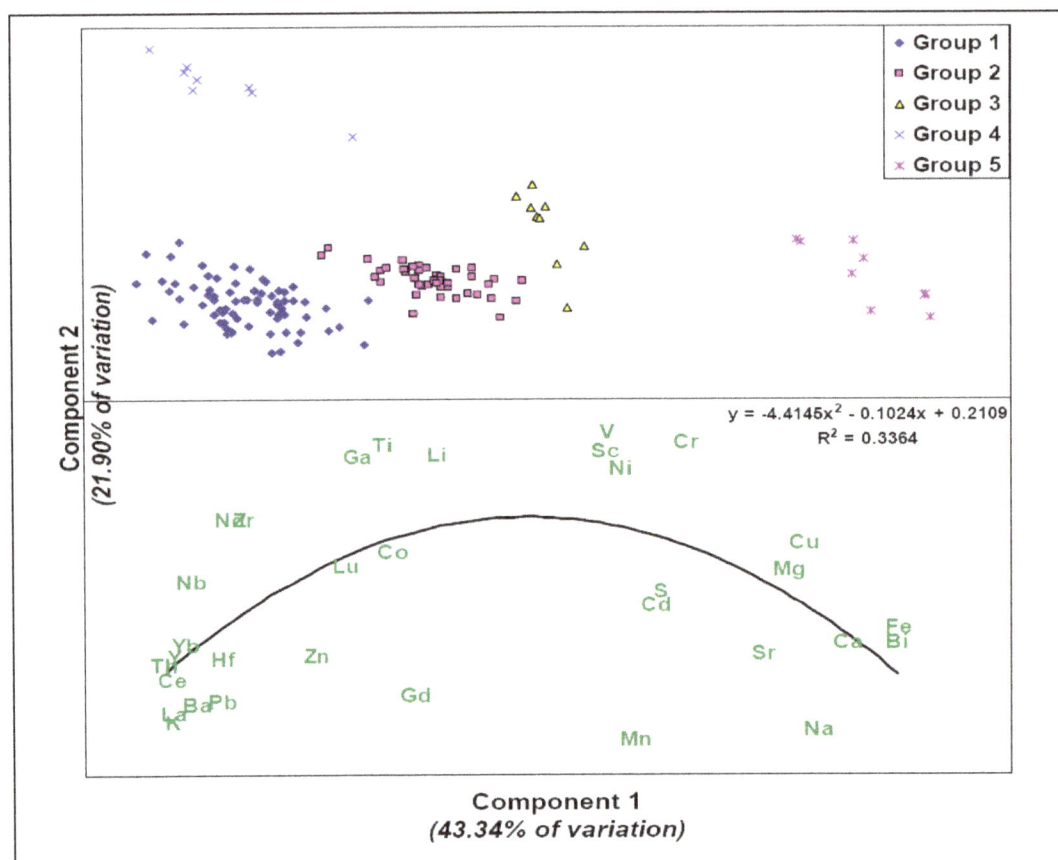

Figure 2. PCA scatterplot of the first two components for the non-optimised dataset showing structure of objects (above) and elements with polynomial regression and R^2 value (below).

Element	Group 1 (n=76)		Group 2 (n=55)		Group 3 (n=9)		Group 4 (n= 8)		Group 5 (n= 9)	
	Average (ppm)	% Stdev	Average (ppm)	% Stdev	Average (ppm)	% Stdev	Average (ppm)	% Stdev	Average (ppm)	% Stdev
Ba	422.02 ± 9.02		259.77 ± 13.67		290.18 ± 22.98		336.48 ± 8.41		96.61 ± 66.76	
Bi	10.64 ± 19.74		15.71 ± 17.06		16.87 ± 11.5		9.95 ± 27.34		22.13 ± 3.57	
Ca	1598.43 ± 39.44		1907.34 ± 86.12		2766.31 ± 25.78		1611.54 ± 33.79		6154.27 ± 16.68	
Cd	0.12 ± 41.67		0.17 ± 58.82		0.23 ± 104.35		0.18 ± 105.56		0.2 ± 50	
Ce	112.91 ± 13.78		75.84 ± 18.54		65.39 ± 13.46		98.62 ± 8.51		14.57 ± 45.16	
Co	36.16 ± 38.61		39.58 ± 87.37		29.36 ± 24.66		64.23 ± 44.89		21.84 ± 39.1	
Cr	18.95 ± 18.58		31.19 ± 15.2		55.16 ± 21.32		78.58 ± 4.16		46.76 ± 39.61	
Cu	10.98 ± 84.97		10.72 ± 29.01		20.3 ± 23.4		18.55 ± 7.98		28.49 ± 17.16	
Fe	17721.7 ± 9.86		22723.89 ± 10.97		24199.6 ± 4.58		17658.06 ± 3.53		28563.38 ± 3.2	
Ga	20.04 ± 11.58		18.92 ± 10.15		19.73 ± 9.38		34.11 ± 5.69		16.11 ± 24.89	
Gd	8.7 ± 19.31		8.79 ± 24.8		7.54 ± 4.77		7.55 ± 4.64		6.62 ± 9.06	
Hf	6.28 ± 18.95		4.32 ± 20.6		3.04 ± 21.05		6.26 ± 14.22		0.44 ± 88.64	
K	21173.59 ± 8.6		14541.7 ± 11.93		9591.28 ± 13.95		16081.24 ± 9.36		2033.69 ± 49.14	
La	60.38 ± 13.66		37.11 ± 15.33		27.56 ± 6.57		42.96 ± 9.57		6.39 ± 32.86	
Li	23.6 ± 12.5		30.63 ± 15.18		35.42 ± 23.94		73.34 ± 11.82		8.24 ± 23.91	
Lu	1.79 ± 20.11		1.73 ± 26.59		1.38 ± 7.97		2.48 ± 46.37		1.19 ± 14.29	
Mg	2923.27 ± 33.21		4653.08 ± 38.36		6802.83 ± 40.01		5231.36 ± 17.86		6326.63 ± 21.92	
Mn	391.38 ± 14.5		269.23 ± 71.17		430.49 ± 29.39		122.31 ± 34		587.61 ± 31.79	
Na	939.81 ± 19.56		1069.87 ± 17.75		700.24 ± 11.57		399.69 ± 8.12		2229.85 ± 29.4	
Nb	13.81 ± 12.45		10.31 ± 12.22		8.53 ± 11.02		18.29 ± 6.94		5.31 ± 21.85	
Nd	18.77 ± 9.27		16.01 ± 11.74		16.24 ± 10.1		26.95 ± 5.42		10.51 ± 27.88	
Ni	10.33 ± 20.52		14 ± 23.5		29.95 ± 20.8		27.21 ± 6.5		11.97 ± 22.81	
Pb	42.13 ± 19.7		30.67 ± 58.59		17.51 ± 7.65		31.36 ± 15.11		7.63 ± 199.87	
S	69.97 ± 80.92		116.93 ± 120.5		322.26 ± 102.23		136.42 ± 45.59		154.08 ± 84.62	
Sc	10.23 ± 8.31		10.4 ± 10.29		12.3 ± 15.12		14.99 ± 6.14		12.58 ± 7.63	
Sr	46.14 ± 12.85		46.3 ± 32.57		42.59 ± 16.29		50.64 ± 7.72		84.27 ± 32.27	
Th	25.97 ± 6.85		22.46 ± 6.68		20.78 ± 3.37		24.95 ± 5.85		14.19 ± 4.72	
Ti	3964.48 ± 10.23		3736.79 ± 11.27		4160.76 ± 11.62		7290.77 ± 5.79		3323.87 ± 28.53	
V	72.37 ± 11.52		81.03 ± 14.18		100.59 ± 19		173.94 ± 5.85		110.65 ± 18.98	
Y	33.93 ± 13.79		25.77 ± 13.74		24.22 ± 6.77		33.57 ± 9.06		11.06 ± 22.6	
Yb	3.69 ± 12.2		2.85 ± 9.82		2.71 ± 6.27		3.68 ± 7.61		1.86 ± 15.59	
Zn	77.4 ± 16.46		79.17 ± 30.83		70.28 ± 35.84		76.51 ± 17.03		34.92 ± 16.15	
Zr	130.05 ± 15.22		107.71 ± 13.75		108.39 ± 9.76		213.24 ± 19.87		52.08 ± 22.37	

Table 2. Summary statistics for the five elemental groups of stoneware jars identified through ICP analysis. Averages based on number of replicates (n=#) with % standard deviations (calculated by taking the standard deviation for the element and dividing it by the element mean X 100).

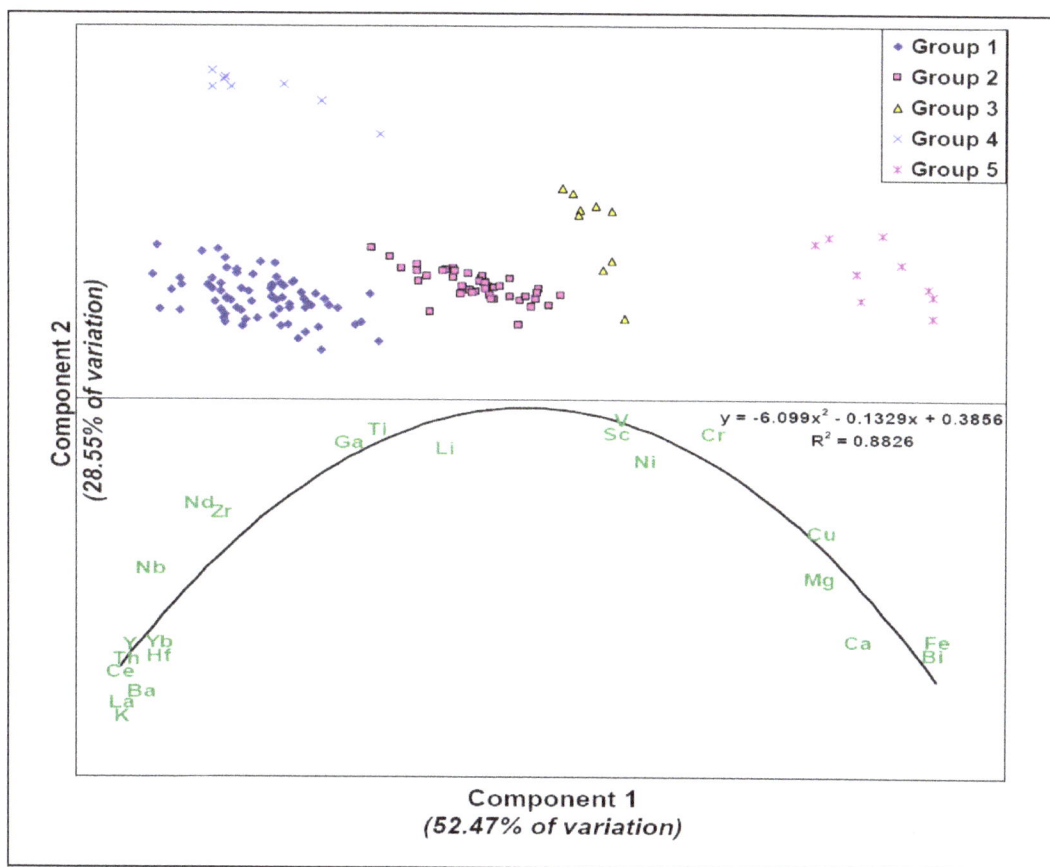

Figure 3. PCA scatterplot of the first two components for the optimised dataset showing structure of objects (above) and elements with polynomial regression and R^2 value (below).

Ten elements were discarded to produce the optimised dataset. These include elements at concentrations well below those of the standards or not represented in them and those that are likely to have poor measurement characteristics (Cd, Gd, Lu, Pb). However, they also include elements in concentrations for which measurement is either precise or both accurate and precise (Co, Mn, Na, S, Sr, Zn). Three of these elements can be discounted because of outlier behaviour (Co: a single sample in group 2 with significantly elevated levels, >250 ppm), S (significantly enhanced, >500 ppm) for several group 2, 3 and 5 samples and Zn (elevated levels, >150 ppm) in three group 1 and 2 samples). The remaining elements Mn, Sr and Na otherwise appear statistically normal (i.e. have good measurement statistics and are not subject to outlier distortion).

Elevated concentrations of these elements characterise group 5. Mn and Na, relatively abundant components of sea water, could be expected as precipitates from post-depositional sea water infiltration (Rodshkin and Ruth 1997). A more complex argument needs to be mounted for Sr, a trace element that normally tracks Ca. Structural similarities allow substitution of Sr into the Ca molecular lattice. This relationship is charge dependant and environmentally sensitive (Porder et al. 2005). The Ca/Sr ratio is substantially higher (Sr depleted) than in the remaining groups (average group 1–4 Ca/Sr ratio: 43.15± 8.72%; group 5 Ca/Sr ratio: 73.03). As Sr-depleted-Ca in group 5 is consistent with a marine rather than terrestrial source, this distinction also suggests possible sea water infiltration. In other words group 5 samples may be more porous, less vitrified and therefore affected by post-depositional precipitates than the remaining groups.

Scanning electron microscopy (SEM)

Low and high magnification SEM and X-ray mapping of fresh carbon-coated fracture surfaces was used to assess the likely extent of post-depositional chemical alteration of a selection of fabrics from each group. All groups exhibit extensive vitrification, closed pore spaces and the development of gas bubbles (Figure 4a). Contrary to expectations from the elemental profiles, Group 5 samples appear to be more vitrified with larger coalescing bubbles than the remainder and have no evidence of post-depositional precipitates (Figure 4b). Group 5 samples also contained numerous fine and coarser Ca- and Na-rich mineral inclusions. These inclusions retained clear (unmelted) grain boundaries and were part of the original pre-fired fabric (Figure 4c and d). No samples from the other groups contained inclusions of comparable composition, size or preservation.

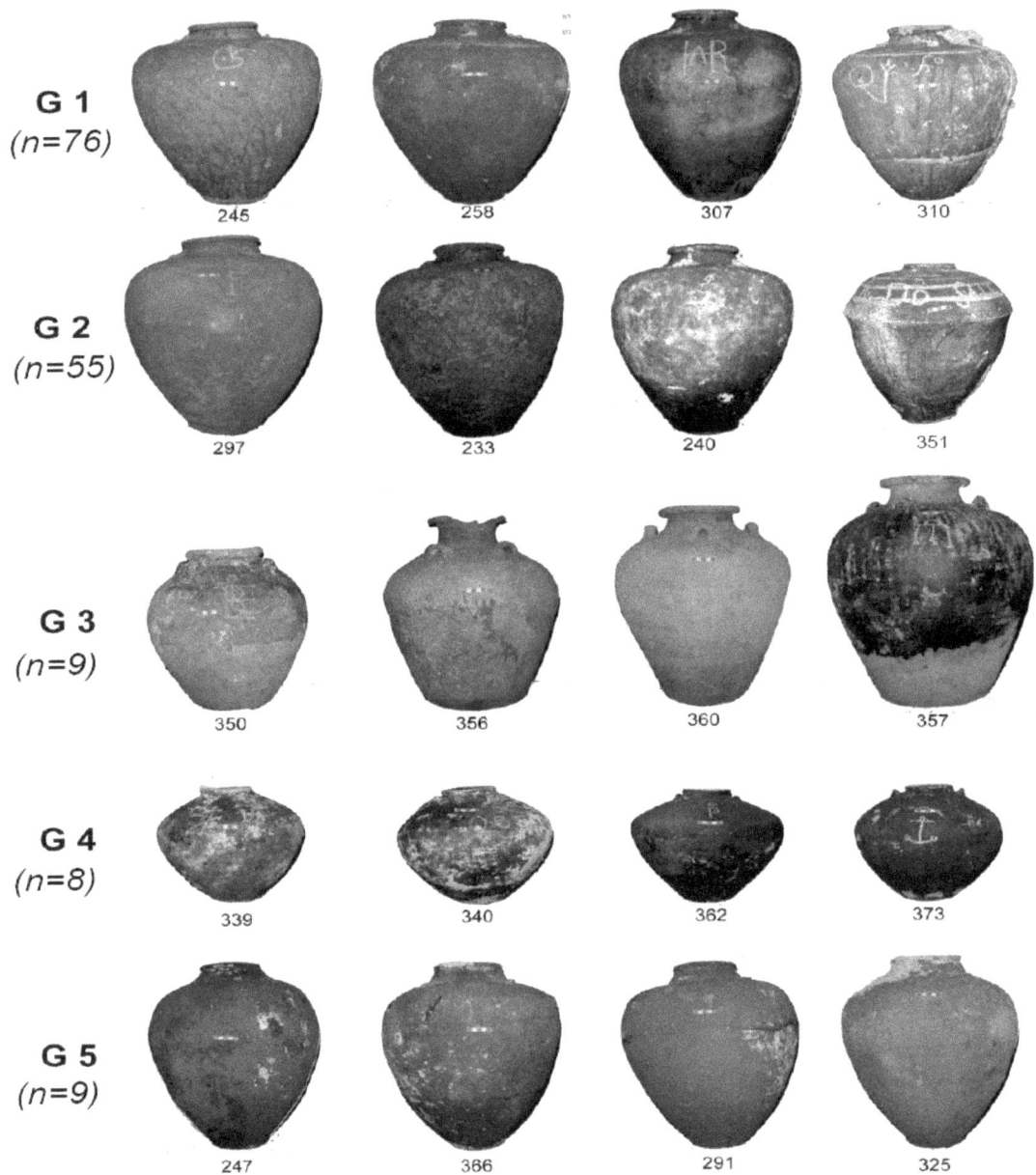

Figure 4. Representative jar types for each of the 5 compositional groups discussed in the text (all to scale).

Discussion

Returning to the three models we can now re-evaluate these results.

Model A (provenance postulate):

For model A, the independent attribute of jar form provides a check on whether the five compositional groups are likely to come from different centres (Figure 5). Groups 1 and 2 share a jar form that is indistinguishable in style and decorative detail (Figure 5, group 1, group 2). This close similarity suggests the same potters rather than emulation of a popular jar form by two centres. The significance of the compositional separation will be returned to below. Group 2 also includes several examples of a second smaller form representing a distinctive fabrication convention (two-piece shoulder joined type (Figure 5, group 2, sample 351). Together, group 1 and 2 account for the majority of the sample population suggesting that these jars were a predominant type available to Spanish ships leaving Manila in the early 17th century. Based on comparison with published data, group 1 and 2 forms have a likely origin in a southeastern Chinese production complex in Fujian or Guandong Province (Hogervost 1982; Southeast Asian Ceramic Society (West Malaysia Chapter) 1985; Harrisson 1986; Long 1992; Valdes et al. 1992; Dupoizat 1996). The remaining groups represent about 5% each of the total sample. The possible spatial significance of this relatively low figure can be gauged from group 3. This group includes a range of types derived from production centres in more distant Thailand (Brown 1988; Brown and Sjostrand 2002), suggesting that the other small groups may also reflect more distant production centres (Figure 5, group 3). Group 4 samples (Figure 5, group 4) are of a distinctive type without sufficiently good parallels to nominate a source area but we can infer a single production region from the relatively homogenous form and the distinctiveness of the titanium-rich composition of this group. For group 5 (Figure 5, group 5) large elemental differences coupled with stylistic anomalies (less well-defined neck and rim; less symmetrical body indicating greater degree of hand finishing) suggest that this group, though broadly similar in form to the most common type shared by groups 1 and 2, is an emulation and does not originate in the same production complex.

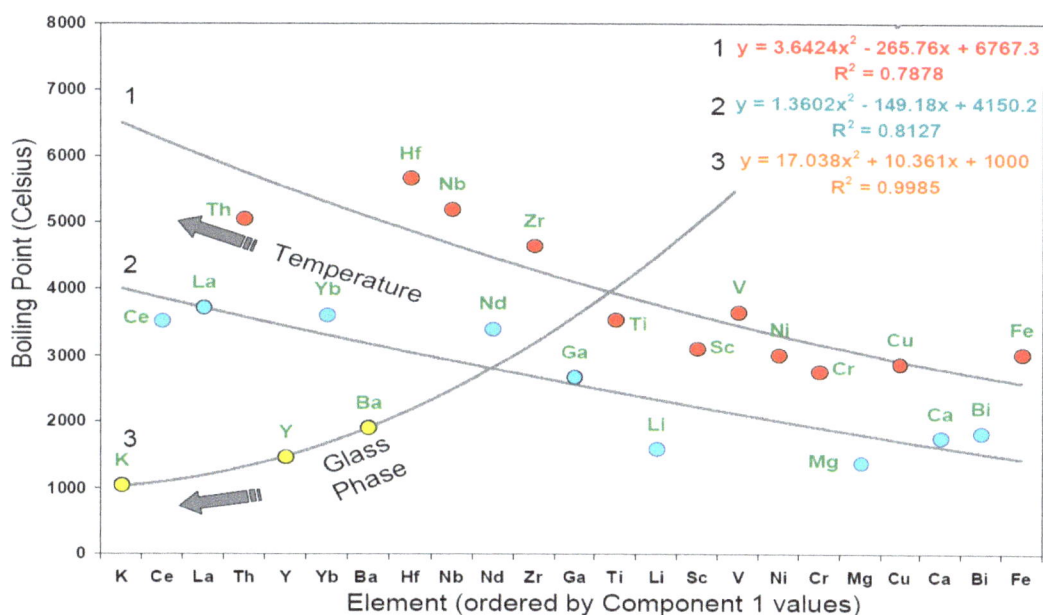

Figure 5. The Boiling Point model: polynomial regression analyses and R^2 values for boiling point data (y axis) compared with order of elements (x axis) on the first component from the PCA of the optimised dataset. Note that the very high value for the 3rd regression reflects the low number of elements and is retained primarily for heuristic value. Contrast this with the high values for regressions 1 and 2, each with ~10 elements extending across the component and boiling point range.

Model B (technology):

The results from the PCA of the optimised dataset were used to determine if a generalised feature was captured in the order of elements along the first component (effectively detrending the PCA 'horseshoe' distribution inherent in the eigenvalues themselves). A strong correlation is evident between component-ordered elements and boiling point data (Aylward and Findlay 1983). Several iterations to improve the polynomial R^2 value result in not one but three regression solutions for the component 1 order of elements against their boiling points (Figure 6). Two of these correlate with elemental order from left to right and an overall decrease in boiling point. The third describes low boiling point elements in what is otherwise the high boiling point side of the plot. While only trace elements are represented at elevated boiling points, the third trend runs counter to this with low boiling point major, minor and trace elements in the otherwise high boiling point end of the first component. This generalised boiling point behaviour is consistent with progressive volatilisation of elements under increasingly high temperature firing regimes. In this scenario the enhancement of a limited suite of low temperature elements in the high temperature end of the component reflects development of an extensive K-Y-Ba-rich felspathic glass phase depleted in other low temperature elements (Kopeikin 1960; Yakovleva and Beznosikova 1960).

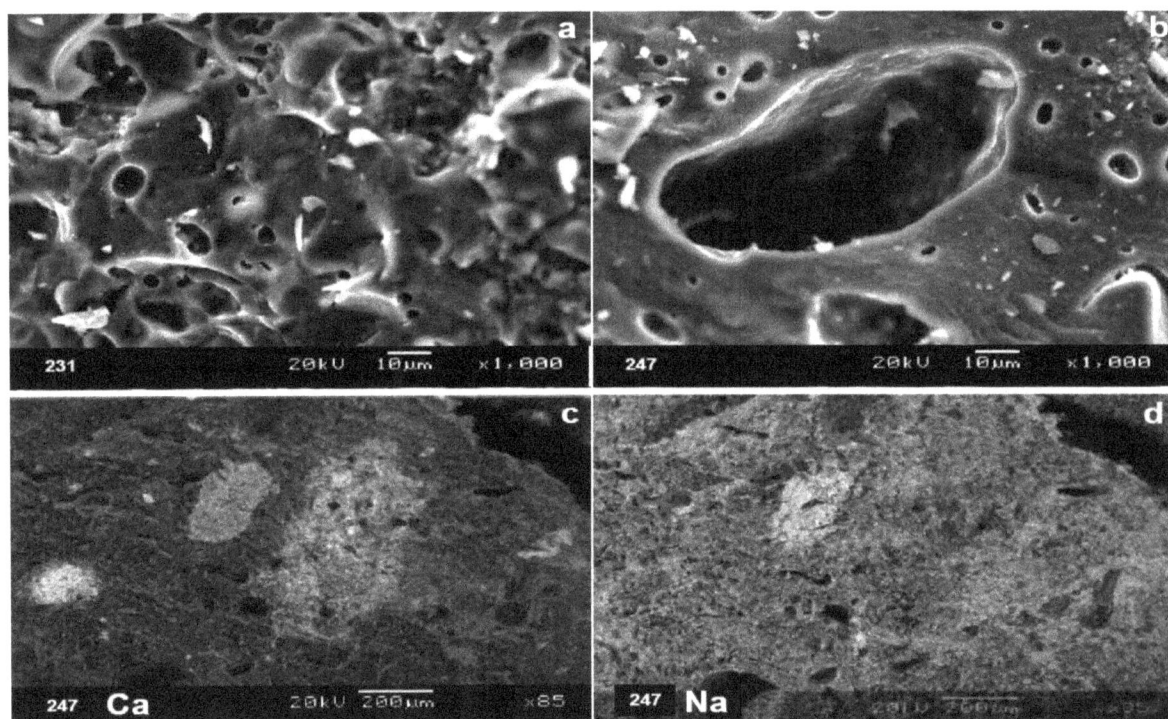

Figure 6. Scanning electron micrographs of representative group fabrics. Above, high magnification stoneware fabric typical of groups 1–4 (a); and fabric typical of group 5 (b). Both show evidence of extensive glass formation with gas bubbles but note the group 5 sample appears more evolved with larger areas of glass and coalesced gas bubbles. Below, low magnification X-ray maps of the same area of a group 5 fabric preserving (c) Ca- and (d) Ca/Na-rich minerals with clear grain boundaries.

Model C (environmental effects):

Post-depositional alteration does not appear to have had appreciable influence on the stoneware fabrics of the sample population. In contrast to relatively porous earthenwares with a greater potential to take up elements mobilised in ground water, the stonewares all exhibited extensive glassy phases (i.e. non-continuous pore structures) that appear impervious to penetration by ground water or salts. Even assuming some degree of alteration undetectable by SEM, the samples of this study, immersed in salt water for almost 400 years, shared the same conditions of temperature and salinity, and have been treated in the same manner for museum storage, such that post-depositional alteration should affect all samples equally.

Summary

For the 'Concepción' stoneware assemblage, the geographic provenance of compositional groups cannot be disengaged from systematic differences produced by firing temperature. The first component of the PCA analysis for the optimised dataset isolated a generalised temperature-related behaviour; polynomial regression analysis highlighted the explanatory efficiency of the boiling point model and underscored the potential pitfalls in optimisation techniques, such as proposed by Baxter and Jackson (2001), that arbitrarily remove ostensibly redundant elements. In this study the elemental end members of the optimised PCA that appeared to contain a high level of redundancy in fact represent a complex relationship between temperature and fabric composition.

Of the three ceramic elemental hypotheses (provenance, environment and technology) two can be accommodated within the dataset (Figure 7). While provenance (Model A) can be demonstrated to account for group distinctions, firing temperature differentials (Model B) represent the most parsimonious explanation of data structure.

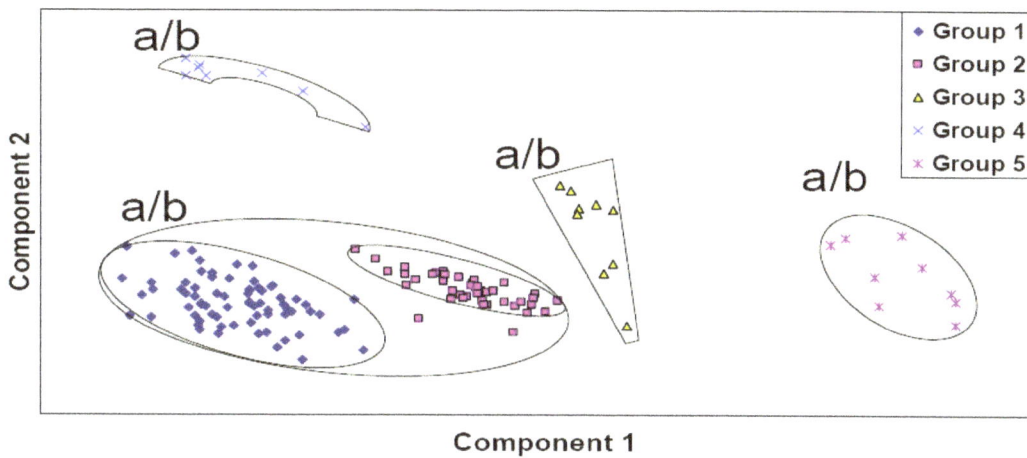

Figure 7. PCA scatterplot of the first two components for the optimised dataset showing structure of objects with hypothetical schemas superimposed (labels follow Fig. 1).

If, as proposed, groups 1 and 2 come from the same production centre then they appear to reflect a complex production environment. The close typological similarity of both groups suggests they were produced by the same potters. However, if the distinction between them primarily relates to differences in firing technology then this suggests that the products of a single pottery workshop were sent to at least two kilns with different firing characteristics. This level of complex organisation has good historical antecedents for the operation of large pottery production centres in Southeastern China from at least as early as the 17th century (Sung 1966).

Group 3 consists of vessels that are of entirely different types and are typical of production centres in Thailand. This group, about the same distance on the first component from group 2 as group 2 is to group 1, also appears to reflect a lower firing temperature technology. Group 4, both compositionally and typologically distinct from the remaining groups, represents a class of jar from yet another, (?) more distant, production complex but that more closely matches the higher temperature range of group 1 and 2 and therefore is more closely related technologically (e.g. northern China, Vietnam).

Group 5, the most compositionally distant of the groups, broadly conforms to the most common jar type of groups 1 and 2 but diverges in typological details. Model C (post-depositional salt water enrichment), suspected to be responsible for this distance, was not sustainable. SEM data indicates levels of vitrification comparable or even more developed than for the remaining groups. In group 5 the low-temperature end member is also richest in fluxing elements (Ca, Mn and Na) that would lower the eutectic point and promote melting at lower temperatures. The presence of Ca/Na rich minerals with distinct boundaries in the group 5 fabrics suggests that while extensive melting had occurred this was curtailed before a full glass phase could develop. The survival of minerals composed of low boiling point elements further supports group 5 as a low temperature end member best explained in Model B. Group 5 wares appear to be the product of a firing technology with significantly lower temperatures than represented, for example, in the distance between the Southern Chinese groups 1, 2 and Thai group 3 samples. As an adaptation of high temperature firing technology to clays with high flux, low melting-point characteristics more typical of earthenware compositions, group 5 appears to represent a stoneware firing technology where stoneware was adapted to a less than optimum clay type.

Conclusions

Temperature plays a role in elemental sequestration of stoneware ceramic datasets. This observation underlines the opportunities for a better quantitative understanding of the differences between both contemporary and successive stoneware firing technologies of a region. It also highlights the potential of this approach as an analytical proxy for complex technological organisation (e.g. the apparent decoupling of potters of a common jar type from multiple firing complexes with different temperature characteristics). These conclusions tend to contradict conventional modelling assumptions for archaeological ceramic elemental datasets with provenance/temperature effects perhaps more widespread than previously supposed. If so, these types of signatures may prove of broader utility in other archaeological settings for identifying systematic techno-cultural differences between production centres or regions.

Acknowledgments

This work was conducted under Australian Research Council Discovery Grant DP0344782. Access to the samples used in this study from the 'Nuestra Señora de la Concepción' was generously given by the CNMI Museum of History and Culture, Garapan, Saipan. Thanks to Lisa Kealhofer who commented on an earlier draft of this paper, Andrew Boulton for input on multivariate assumptions, and anonymous reviewers for their constructive suggestions.

References

Adriaens, D. and W. Verraes. 2002. An empirical approach to study the relation between ontogeny, size and age using geometric morphometrics. In P. Aerts, K. D'Août, A. Herrel and R. V. Damme (eds), *Topics in functional and ecological vertebrate morphology*. Maastricht: Shaker Publishing.

Aylward, G. H. and T. J. V. Findlay. 1983. *SI chemical data*. Milton, Queensland: Wiley.

Baxter, M. J. and C. M. Jackson. 2001. Variable selection in artefact compositional studies. *Archaeometry* 43(2):253–268.

Bronson, B. 1990. Export porcelains in economic perspective: The Asian ceramic trade in the 17th century. In Ho Chuimei (ed.), *Ancient ceramic kiln technology in Asia*, pp 126–151. Hong Kong: Centre of Asian Studies.

Brown, R. M. 1988. *The ceramics of Southeast Asia, their dating and identification*. 2nd ed. Singapore, Oxford University Press.

Brown, R. and S. Sjostrand. 2002. *Maritime archaeology and shipwreck ceramics in Malaysia*. Kuala Lumpur: Department of Museums and Antiquities.

Bruner, E. and G. Manzi. 2004. Variability in facial size and shape among North and East African human populations. *Italian Journal of Zoology* 71:51–56.

Buxeda i Garrigos, J. 1999. Alteration and contamination of archaeological ceramics: The perturbation problem. *Journal of Archaeological Science* 26:295–313.

Buxeda i Garrigos, J., V. Kilikoglou and P. M. Day. 2001. Chemical and mineralogical alteration of ceramics from a late Bronze Age kiln at Kommos, Crete: The effect on the formation of a reference group. *Archaeometry* 43(3):349–371.

Chen, T., J. G. Rapp Jr, Z. Jing and N. He. 1999. Provenance studies of the earliest Chinese protoporcelain using instrumental neutron activation analysis. *Journal of Archaeological Science* 26:1003–1015.

Cogswell, J. W., H. Neff and M. D. Glascock. 1996. The effect of firing temperature on the elemental characterization of pottery. *Journal of Archaeological Science* 23:283–287.

Descantes, C., H. Neff and M. D. Glascock. 2002. Yapese prestige goods: The INAA evidence for an Asian dragon jar. In M. D. Glascock (ed.), *Geochemical evidence for long-distance exchange*, pp 229–256. Westport, Conn.: Greenwood Publishing Group.

Dupoizat, M.-F. 1996. Asian stoneware jars. In J.-P. Desroches, F. G. Casal and F. Goddio (eds), *The treasures of the San Diego*, pp 222–251. New York: Elf Aquitaine International Foundation/ National Museum of The Philippines.

Fan, J. and I. Gijbels. 1996. *Local polynomial modelling and its applications*. London: Chapman and Hall/CRC.

Ferring, C. R. and T. K. Perttula. 1987. Defining the provenance of red slipped pottery from Texas and Oklahoma by petrographic methods. *Journal of Archaeological Science* 14:437–456.

Garnsey, W. and R. Alley. 1983. China: *Ancient kilns and modern ceramics: A guide to the potteries*. Canberra: Australian National University Press.

Gomez, B., H. Neff, M. L. Rautman, M. D. Glascock and S. J. Vaughan. 2002. The source provenance of Bronze Age and Roman pottery from Cyprus. *Archaeometry* 44(1):23–36.

Gosselain, O. P. 1992. Bonfire of the enquiries. Pottery firing temperatures in archaeology: What for? *Journal of Archaeological Science* 19:243–259.

Grave, P. 2004. Elemental characterisation of Tepe Yahya Iron Age ceramics. The Archaeology of Resistance and Integration. In P. Magee (ed.), *Excavations at Tepe Yahya, Iran, 1967–1975, Volume IV*: The Iron Age Settlement, pp 62–80. American School of Prehistoric Research Bulletin 46. Cambridge, Mass.: Peabody Museum, Harvard University.

Grave, P., M. Barbetti, M. Hotchkis and R. Bird. 2000. The stoneware kilns of Sisatchanalai and Early Modern Thailand. *Journal of Field Archeology* 27(2):169–182.

Grave, P., L. Lisle and M. Maccheroni. 2005a. Multivariate comparison of ICP-OES and PIXE-PIGE analysis of East Asian storage jars. *Journal of Archaeological Science* 32(6):885–896.

Grave, P., M. Maccheroni and L. Lisle. 2005b. The transition to the early modern economy in East and Southeast Asia, 1400-1750 AD: Production and exchange of storage jars. In Cheng Pei-kai (ed.), *Proceedings of the international conference: Chinese export ceramics and the maritime trade, 12th–15th centuries*, pp 163–186. Hong Kong: Chinese Civilisation Centre, City University of Hong Kong.

Guo Yanyi and Li Guozhen 1986. Song Dynasty Ru and Yaozhou green glazed wares. *In Scientific and technological insights on ancient Chinese pottery and porcelain, Proceedings of the International Conference on ancient Chinese pottery and porcelain, Shanghai Institute of Ceramics*, pp 153–160. Beijing: Science Press.

Harrisson, B. 1986. *Pusaka: Heirloom jars of Borneo*. Singapore: Oxford University Press.

Hayashiya, S. and G. Hasebe. 1966. *Chinese Ceramics*. Rutland, Vermont and Tokyo: Charles E. Tuttle.

Hein, A., A. Tsolakidou and H. Mommsen. 2002. Mycenaean pottery from the Argolid and Achaia: A mineralogical approach where chemistry leaves unanswered questions. *Archaeometry* 44(2): 177–186.

Hein, D. and M. Barbetti. 1988. Sisatchanalai and the development of glazed stoneware in Southeast Asia. *Siam Society Newsletter* 4(3):8–18.

Hennessy, R. J. and J. P. Moss. 2001. Facial growth: Separating shape from size. *European Journal of Orthodontics* 23(3):275–285.

Ho Chuimei (ed.), 1990. *Ancient ceramic kiln technology in Asia*. Hong Kong: Centre of Asian Studies, Hong Kong University.

Hogervost, A. 1982. Stoneware storage jars (Martavans). In C. L. van der Pijl-Ketel and J. B. Kist (eds), *The ceramic load of the 'Witte Leeuw' 1613*, pp 220–245. Amsterdam: Rijksmuseum.

Hughes-Stanton, P. and R. Kerr. 1980. *Kiln sites of ancient China: An exhibition lent by the People's Republic of China*. London: Oriental Ceramic Society.

Hughes, M. J., K. J. Matthews and J. Portal. 1999. Provenance studies of Korean celadons of the Koryo Period by neutron activation analysis. *Archaeometry* 41(2):287–310.

Impey, O. 1972. *A tentative classification of the Arita kilns. In International symposium on Japanese ceramics*. Seattle.

Impey, O. 1996. The early porcelain kilns of Japan: Arita in the first half of the seventeenth century. Oxford: Clarendon Press.

Impey, O. and M. Tregear. 1983. Provenance studies of Tanqui porcelain. *Trade Ceramic Studies* 3: 102–118.

Jones, R. E. 1986. Techniques and methodology in characterisation and provenance work. In R. E. Jones, *Greek and Cypriot pottery: A review of scientific studies*, pp 15–56. Athens: The British School at Athens.

Kerr, R. and N. Wood. 2004. *Chemistry and chemical technology*. Cambridge, Mass.: Cambridge University Press.

Kilikoglou, V., Y. Maniatis and A. P. Griminas. 1988. The effect of purification and firing of clays on trace element provenance studies. *Archaeometry* 30(1):37–46.

Kolb, C. 1982. Ceramic technology and problems and prospects of provenience in specific ceramics from Mexico and Afghanistan. In J. S. Olin and A. D.Franklin (eds), *Archaeological ceramics*, pp 193 – 208. Washington: Smithsonian Institution Press.

Kopeikin, A. A. 1960. The effect of fluxing additions on the structure and properties of semiporcelain. *Glass and Ceramics (Historical Archive)* 15(10):531–536.

Krzanowski, W. J. 1987. Selection of variables to preserve multivariate data structure, using principal components. *Applied Statistics* 36(1):22–33.

Li Jiazhu. 1990. The evolution of Chinese pottery and porcelain technology. In W. D. Kingery (ed.), *Ceramics and civilisation: Ancient technology to modern science*, Volume 1, pp 135–62. Westerville, Ohio: The American Ceramic Society.

Liu Zhenqun. 1986. Investigations of the kilns and firing processes of pottery and porcelain in Chinese past dynasties. In *Scientific and technological insights on ancient Chinese pottery and porcelain, Proceedings of the International Conference on ancient Chinese pottery and porcelain, Shanghai Institute of Ceramics*, pp 295–300. Beijing: Science Press.

Long, K. N. 1992. Typology and classification of jars. In C. O. Valdes, K. N. Long and A. C. Barbosa (eds), *A thousand years of stoneware jars in the Philippines*, pp 168–199. Manila: Jar Collectors (Philippines).

Maggetti, M., H. Westley and J. S. Ofin. 1984. Provenance and technical studies of Mexican majolica using elemental and phase analysis. In J. B. Lambert, *Archaeological chemistry III, advances in chemistry series* 205, pp 151–191. Washington DC: American Chemical Society.

Mallory-Greenough, L. M., J. D. Greenough and J. V. Owen. 1998. New data for old pots: Trace-element characterization of ancient Egyptian pottery using ICP-MS. *Journal of Archaeological Science* 25:85–97.

Mason, R. B. and E. J. Keall. 1988. Provenance of local ceramic industry and the characterization of imports: Petrography of pottery from medieval Yemen. *Antiquity* 62(236):452–463.

Mathers, W. M., H. S. Parker III and K. A. Copus (eds), 1990. *Archaeological report: The recovery of the Manila galleon 'Nuestra Señora de la Concepción'*. Sutton, Vermont, USA: Pacific Sea Resources.

Matson, F. R. 1971. A study of temperature used in firing ancient Mesopotamian pottery. In R. H. Brill (ed.), *Science and Archaeology*, pp 65–79. Cambridge, Mass.: The MIT Press.

Matsunaga, M. and I. Nikai. 2004. A study of the firing technique of pottery from Kaman-Kaleoyuk, Turkey, by synchrotron radiation-induced fluorescence X-ray absorption near-edge structure (XANES) analysis. *Archaeometry* 46(1):103–114.

Medley, M. 1976. *The Chinese potter: A practical history of Chinese ceramics*. Oxford: Phaidon.

Middleton, A. P., M. R. Cowell and E. W. Black. 1992. Romano-British relief-patterned flue tiles: A study of provenance using petrography and neutron activation analysis. In *Sciences de la Terre et Ceramiques Archeologiques, Experimentations, Applications* 16, pp 49–59. Saint Louis a Cergy: Institut Geologique Albert de Lapparent.

Mikami, T. 1972. *The art of Japanese ceramics*. New York: Weatherhill.

Mommsen, H., T. Beier, U. Diehl and C. Podzuweit. 1992. Provenance determination of Mycenaean sherds found in Tell el Amarna by neutron activation analysis. *Journal of Archaeological Science* 19: 295–302.

Neff, H. 1994. RQ-mode principal component analysis of ceramic compositional data. *Archaeometry* 36: 115–130.

Neff, H. 2000. Neutron activation analysis for provenance determination in archaeology. In E. Ciliberto and G. Spoto (eds), *Modern analytical methods in art and archaeology, chemical analyses series* 155, pp 81–134. New York: John Wiley and Sons.

Olin, J. S., G. Harbottle and E. V. Sayre. 1978. Elemental compositions of Spanish and Spanish-colonial majolica ceramics in the identification of provenience. In G. F. Carter (ed.), *Archaeological chemistry II, advances in chemistry series* 171, pp 200–229. Washington DC: American Chemical Society.

Pollard, A. M. 1986. Multivariate methods of data analysis. In R. E. Jones, *Greek and Cypriot pottery: A review of scientific studies*, pp 56–83. Athens: The British School at Athens.

Pollard, A. M. and E. T. Hall. 1986. Provenance studies of oriental porcelain by chemical analysis. In *Scientific and technological insights on ancient Chinese pottery and porcelain, Proceedings of the International Conference on ancient Chinese pottery and porcelain, Shanghai Institute of Ceramics*, pp 377–381. Beijing: Science Press.

Pollard, A. M. and H. Hatcher. 1986. The chemical analysis of oriental ceramic body compositions: Part 2 — greenwares. *Journal of Archaeological Science* 13:261–287.

Pollard, A. M. and H. Hatcher 1994. The chemical analysis of oriental ceramic body compositions: Part 1: Wares from North China. *Archaeometry* 36(1):41–62.

Porder, S., A. Paytan and P. M. Vitousek. 2005. Erosion and landscape development affect plant nutrient status in the Hawaiian Islands. *Oecologia* 142:440–449.

Rinaldi, M. 1990. The ceramic cargo of the 'Concepción'. In W. M. Mathers, H. S. Parker III and K. A. Copus (eds), *Archaeological report: The recovery of the Manila galleon 'Nuestra Señora de la Concepción'*, pp 398–528. Sutton, Vermont USA: Pacific Sea Resources.

Rodshkin, I. and T. Ruth. 1997. Determination of trace metals in estuarine and sea-water reference materials by high resolution inductively coupled plasma mass spectrometry. *Journal of Analytical Atomic Spectrometry* 12:1181–1185.

Schwedt, A., H. Mommsen and N. Zacharias. 2004. Post-depositional elemental alterations in pottery: Neutron activation analyses of surface and core samples. *Archaeometry* 46(1):85–101.

Shaw, J. C. 1985. The kilns of Lan Na or northern Thai ceramics. *In Technical workshop on ceramics (T-W4), Bangkok and Chiang Mai, Thailand, December 1–12, 1985.* Bangkok: Southeast Asian Ministers of Education Organisation, SEAMEO Project in Archaeology and Fine Arts.

Southeast Asian Ceramic Society (West Malaysia Chapter) (ed). 1985. *A ceramics legacy of Asia's maritime trade: Song Dynasty Guangdong wares and other 11th to 19th century trade ceramics found on Tioman Island, Malaysia.* Singapore: Oxford University Press.

Stenger, A. 1992. Sourcing and dating of Asian porcelains by elemental analysis. In P. Wegars (ed.), *Hidden heritage: Historical archaeology of the overseas Chinese*, pp 315–331. Amityville, New York: Baywood Publishing Co.

Stevenson, J. and J. Guy. 1997. *Vietnamese ceramics: A separate tradition.* Chicago: Art Media Resources/ Avery Press.

Sung Ying-Hsing. 1966. *Chinese technology in the seventeenth century* (trans. E.-T. Zen Sun and S.-C. Sun). Mineola, New York: Dover Publications.

Tangri, D. and R. Wright. 1993. Multivariate analysis of compositional data: Applied comparisons favour standard principal components analysis over Aitchison's loglinear contrast. *Archaeometry* 35(1): 103–15.

Tufte, E. 1983. *The visual display of quantitative data.* Cheshire, Conn.: Graphics Press.

Valdes, C. O., K. N. Long and A. C. Barbosa. 1992. *A thousand years of stoneware jars in the Philippines.* Manila: Jar Collectors (Philippines).

Vallibhotama, S. 1974. The Khmer ceramic kilns of Ban Kruat and their preservation. *Our Future* 2:30–33.

Vitali, V. and U. M. Franklin. 1986. New approaches to the characterization and classification of ceramics on the basis of their elemental composition. *Journal of Archaeological Science* 13:161–170.

Wilson, L. and A. M. Pollard. 2001. The provenance hypothesis. In D. R. Brothwell and A. M. Pollard (eds), *Handbook of archaeological sciences*, pp 507–517. Chichester: John Wiley.

Yakovleva, M. E. and A. V. Beznosikova. 1960. The microstructure of earthenware and semiporcelain fired at 1100–1300. *Glass and Ceramics* (Historical Archive) 15(9):483–490.

Yang Wenxian and Wang Yuxi. 1986. A study of ancient Jun ware and modern copper red glaze from the chemical compositions. *In Scientific and technological insights on ancient Chinese pottery and porcelain, Proceedings of the International Conference on ancient Chinese pottery and porcelain, Shanghai Institute of Ceramics*, pp 204–210. Beijing: Science Press.

Yap, C. T. and Y. Hua 1992. Principal component analysis of Chinese porcelains from the Five Dynasties to the Qing Dynasty. *Zeitschrift für Naturforschung* 49a:1029–1033.

16

New approaches for integrating palaeomagnetic and mineral magnetic methods to answer archaeological and geological questions on Stone Age sites

Andy I. R. Herries

Human Origins Group and Primate Origins Program
School of Medical Sciences
University of New South Wales
Sydney, Australia.
a.herries@unsw.edu.au

Geomagnetism Laboratory
Oliver Lodge, University of Liverpool, UK
andyherries@yahoo.co.uk

Introduction and aims

Archaeomagnetism as defined here is the use of magnetic methods of analysis on archaeological materials and deposits, although in its widest context it refers to the magnetisation of any materials relating to archaeological times. It is most widely known for its use in dating, but more recently it has been utilised for other purposes including site survey, sourcing and palaeoclimatic reconstruction. These applications have different site requirements, as discussed below. Two main methods of analysis exist: those that look at the direction and intensity of fossil remnant magnetisations, as in palaeomagnetism; and those related to looking at the mineralogy, grain size and concentration of minerals within a rock or sediment, as in mineral (rock, environmental) magnetism. In the later case, identification of these parameters is achieved by different types and strengths of laboratory-induced remnant magnetisations and/or heat into samples to see how they react or alter.

Magnetic methods have, over the last 10 years, been increasingly used as a Quaternary method of analysis for a variety of applications including dating, sediment-source tracing, and palaeo-environmental/climatic reconstruction. While these methods have been used on some archaeological sites (e.g. Ellwood et al., 1997; Dalan and Banerjee 1998; Moringa et al. 1999; Gose 2000; Peters et al. 2001), their application has been sporadic and their potential as a major tool for reconstructing archaeological data remains underutilised. This paper provides a review of methods of archaeomagnetic analysis to show the potential for recovering various types of primary data that can be integrated to form a powerful tool for reconstructing archaeological site evolution and behavioural patterns. The paper also provides preliminary data from a number of Stone Age cave sites in South Africa and Spain.

Mineral magnetism

Sediment sourcing, input, alteration and palaeoclimatic modelling

Mineral magnetic studies examine the magnetic nature (minerals, grain sizes and concentration) of the sedimentary input into a depositional system and the various processes that have acted on that material both before and after its deposition in that system. Soil formed on volcanic soils will have a very different magnetic character to soils formed on karstic landscapes. A study of the magnetic mineralogy of cave sediments can therefore show changes in sedimentary input, such as a change from locally derived material to material derived from further afield. This particularly occurs for sites where sediments are deposited by aeolian activity due to changes in wind speed and direction, which can change between glacial and interglacial periods (Begét 2001). In fluvially derived deposits, changes generally tend to reflect more locally changing conditions, such as changes in pedogenic activity, soil cover, weathering and erosion. A study of the magnetic mineralogy of sediment sequences can often be used as a palaeoclimatic proxy record because these processes are governed by climatic factors such as rainfall and temperature variations. These are in turn driven by glacial/interglacial cycles. Such palaeoclimatic proxy records have primarily been undertaken on lake and marine sequences (for example, Peck et al. 2004; Williamson et al. 1998), but work has shown that cave sediments are also suitable, as they are relatively protected from external climatic processes after deposition (Herries 2006).

Mineral magnetic analysis of sediment sequences can reveal a climatically-driven signal for a number of reasons. Primary and secondary iron phases in rocks and sediments are transformed by weathering and various other climatically driven processes into secondary iron oxides (such as magnetite, maghaemite and haematite), hydroxides (such as goethite) and sulphides (such as pyrrhotite). Most often, these pedogenic processes convert primary and secondary iron into magnetite, with later oxidisation to maghaemite through oxidisation/reduction cycles (Maher 1998). This is known as pedogenic enhancement and it produces a dominance of these ferrimagnetic minerals. With prolonged weathering, these processes may also cause the formation of anti-ferromagnetic minerals, haematite and goethite. The degree of pedogenic enhancement is not only determined by climatic and environmental processes but also by the local lithology, which provides the initial magnetic input into the pedogenic system through weathering of underlying rock strata. In certain rocks, iron oxides, hydroxides and sulphides may already coexist with amorphous iron. In igneous landscapes magnetite is the dominant iron oxide, whereas sedimentary rocks may contain significant haematite, especially in sandstones. As large amounts of highly magnetic phases are weathered directly from the local rock strata in volcanic landscapes, the soils tend to show little pedogenic enhancement. The greatest enhancement is seen in rock types that do not contain large amounts of primary iron oxides, such as limestone (Maher 1998).

Pedogenic formation of ferrimagnets is driven by temperature and moisture. It appears to be favoured in well-drained, poorly-acidic soils on weatherable, Fe-bearing (but often not Fe-rich) substrates in a climate that produces wetting/drying cycles (Maher 1998). Excessively arid, waterlogged or acidic soils display little pedogenic enhancement. Maher and Thompson (1995) suggest that in many cases maximum values of strong ferrimagnetic mineral phases, and so greatest pedogenic enhancement, correlate well with absolute rainfall. However, in cases where a certain rainfall threshold is exceeded, water-logging occurs and can result in the depletion of magnetic minerals by processes such as gleying (Liu et al. 1999). Detailed in situ magnetic susceptibility (MS) measurements of multiple sections can quickly identify the potential of a site for palaeoclimatic analysis by identifying bulk changes in iron-bearing minerals. However, to understand fully the magnetic mineralogical changes, a suite of mineral magnetic measurements needs to be taken. Other analytical techniques for characterising mineralogy (e.g. XRF, XRD, FTIR) can most often not be used due to the small amounts of magnetic material within the samples.

The survey and identification of archaeological sites using archaeomagnetism is based on locating strong magnetic anomalies that are normally related to the use of fire by humans (Mullins 1977). These anomalies can be strongly magnetised rocks that once made up a fireplace or sediments that have been anthropogenically altered. Palaeoclimatic analysis on archaeological sites is complicated by anthropogenic alteration of the deposits after deposition. It is therefore of primary importance to understand how the heating of sediments affects the magnetic mineralogy of the deposits. Magnetic measurements are sensitive to fire histories because burning causes the transformation of trace iron within the fuel source itself or/and within sediments associated with the heating (Peters et al. 2001, 2002). Changes between different minerals and different grain sizes can occur; these are dependent on the temperature, longevity and atmosphere of heating (Herries and Kovacheva 2007). Oxidising conditions cause the formation of fine- to ultra-fine-grained single domain ferrimagnetic minerals and at higher temperatures, haematite (Herries et al. 2007a). In contrast, reducing conditions cause the formation of larger stable single-domain grained magnetite (Herries et al. 2007a). While heating of sediments in kilns and ovens might produce a purely reducing environment (Herries et al. 2007a), campfires, as seen on Stone Age sites, almost always produce an oxidising or mixed environment of heating. As with magnetic enhancement in soils, the effect of heat on sediments is determined primarily by their initial magnetic makeup (Herries et al. 2007a). A number of factors can affect the thermo-magnetic enhancement of burnt deposits, including the type of combustion process used, contaminant mineralogy and fuel chemistry (Peters et al. 2002). Different fuel sources can often be determined from the magnetic mineralogy of ash residues (Peters et al. 2002). Mineralogically complex fuel-ash is normally confined to fuels such as coals and peat. Pure wood-ash, which would be expected to dominate Stone Age and Palaeolithic hearths, should cause no thermo-magnetic enhancement because wood itself is non-magnetic, but magnetic enhancement does occur due to small amounts of burnt sediment within the ash.

South African soils and burning

Over the last 10 years a series of samples have been collected from modern and ancient campfires in South Africa (Figure 1), to come to an understanding of the effect of fire on South African soils and sediments at different sites. Most often, heating causes the formation of fine-grained ferrimagnetic particles, which produces magnetic enhancement and high MS (Peters et al. 2001). The viscous grain sizes formed by heating can normally be detected by frequency dependence of MS (FD-MS [XFD%]) (Thompson and Oldfield 1986). However, South African cave sediments already contain large amounts of these viscous magnetic grains due to pedogenesis and natural burning in the open landscape (Table 1; Figure 2 colluvium). Further burning causes a high concentration of ultra-fine (superparamagnetic) grains that are smaller than can be detected by this method using the standard Bartington (Ltd.) MS equipment. An increase in MS still occurs as these grain sizes are still easily magnetisable. However, rather than seeing an increase in FD-MS with heating, as seen at many Eurasian sites (Morinaga et al. 1999; Peters et al. 2001; Herries et al. 2007a), no change, or even a decrease in FD-MS is seen (Table 1).

In South Africa, the variation in this parameter is seemingly partly geographically dependant. At more coastal sites, such as Pinnacle Point and Sibudu Cave, a decrease in FD-MS is seen in conjunction with increased MS (Table 1; Figure 2 fire). At most inland sites, such as Grand Canyon Rockshelter (Limpopo Province), Rose Cottage Cave (Free State) and Molony's Kloof (Northern Cape) only a slight decrease or no change in FD-MS can be noted, although an increase in MS is still seen (Table 1). While a comparison of MS and FD-MS can still be used to identify burnt sediments in South Africa, not all burnt sediments can unequivocally be identified with this method unless detailed comparative mineralogical studies are undertaken.

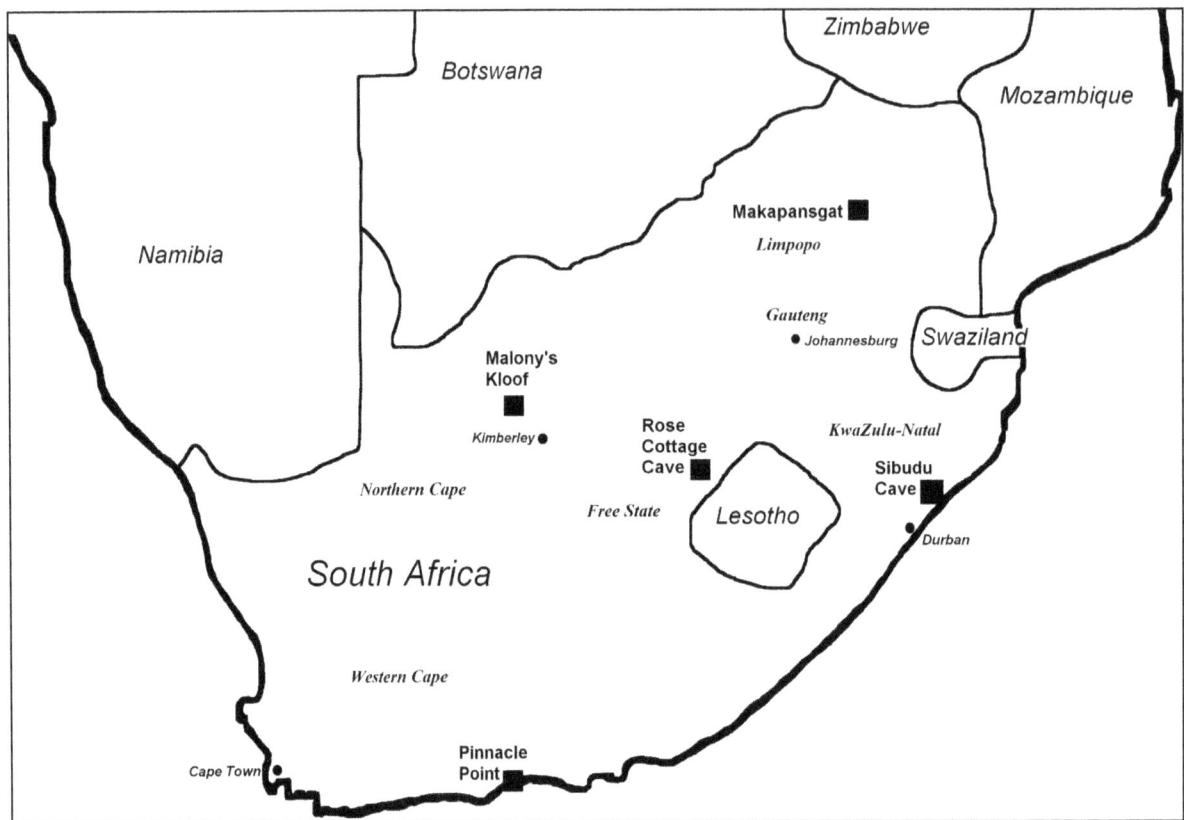

Figure 1. Location of archaeological sites studied in South Africa.

Site	Location	State	MS	FD-MS
Grand Canyon Rockshelter	Limpopo	unburnt	278.4	11.4
Grand Canyon Rockshelter	Limpopo	mixed	544.0	10.6
Grand Canyon Rockshelter	Limpopo	burnt	1876	9.37
Rose Cottage Cave	Free State	unburnt	125.0	9.8
Rose Cottage Cave	Free State	mixed	200	10.1
Rose Cottage Cave	Free State	burnt	285.0	10.7
Molony's Kloof	Northern Cape	unburnt	35.8	5
Molony's Kloof	Northern Cape	mixed	160.0	6.6
Molony's Kloof	Northern Cape	burnt	231.7	8.9
Pinnacle Point	Western Cape	unburnt	1.1	0.0
Pinnacle Point	Western Cape	mixed	41.0	8.8
Pinnacle Point	Western Cape	burnt	138.0	2
KwaZulu Natal recent	KwaZulu Natal	unburnt	750.0	12.5
KwaZulu Natal recent	KwaZulu Natal	natural burning	900.0	12.0
Sibudu Cave	KwaZulu Natal	unburnt	92	8.7
Sibudu Cave	KwaZulu Natal	mixed	362	5.5
Sibudu Cave	KwaZulu Natal	burnt	601	3

Table 1. Magnetic susceptibility (MS) and frequency dependence of MS (FD-MS) for burnt, mixed and unburnt sediments from sites in different areas of South Africa.

Low temperature magnetic susceptibility (LT-MS) provides another potential method for identifying burnt sediments as it can be used for identifying ultra-fine superparamagnetic (SP) grains that are formed by the heating process (Peters et al. 2002). SP grains do not hold a magnetic remanence, are slightly smaller than those detected by FD-MS and are the grain size formed by heating of sediments in South Africa. Larger stable single-domain (SSD) ferrimagnetic grains show little change in MS down to the temperature of liquid nitrogen ($-196°C$). In contrast SP grains show a large drop in MS down to $-196°C$. The RS ratio is the ratio of MS at $+25°C$ to MS at $-196°C$. As such, SSD ferrimagnetic grains have an RS value close to 1 and SP values are much lower. The shape of the LT-MS can also be used to identify the mineralogy of the sample. SP magnetite will have a low RS ratio that is the same between -196 and $-150°C$ (the isotopic point of magnetite), causing the formation of a low temperature tail. Maghaemite does not have such a low temperature tail. Haematite can cause a variety of behaviours depending on its grain size and its MS behaviour at low temperatures is generally less well understood than for magnetite.

Burnt clays from Bulgarian clay-built pottery kilns and LSA hearths from Rose Cottage Cave in South Africa show an LT-MS behaviour that is quite distinctive and suggests that a particular mineralogy

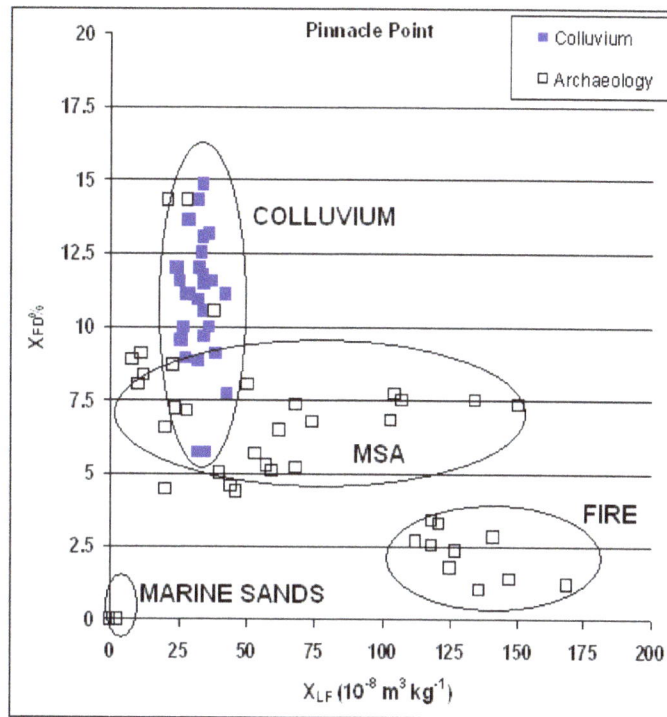

Figure 2. Frequency dependence of magnetic susceptibility (FD-MS; XFD%) versus low frequency magnetic susceptibility (MS; XLF) data for modern soils, unaltered sand and archaeological layers and in situ hearths from MSA bearing Cave 13b at Pinnacle Point.

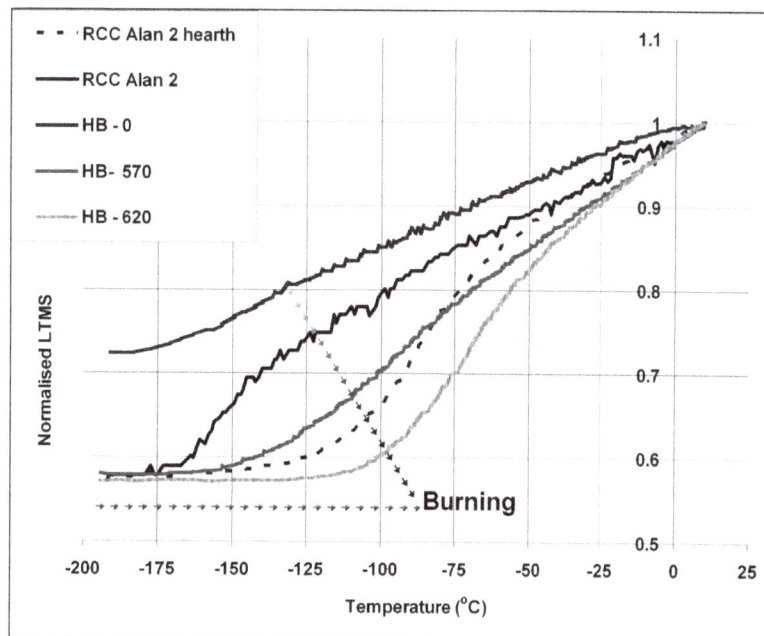

Figure 3. Low temperature magnetic susceptibility (LT-MS) data for unburnt and burnt LSA layer Alan 2 from Rose Cottage Cave and experimentally burnt clay samples from the Thracian site of Halka Bunar, Bulgaria.

prevails, at least in more clay based sediments (Figure 3). Unburnt sediments have a typical SP character. Burnt sediments have a curve with a lower RS, a low temperature tail that prevails to temperatures above –150°C and overall the cure has a more S-shaped nature. This is thought to be in part due to the formation of coarser grained haematite in the samples. Further work is need to clarify this but these data along with the work of Peters et al (2002) has shown the power of using LT-MS in identifying burnt deposits. Bellomo (1994) has also suggested certain magnetic parameters for identifying human/hominin burnt material. However, many of these will be dependent on the location of the site, lithology and climate among other things. As such, the use of archaeomagnetic methods to determine burning must always be on a case by case basis and use as many parameters and methods as possible. Magnetic enhancement will only be relative to the base mineralogy of the layer as it was before it was altered.

In the past, the colouration of sediments was taken as an indicator of burnt archaeological deposits. However, recent work has shown that reddening of sediments by burning only happens occasionally in campfires (Canti and Lindford 2000). Reddening of deposits is caused by the formation of haematite. While haematite can form at low temperatures in a heavily oxidising environment due to the transformation of maghaemite (Herries and Kovacheva 2007), it is fine grained and so does not affect many remanence properties. Courser grained haematite is often not produced until temperatures in excess of 600°C (Herries et al. 2007a). In most instances the temperature beneath most fires remains below 500°C and reddening of the soil happens only rarely (Canti and Lindford 2000). Moreover, more reducing conditions occur in sediments beneath the campfire and a reddening of deposits would be unlikely to occur as haematite would not be formed by oxidisation processes.

The identification of early fire use by hominins in the open landscape is difficult due to the occurrence of natural fires, particularly in Africa. Often coloured sediments in caves, where fires do not occur naturally, are considered to be more reliable indicators of the hominin/human use of fire. However, the work of Weiner et al. (1998) has also shown that coloured sediments at the Palaeolithic cave site of Zhoukoudian are due to entirely natural processes. At the Cave of Hearths, at Makapansgat, a series of coloured sediments and circular bone accumulations occurs in Early Stone Age (ESA) layers that have also been interpreted as early fire use (Mason 1988; Maguire 1998). Unaltered alluvial and colluvial sediments from Makapansgat (ZKL-003; GKC; Figure 4), and from the Cave of Hearths (PM, WWS; Figure 4) have high proportions of fine to ultra-fine VSD grains. Burnt sediments from the nearby Grand Canyon Rockshelter fall into two groups. The lower MS group represents unaltered colluvially-derived sediments. The higher MS group represent burnt sediments from a hearth at the site and show an increase in MS but a similar to lower FD-MS. The increase in MS is due to an increase in ultra-fine, SP magnetite during the process of burning. In contrast, the purported burnt ESA layers from the Cave of Hearths have low MS and FD-MS due to a lack of fine to ultra-fine VSD and SP grains (HWS; Figure 4). The layers therefore show no form of magnetic enhancement as would be expected from fire use or a mineralogy consistent with any documented burnt sediments. The magnetic mineralogy indicates that the fine-grained magnetite fraction, present in surrounding deposits (WWS, PM), is absent and has most likely been altered or destroyed by some other process. It is suggested that the coloured horizons actually represent waterlogged pool deposits, where complex mineralogical changes have occurred due to the presence of owl and bat guano. White ashy-looking horizons within the sequence consist of un-burnt and crushed rodent bones. This research has shown that there is no evidence of fire in the ESA horizons at the site.

In theory, the identification of heavily anthropogenically altered layers in a cave sequence is possible using these magnetic methods and may be used to identify heavily occupied layers or zones of an archaeological cave. If the base material is particularly unmagnetic, in the case of more sand- or clay-based sediments, the identification of anthropogenic alteration, and therefore heavy occupation should be simpler. Figure 5 shows a section through the test trench of Sopeña Cave in Asturias, northern Spain. The sequence is dated to between the end of the Last Glacial Maximum ~18–20,000 years and >38,000 years, and records the Mousterian to Upper Palaeolithic transition. The unaltered base sediment is clay-, sand- and silt-rich and has a low mean value of MS (X mean 0.24 x 10^{-6} m^{3}kg^{-1}). Little change in the MS is seen for the section that would suggest a climatically determined signal from variation in sedimentary input. However, there are four distinct layers that have enhanced MS values. Unaltered layers are dominated by more coarse-grained magnetite and some maghaemite. When these deposits are heated they convert to fine-grained magnetite and a large increase in magnetisation and MS is recorded. In contrast, burnt layers are dominated by ultra-fine-grained magnetite and do not alter on heating to 700°C. The magnetic sequence at Sopeña Cave identifies four main levels of occupation at the site (Levels 3, 5, 7 and 13). However, a section through different parts of the cave may alter this picture as occupation may have taken place in different lateral areas of the cave at different periods.

Therefore, the task now is to try and recreate these data multi-dimensionally to create a three dimensional picture of anthropogenic alteration that could ultimately be used to identify spatial patterning of occupation within archaeological cave sites. Figure 6 shows a 3D GIS image of magnetic measurements undertaken at a Middle Stone Age (MSA) cave from Mossel Bay in South Africa (Marean et al.. 2007). Over 600 samples were measured from bulk samples taken during excavation of the site. The base sediment in the caves consists of magnetically weak dune sand and quartzitic roof spall. When the sediments are heated magnetite is formed. High MS values show close relationships

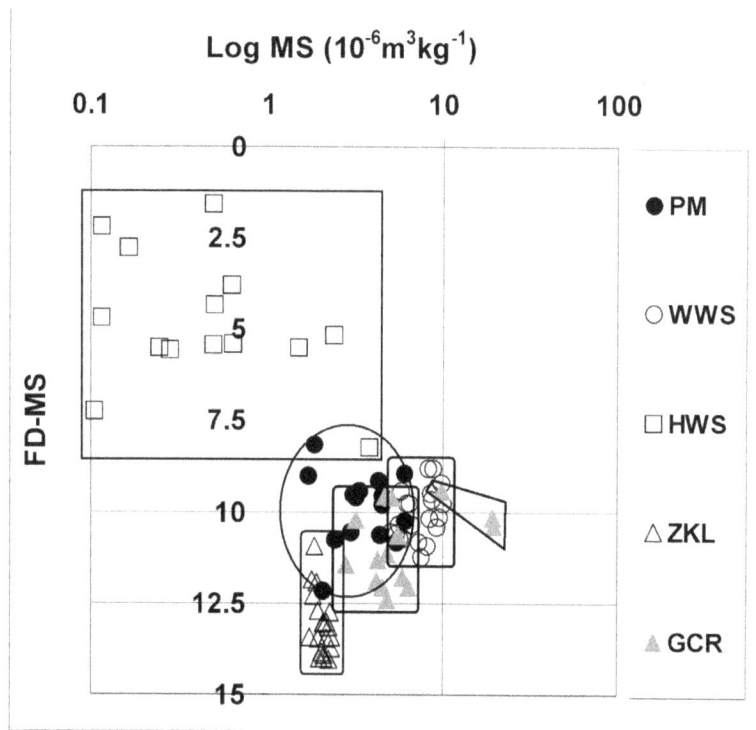

Figure 4. A comparison of FD-MS versus log MS for sites from Makapansgat, South Africa. Samples from site ZKL-003 represent alluvial sediments. Samples from Grand Canyon Rockshelter (GKC) represent colluvial sediments. The second grouping with higher MS and similar to low FD-MS represent hearth sediments from the site. WWS samples come from lightly calcified unaltered ESA layers of the Cave of Hearths. PM samples come from calcified ESA deposits of the Cave of Hearths and have lower MS values than the WWS due to the diluting influence of higher calcite content. HWS samples represent coloured horizons suggested to be evidence for hominin fire use from the ESA layers at the Cave of Hearths.

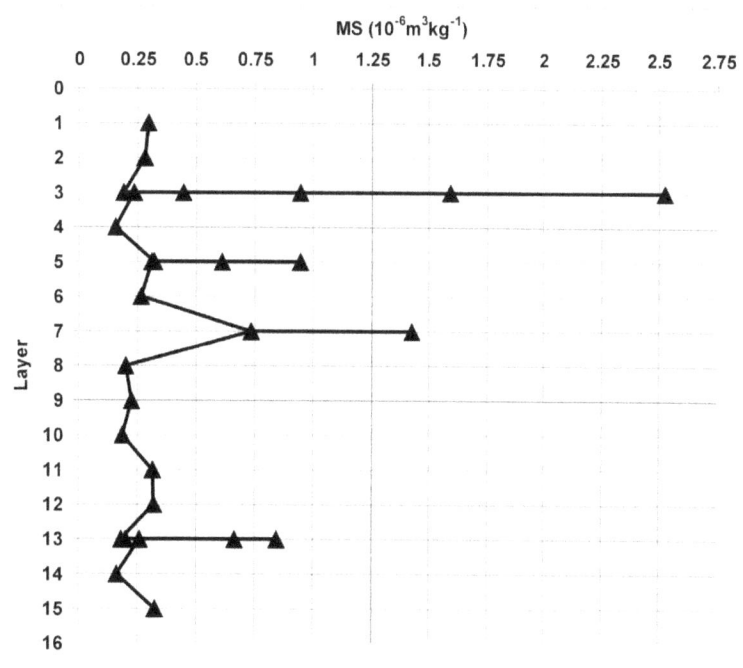

Figure 5. Increase in MS for heavily occupied layers from Sopeña Cave (Asturias, Spain) caused by anthropogenic alteration due to the use of fire.

to 'burnt' stratigraphic units identified during excavation. Between these burnt units, lower MS values are recorded that represent the mixing of magnetically strong phases into magnetically weak sediments. This will be in part due to the movement of humans around the floor of the cave and can be used to delineate movement and occupation zones within the cave.

Palaeoclimatic reconstruction

Once an understanding of the effect of anthropogenic alteration on cave deposits has been established it is possible in some cases to recover palaeoclimatic information from Stone Age cave sites. In situ MS measurements were made for the entire excavated sequence at Rose Cottage Cave in the Free State of South Africa (Figure 7).

Figure 6. Three-dimensional GIS model of MS at the MSA bearing Pinnacle Point Cave 13B, Western Cape, South Africa. The spheres represent MS values (large sphere = high MS) plotted against stratigraphic units defined as burnt during excavation. (image by Erich Fisher, University of Florida).

The sequence covers the pre-Howiesonspoort (HP) MSA, HP, post-HP MSA, transitional MSA/Late Stone Age (LSA) and LSA period. Samples were additionally taken from each major layer within the sequence for laboratory analysis. A series of standard mineral magnetic tests (as per Walden et al. 1999) were undertaken on these laboratory samples to assess the reason for MS changes seen in the in situ sequence. Hearths are easily identified as they have very high MS values. Moreover, they are often well delineated in the upper sequence and so can be avoided during sampling.

Until recently, the older MSA sequence (>40ka) had no reliable age estimates, but is important as it contains a HP occupation. The base of the sequence has been dated to 70 ka using thermoluminescence (Valladas et al. 2005). Magnetic susceptibility variations of these deposits also suggest that the base of the MSA deposits were deposited at the transition from OIS 5a to OIS 4 (~68 ka). During the OIS 4 period (~58–68 ka) lower magnetic susceptibility values are recorded due to a decrease in the input of fine-grained ferrimagnetic minerals and increased influence from diamagnetic and paramagnetic weathered sandstone fractions. The layers represented by the beginning of OIS 4 (~68 ka) contain the HP industry and are associated with spikes in MS related to in situ hearths. In situ hearths are easily identifiable in all layers of the sampled sequence in which they occur, as the MS values are between two and three times the value of the background values (Figure 7). A slow increase in MS represents the change from OIS 4 to OIS 3 with peak MS occurring around 50 ka. After this a fluctuating climate occurs before the MS values reach their lowest values around 14 to 20 ka during the Last Glacial Maximum. MS values then increase again to a maximum during the Holocene, around 7–5 ka. The MS sequence at Rose Cottage Cave shows up not only major changes in climate, such as the transition from more glacial to interglacial stages, but also small-scale fluctuations within the Holocene including an event at around 9.2-9.6 ka.

At Sibudu Cave, in KwaZulu-Natal, a palaeoclimatic sequence has also been recovered. However, the processes and mechanisms behind the palaeoclimatic system are different (Herries 2006). The upper sequence covers the period between >60 ka and 40 ka (Wadley and Jacobs 2006). Towards the base of this sequence a distinct mineral magnetic transition is noted with a change from sediments dominated by derived wind-blown soils containing high proportions of ultra-fine-grained magnetite and maghaemite in the top section to sediments in the base with little derived soil and an increased amount of haematite from weathering of the sandstone rockshelter. This

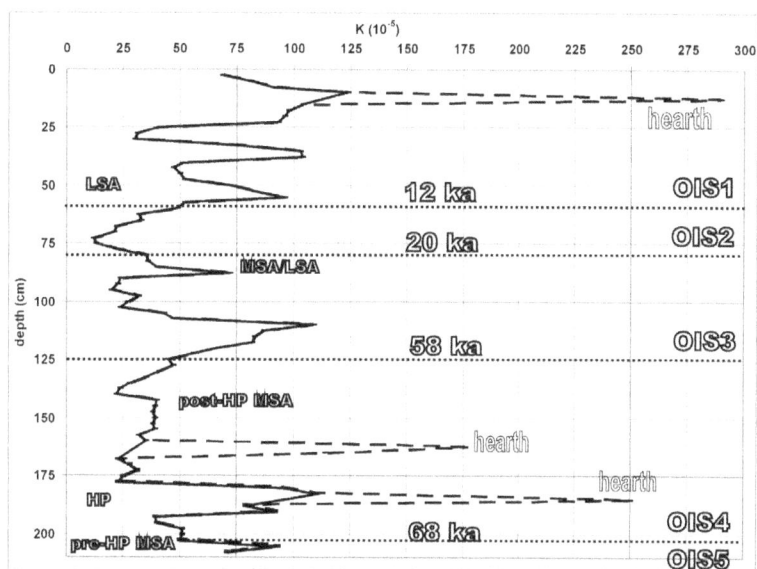

Figure 7. In situ magnetic susceptibility (K x10-5) curve versus depth (cm), age and stone tool industries from an excavated section at Rose Cottage Cave over the last 70 ka. Spikes in MS that are related to in situ hearths are also shown. The cave has a pre-Howiesoon-spoort MSA (pre-HP MSA), Howiesoonspoort (HP), post-Howiesoon-spoort MSA (post-HP MSA), MSA to LSA transitional industry (MSA/LSA) and a number of LSA industries.

transition occurs at around 60 ka (Herries 2006) and is linked to changes in pedogenesis, wind patterns and deposition of derived soils within the cave between the colder, more arid, low MS, OIS 4 and the warmer, high MS OIS 3. Ongoing analysis on the base of the MSA sequence again suggests that a Howiesonspoort industry occurs within OIS 4 (Wadley and Jacobs 2006). Because this site was heavily occupied by humans the sediments have undergone extensive anthropogenic alteration in certain areas due to fire use. As such, small changes and spikes in MS are not interpreted as climatically determined (Herries 2006). However, major climatic events such as the OIS 4 to 3 transition are still noted in multiple sections and show that the climatic signal is not entirely overprinted (Herries 2006).

Sourcing and behaviour

Ochre is found in MSA sites in South Africa (Marean et al.. 2007). As this material is highly magnetic it is also important to assess its contribution to the magnetic signal when conducting detailed archaeomagnetic studies of an archaeological cave and its deposits. Hysteresis loops, isothermal remanent magnetisation (IRM) acquisition curves and thermomagnetic curves were conducted on archaeological ochre from a MSA cave site near Mossel Bay using a variable field translation balance (VFTB). The samples also underwent X-ray diffraction experiments at Arizona State University. In most cases the mineral magnetic character is much more distinct than X-ray diffraction spectra and allowed a number of distinct magnetic ochre types to be distinguished. Figure 8 shows thermomagnetic curves for three ochre samples from the site. One sample has a Curie point (Tc) of 680°C and indicates the dominance of haematite. Another sample has a Tc of 575°C that indicates the dominance of magnetite. The final sample has a Tc of 620°C and a drop in magnetisation on heating that suggests the dominance of maghaemite. IRM acquisition curves and hysteresis loops indicate that while most red ochre samples contain haematite, some also contain magnetite and/or maghaemite. Further analysis utilising hysteresis loops and IRM acquisition curves can provide distinct mineralogies that can further characterise the ochre deposits. Mooney et al. (2003) conducted successful sourcing of ochre from Australia using combined

X-ray diffraction and mineral magnetic methods to help characterise ochre. This work shows the great potential for sourcing ochre from Stone Age sites and therefore recovering valuable behavioural data on group ranges or even trading networks.

Palaeomagnetism

Geomagnetic field variation

Geomagnetic data consists of palaeo-directional measurements in the vertical (inclination) and horizontal (declination) plain as well as the intensity of the Earth's field. Unlike mineral magnetic studies, palaeomagnetic studies need to be undertaken on oriented samples if palaeo-directional or taphonomic data are to be recovered. However, for palaeo-intensity studies, samples need not be oriented. Perhaps the best known application of palaeo-magnetic studies to archaeology is the dating of hominin bearing deposits using magnetostrati-graphic analysis (Tamrat et al.

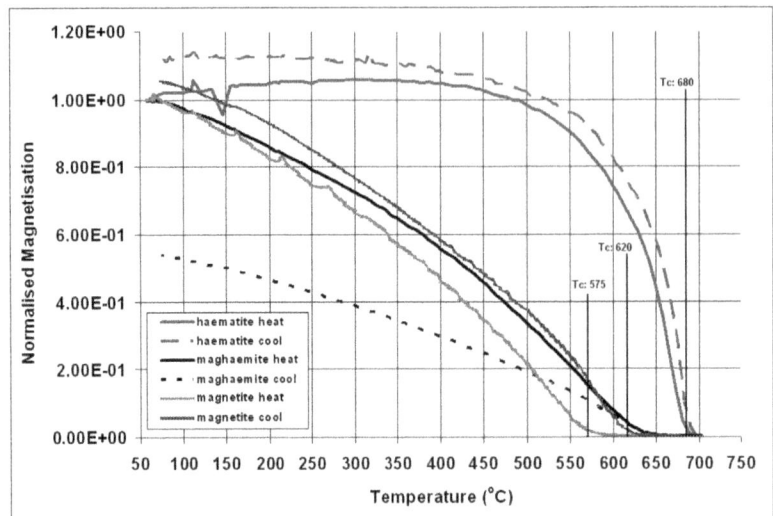

Figure 8. Thermomagnetic curves for three ochre samples from Pinnacle Point Cave 13B. One sample has a Curie point (Tc) of 680°C and indicates the dominance of haematite. One sample has a Tc of 575°C that indicates the dominance of magnetite. The final sample has a Tc of 620°C and a drop in magnetisation on heating that suggests the dominance of maghaemite.

1995; Lacruz et al. 2002; Zhu et al. 2003; Herries et al. 2006a, b). Reversal magnetostratigraphy works on the basis of comparing the fossil polarity of sediments on archaeological and fossil sites with the globally known variation in the Earth's magnetic field (Cande and Kent 1995; Ogg and Smith 2004).

Even when the dipole field has a steady polarity it undergoes swings in direction and variation in its intensity (Ogg and Smith 2004). This fluctuation, or 'secular variation', of the geomagnetic field over stable polarity time periods is also much less well defined, especially for earlier archaeological time periods and outside of Europe and the United States. Secular variation studies can be done using either the palaeo-direction and/or the palaeo-intensity of magnetisation of sediments, speleothems, lavas, or more normally in archaeological studies, burnt materials (Openshaw et al. 1993, Herrero-Bervera and Valet 2002; Herries et al. 2007a, b). When these were deposited or cooled they acquired a magnetic remanence (depositional remanent magnetisation or thermoremanent magnetisation) in the same direction as the Earth's ambient magnetic field. This remanence also has an intensity of magnetisation that is proportional to the strength of the magnetic field at that time. As the Earth's magnetic field gradually changes both direction and intensity, the direction and intensity of samples from the site can be compared and hence dated by comparison with the recorded direction and field intensity for past times at that locality, if they exist. In Europe, archaeomagnetic studies have been undertaken to develop regional archaeomagnetic curves that can be used to date archaeological sites and features (e.g. Kovacheva 1997; Herries et al. 2007a, b). As a dating method 'archaeomagnetic dating', as it is traditionally termed, has a typical accuracy of a few hundred years and an age range back to around 8 ka (Kovacheva 1997), depending on the length and completeness of the archaeomagnetic record for the region under study. Unfortunately, for much of the world, especially the southern hemisphere and over most of the Stone Age time period, no comparative geomagnetic secular variation curve currently exists. The best records for this time period come from lake or marine sediments, where many post-depositional effects occur, or from lava sequences (Herrero-Bervera and Valet 2002; Roberts 2006).

Archaeological sites provide a prime resource for the recovery of geomagnetic field data. In the southern hemisphere, particularly South Africa, the lack of a long term record of kiln and stone hearth construction has led to an inability to create an archaeomagnetic reference curve. The recent development of the microwave palaeointensity technique (Böhnel et al. 2003) enables much smaller samples, such as pottery, to be utilised to record geomagnetic field variation. As such, there is the potential in South Africa to create an archaeointensity record that covers at least the last 2000 years. To extend beyond this time period the only potential data are from Stone Age hearths. Heat-retainer hearths, as used in Australia, are ideal for this type of analysis and preliminary work in this area has been conducted by Barbetti (1977) and Barton and Barbetti (1983). However, few built hearths have been noted from southern Africa. Recent research has been undertaken on burnt rocks from South African Stone Age sites. These studies have a number of values: 1) the date recovered from in situ burnt rocks provides evidence for geomagnetic field changes over a much longer time range; 2) the method can identify burnt rocks that are in situ and their maximum temperature of heatinng; 3) this can be used along with magnetic mineralogical studies of burnt sediments to create a powerful tool for reconstructing site taphonomy.

Palaeomagnetism and fire identification

When sediments are deposited the grains orient themselves in the direction of the Earth's magnetic field and produce a depositional remanent magnetisation (DRM). This is retained after lithification with small changes occurring on dewatering and compaction, to form a post-depositional DRM (pDRM). Sand is not a good recorder of DRM due to the weakness of quartz and also its susceptibility to movement. However, sandstones can preserve a secondary chemical remanent magnetisation (CRM) due to the formation of haematite during lithification and weathering. This CRM will record the direction of the Earth's magnetic field at the time of its alteration. When blocks fall off the wall of a rockshelter they lose their orientation and their direction of remanence now lies in a random direction. When the rocks are incorporated into a fireplace, either deliberately or accidentally, and are heated they will acquire a thermoremanent magnetisation (TRM) in the direction of the ambient geomagnetic field (Figure 9).

Figure 9. A rock-bound hearth from Western Cape. Before the rocks are heated their geological palaeodirection is in independent directions. When the rocks are heated a TRM is induced in the sample and their palaeodirection should be in the same direction. A sample outside the hearth will retain its original geological palaeodirection.

If heated to above the Curie temperature (Tc) of the main remanence-carrying mineral, haematite (680°C) or magnetite (575°C), the CRM will be completely overprinted with a TRM and the rock will retain only a single field direction. An unheated rock will also record only a single field direction, however, it will have a weak CRM and its direction will be random. In contrast, a heated rock will retain a strong, single component in a normal (towards North) magnetic field direction. If the rock is not in situ it can still be distinguished from an unheated rock by its high intensity, as the strong ferrimagnetic mineral phase, magnetite (Tc = 575°C), is normally formed during heating. No heating temperature information can be recovered other than that it was heated to above the Curie point. If heated to below the Tc, a partial thermomagnetic remanence (pTRM) is induced. Heating to temperatures below the ancient temperature of heating (ATc) will thermally demagnetise the pTRM. The temperature step at which the pTRM is removed and the magnetic vector changes to the randomised CRM of rock is the estimated last temperature of heating. Therefore, subsequent heating of the rock in the laboratory in small temperature increments (25–50°C) and measurement between these temperature steps can help determine the maximum temperature of heating that the rock or stone artefact has experienced.

A series of experiments has been undertaken to determine the suitability of the palaeomagnetic vector method to different rock types and situations. Gose (2000) showed the potential of this method to stone-bound hearths during more recent periods in the United States. A series of bedrock sandstone samples from Carden Park Rockshelter was burnt in a campfire at the University of Liverpool to see if this material retained a measurable, stable TRM from heating in a hearth. The sandstone samples retained a normal geological magnetic polarity. The samples were placed into the fire in various positions and the campfire was stoked for one and a half hours. The maximum temperature of heating of various parts of the hearth was monitored with thermocouples to confirm the accuracy of the maximum temperature of heating method (Figure 10). Sample BSSB1A came from the surface of the sample and was placed at 90 degrees to the normal geological remanence of the sample and in the same vertical plain. The mean modern overprint that is recorded in the samples was 354 degrees declination by +65.8 degrees inclination. The modern field at the location was 355.5 degrees by +67.9. This suggests that sandstone fulfils the first requirement of recording the current Earth's field accurately. Thermocouples gave a maximum temperature for the fire above the burnt rock of 395°C. The base of the rock gave a maximum temperature of 315°C. The estimate for the maximum temperature of heating using the palaeomagnetic vectors method was between 300 and 400°C for all the samples (these are the temperatures between which the magnetic components switch direction; Figure 10). This suggests that this method is reliable for determining both the direction of the ancient magnetic field and the approximate temperature of last heating that has occurred. By undertaking thermal demagnetisation in smaller temperature stages and by paying careful attention to the location of the rocks and samples within the morphology of the fireplace, this estimate could be refined.

Studies of a modern hearth from Makapansgat, in the Limpopo Province of South Africa, have shown that burnt quartzite can successfully record palaeo-directional data , although the samples are often so weak that a very sensitive magnetometer is needed. The modern field direction at Makapansgat is 346.2 degrees declination and −54.7degrees inclination (IGRF reference field accessed through the British Geological Survey). Burnt rocks from the modern fireplace at Makapansgat record a mean direction of 345.9 degrees declination and −55.1 degrees inclination (Table 2). A third rock (MAKMOD3; Table 2) from the fireplace was not included in the mean as it had obviously been moved since heating. A vector change occurs between 350 and 400°C and indicates that the rocks were heated to a maximum temperature of 350°C (Figure 11). The rocks therefore record both a pTRM from the modern heating and the original geological remanence. These data further support the ability of this method to record reliable palaeodirections and palaeotemperatures in different rocks in different parts of the world.

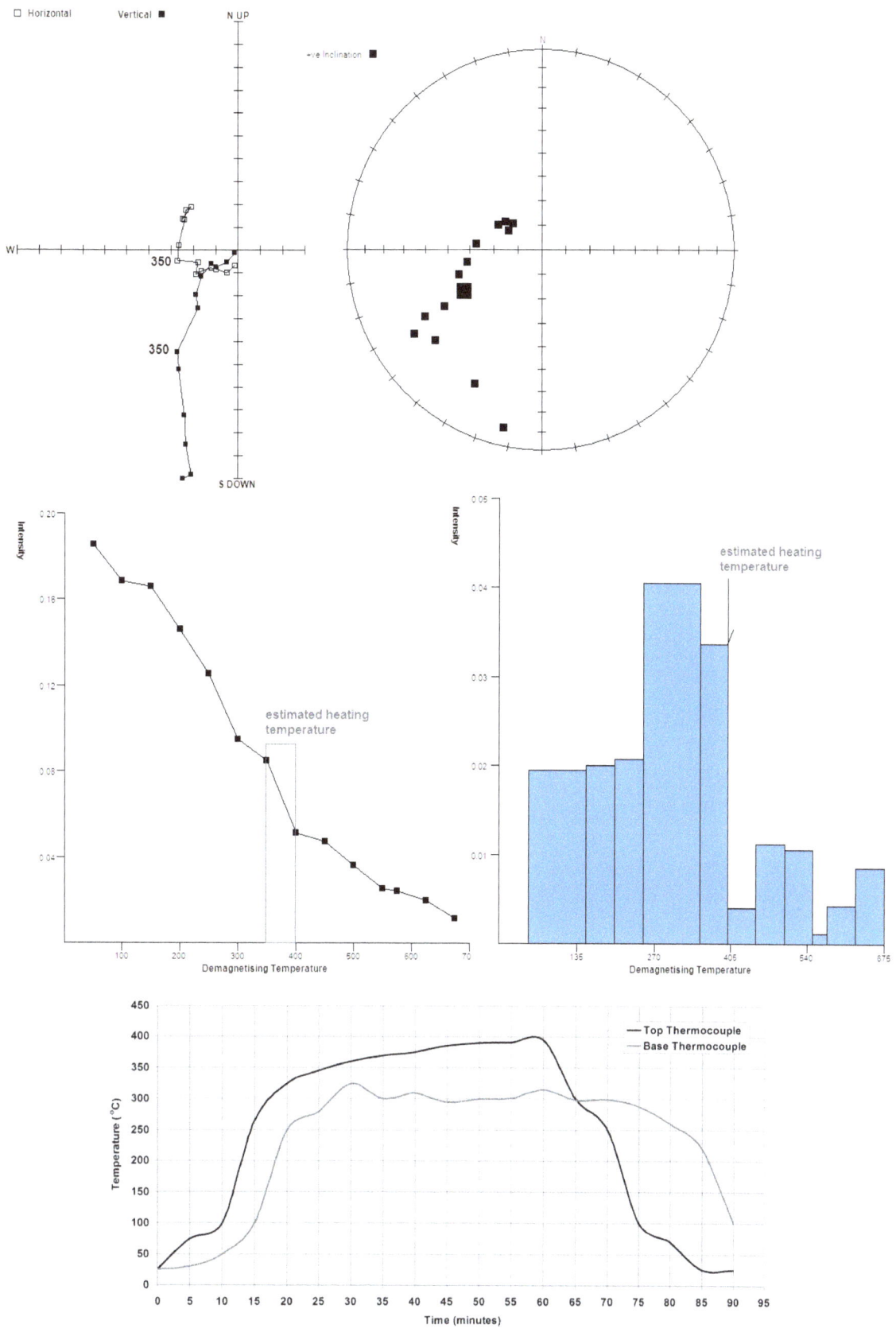

Figure 10. Magnetic data for a sample from an experimental hearth at the University of Liverpool, U.K: a) orthogonal plot; b) stereo plot; c) intensity plot; d) spectrum plot; and e) a plot of time versus temperature for thermocouples associated with the burnt rock (above and below).

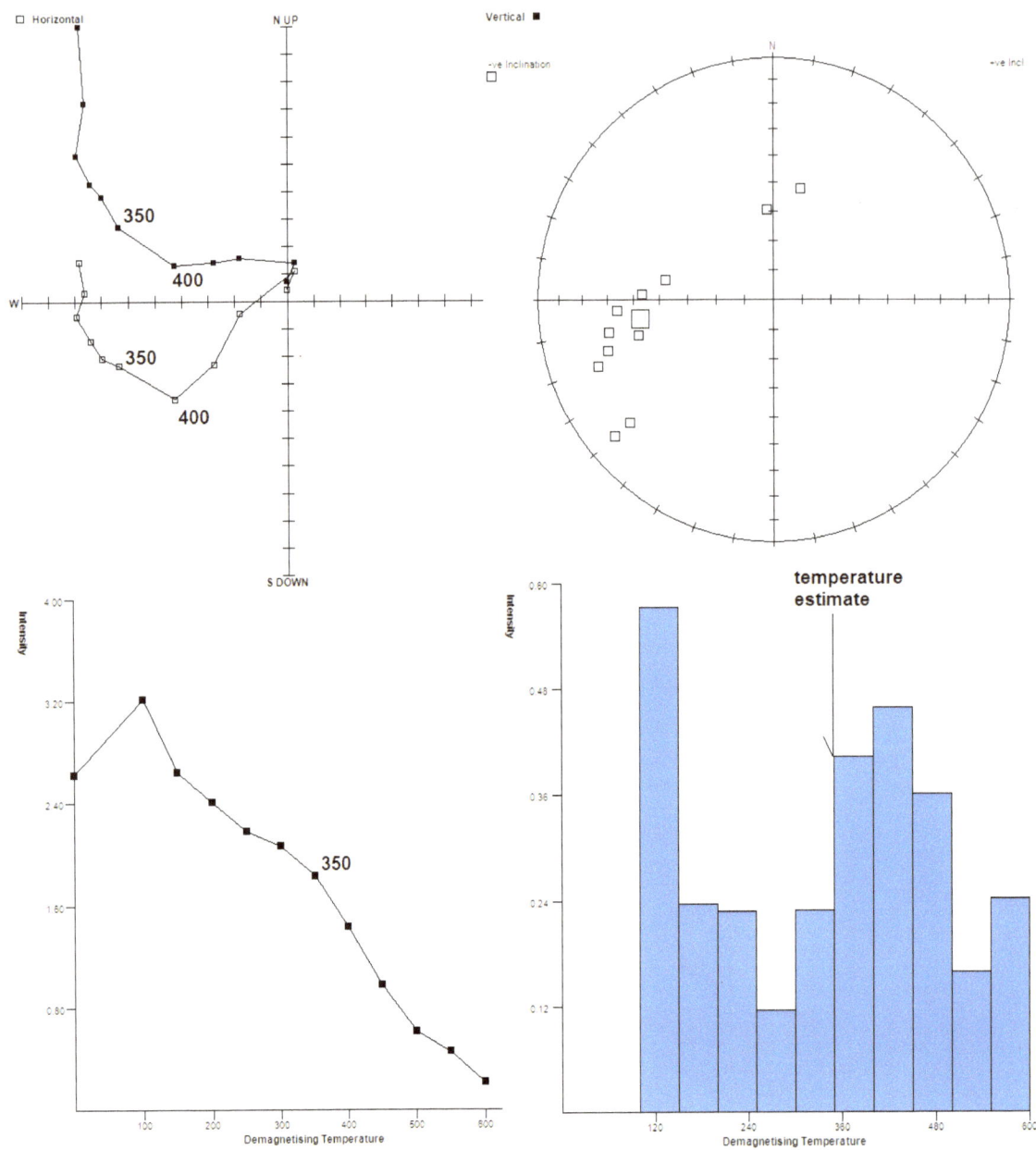

Figure 11. Magnetic data for a sample for a modern hearth at Makapansgat in Limpopo Province, South Africa.

However, recovering reliable data from Stone Age sites is dependent on a number of factors. How successfully the geomagnetic field is recorded will be partially dependent on the morphology of the fireplace. Rocks that surround fireplaces have the tendency to be moved after or between heatings and so many do not retain the correct modern field direction. This is shown by the unreliable directions from rock MAKMOD3 (Table 2). Those that lie within the fire appear to generally be more reliable (MAKMOD1; Table 2) but if too small may be moved during restocking of the fire. Those that are embedded in sediments underneath fireplaces, either as heat retainers or accidentally heated, will generally record the field more accurately as they are not subjected to movement, but they will be heated to a lower temperature than those directly in the hearth (MAKMOD2). Thus, the biggest obstacle to recovering secular variation data from Stone Age sites is normally the size of the burnt rocks along with the weakness of the host rock, the morphology of the hearth and the rocks location within it. However, while palaeodirectional data may not be recoverable, it is still possible to identify a number of things from the burnt rocks. If the samples have been heated to less than the Tc of the remanence-carrying mineral (normally magnetite, 575°C), then both the pTRM induced in the sample from heating and the original geological remanence can be separately identified as with sample MAKMOD3 (Table 2). Also, not all rock types are amenable to this method, mainly due to the lack of magnetic material or the inability to measure such weak samples. Studies of flint from the Middle Pleistocene site of Beeches Pit in the U.K (Haritou 1996; Gowlett 2006), and limestone from the late Pleistocene site of Sopeña in Spain have shown that not all rocks are ideal for this method of analysis.

At Sibudu Cave, in the Kwazulu-Natal, a number of firing events have been discovered throughout the MSA layers of the site (Herries 2006; Wadley and Jacobs 2006). These occur as both discrete, shallow fire pits and as simple scatters of ash and coloured sediment. By looking at the fossil directions of magnetisation within a number of sandstone blocks recovered from these features and horizons, it was hoped to identify the primary heating contexts from secondary ash scatters. A number of rocks taken from 'ashy layers' in the trial trench showed no evidence of burning with a weak NRM and a weak single component of magnetisation. FTIR analysis of the ashy material identified it as gypsum. Naturally forming gypsum nodules were then noted in a number of layers at the site. These gypsum nodules formed during periods of hiatus at the site, as shown by rapid jumps in magnetic susceptibility and OSL dates (Herries 2006; Wadley and Jacobs 2006). However, thermal demagnetisation of sandstone from a number of fire pits at the site gave meaningful directions that estimated the maximum temperatures of heating to be around 450–500°C in some cases (Figure 12). Two separate samples from the same fireplace also gave relatively reliable palaeodirection results (Table 2: SIBBGY1 and SIBBGY2; Figure 12a, 12b) A burnt dolerite tool (SIBBDT1; Table 2) from the site was also sampled and while it did not give a reliable palaeodirection, it did give a consistent temperature of heating around 500°C (Figure 12c). As such, the method can also be used to look at the heat treatment of stone tools (Brown et al., 2008).

Sample	Dec.	Inc.	MAD	Palaeotemp
CARP1a	353.6	65.3	5.7	350-400
CARP1b	352.4	69.3	6.9	350-400
CARP2a	356.5	63.4	8.3	300-350
CARP2b	349.9	65.4	7.2	300-350
CARP3a	357.4	66.3	5.4	350-400
CARP3b	354.2	65.2	4.4	350-400
mean	354.0	65.8		
Modern	355.5	67.9		
MAKMOD1a	344.4	-54.6	7.6	350-400
MAKMOD1b	346.4	-58.2	5.6	350-400
MAKMOD2a	341.5	-52.3	6.7	300-350
MAKMOD2b	351.3	-55.4	4.2	300-350
MAKMOD3a	312.2	-44.3	6.7	300-350
MAKMOD3b	302.9	-40.5	8.2	300-350
mean	345.9	-55.1		
Modern	346.2	-54.7		
SIBBDT1	351.7	-3.9	3.4	450-500
SIBBRSP1	30.1	-43.8	5.6	400-450
SIBBRSP2	31.6	-39.6	6.7	400-450
mean	30.35	-41.7		

Table 2. Palaeodirectional data for burnt rock samples from an experimental fire at Carden Park (CARP), a modern fire at Makapansgat (MAKMOD) and MSA samples from Sibudu Cave (SIBB).

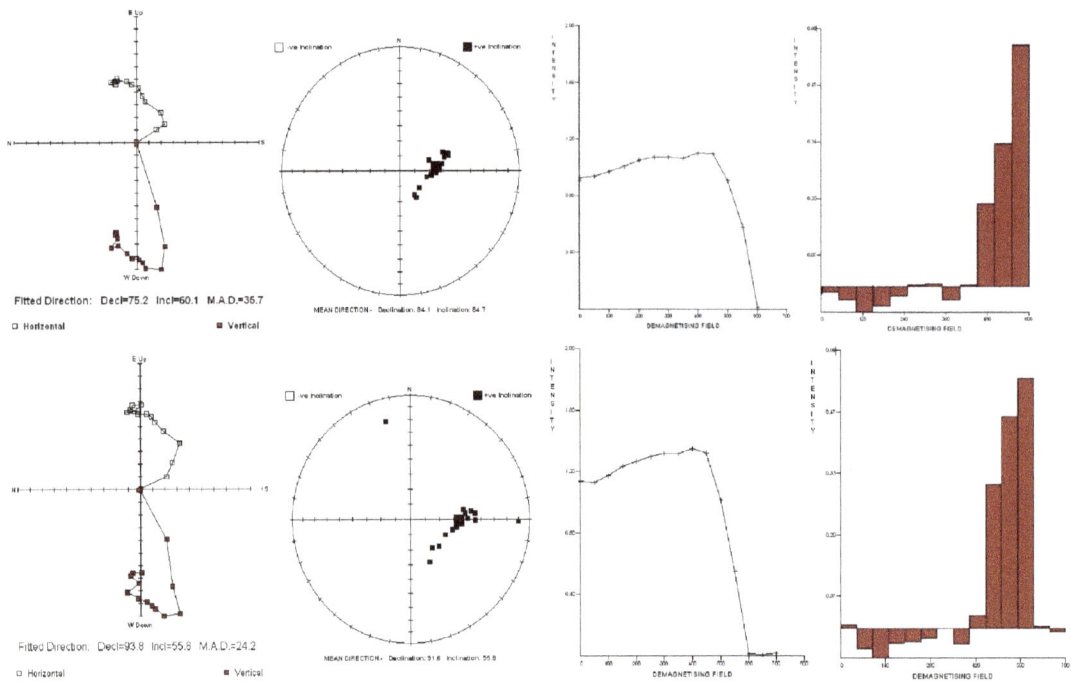

Figure 12. Magnetic data for two burnt rocks from a MSA hearth (~50 ka) at Sibudu Cave, KwaZulu-Natal, South Africa.

Conclusions

This paper illustrates the potential of magnetic methods of analysis, subsumed under the term archaeomagnetism, for recovering information useful for archaeologists including:

- dating
- palaeoclimatic reconstruction
- behaviour:
    1. evidence for fire use
    2. spatial patterning of occupation
    3. heating temperature and taphonomy of burnt rocks and stone tools
    4. sourcing of artefacts.

Acknowledgments

All archaeomagnetic work was undertaken at the Liverpool University Geomagnetism Laboratory (U.K.) with the support of Alf Latham (Archaeology) and John Shaw (Earth Sciences). Work on the South African hominin sites and work at Sibudu and Rose Cottage caves and the Cave of Hearths was funded by the Arts and Humanities Research Board (U.K.) studentship number 99/3050. The work at Pinnacle Point was conducted as part of the Mossel Bay Archaeological (MAP) and SACP4 projects, run by Curtis Marean of the Institute of Human Origins, Arizona State University and funded by the National Science Foundation (USA) (grants # BCS-9912465, BCS-0130713, and BCS-0524087 to Marean), the Hyde Family Trust, the Institute of Human Origins, and Arizona State University. Thanks to the South African Heritage Resources Agency (SAHRA) and Heritage Western Cape (HWC) for providing permits to conduct excavations at the selected sites and export specimens for analysis. Thanks to Kevin Kuykendall, Lyn Wadley, Curtis Marean, Anthony Sinclair and Darren Curnoe for support and access to archaeological sites.

References

Barbetti, M. 1977. Measurements of recent geomagnetic secular variation in southeastern Australia and the question of dipole wobble. *Earth and Planetary Science Letters* 36:207–218.

Barton, C. E. and M. Barbetti. 1982. Geomagnetic secular variation from recent lake sediments, ancient fireplaces and historical measurements in southeastern Australia. *Earth and Planetary Science Letters* 59:375–387.

Bellomo, R. V. 1994. A methodological approach for identifying archaeological evidence of fire resulting from human activities. *Journal of Archaeological Science* 20:525–555.

Begét, J. E. 2001. Continuous Late Quaternary proxy climate records from loess in Beringia. *Quaternary Science Reviews* 20:499–507.

Böhnel, H., A. J. Biggin, D. Walton, J. Shaw and J. A. Share. 2003. Microwave palaeointensities from a recent Mexican lava flow, baked sediments and reheated pottery. *Planetary Science Letters* 214:221–236. Brown, K., Marean, C. Robert, D., Herries, A., Braun, D. Tribolo, C. Jacobs, Z.. Atanasova, M. (2008) First documentation of heat treatment of silcrete in the Howieson's Poort and still Bay Middle Stone Age industries, Southern Cape, South Africa. Paleoanthropology 2008: A5.

Brown, K., Marean, C. Roberts, D., Herries, A., Braun, D. Tribolo, C. Jacobs, Z.. Atanasova, M. (2008) First documentation of heat treatment of silcrete in the Howieson's Poort and Still Bay Middle Stone Age industries, Southern Cape, South Africa. *Paleoanthropology* 2008: A5.

Cande, S. C. and D. V. Kent. 1995. Revised calibration of the geomagnetic polarity timescale for the Late Cretaceous and Cenozoic. *Journal of Geophysical Research* 100:6093–6095.

Canti, M. G. and N. Linford. 2000. The effects of fire on archaeological soils and sediments: Temperature and colour relationships. *Proceedings of the Prehistoric Society* 66:385–395.

Dalan, R. A. and S. K. Banerjee. 1998. Solving archaeological problems using techniques of soil magnetism. *Geoarchaeology* 13:3–36.

Ellwood, B. B., K. M. Petruso and F. B. Harrold. 1997. High-resolution paleoclimatic trends for the Holocene identified using magnetic susceptibility data from archaeological excavations in caves. *Journal of Archaeological Science* 24:569–573.

Gose, W. A. 2000. Palaeomagnetic studies of burned rocks. *Journal of Archaeological Science* 27:409–421.

Gowlett, J. A. J. 2006. The early settlement of northern Europe: Fire history in the context of climate change and the social brain. *Comptes Rendus Palevol* 5:299–310.

Haritou, S. J. A. 1996. *Palaeomagnetism as an indication of hominid fire use: Analysis of flint and clay from Beeches Pit excavation 1996, West Stow, Suffolk, England.* Unpublished M.Sc. thesis, University of Liverpool, U.K.

Herrero-Bervera, E. and J.-P. Valet. 2002. Paleomagnetic secular variation of the Honolulu Volcanic Series (33–700 ka), O'ahu (Hawaii). *Physics of the Earth and Planetary Interiors* 133:83–97.

Herries, A. I. R. 2006. Archaeomagnetic evidence for climate change at Sibudu Cave. *Southern African Humanities* 18:131–147.

Herries, A. I. R., J. W. Adams, K. L. Kuykendall and J. Shaw. 2006. Speleology and magnetobiostratigraphic chronology of the GD 2 locality of the Gondolin hominin-bearing paleocave deposits, North West Province, South Africa. *Journal of Human Evolution* 51:617–631.

Herries, A. I. R and M. Kovacheva. 2007. Using archaeomagnetism to answer archaeological questions at the Thracian site of Halka Bunar, Bulgaria. *Archaeologia Bulgarica* 11, in 2007/3, 25-46.

Herries, A. I. R., M. Kovacheva, M. Kostadinova and J. Shaw. 2007. Archaeo-directional and -intensity data from burnt structures at the Thracian site of Halka Bunar (Bulgaria): The effect of magnetic mineralogy, temperature and atmosphere of heating in antiquity. *Physics of the Earth and Planetary Interiors* 162, 199-216.

Herries, A. I. R. and A. G. Latham. 2003. 'Environmental archaeomagnetism': Evidence for climatic change during the later Stone Age using the magnetic susceptibility of cave sediments from Rose Cottage Cave, South Africa. In P. Mitchell, A. Haour and J. Hobart (eds), *Researching Africa's past: New contributions from British archaeologists, pp* 25–35. *School of Archaeology Monograph* 57. Oxford: Oxford University.

Herries, A. I. R, K. Reed, A. G. Latham and K. L. Kuykendall. 2006. Speleology and magnetobiostratigraphic chronology of the Buffalo Cave fossil bearing palaeodeposits, Makapansgat, South Africa. *Quaternary Research* 66:233–245.

Kovacheva, M. 1997. Archaeomagnetic database from Bulgaria: The last 8000 years. *Physics of the Earth and Planetary Interiors* 102:145–151.

Lacruz, R.S, J. S. Brink, J. Hancox, A. S. Skinner, A. Herries, P. Schmidt and L. R. Berger. 2002. Palaeontology, geological context and palaeoenvironmental implications of a Middle Pleistocene faunal assemblage from the Gladysvale Cave, South Africa. *Palaeontologia Africana* 38:99–114.

Liu, X. M, P. Hesse, T. Rolph and J. E. Begét. 1999. Properties of magnetic mineralogy of Alaskan loess: Evidence for pedogenesis. *Quaternary International* 93:93–102.

Maguire, J. M. 1998. Makapansgat: A guide to the palaeontological and archaeological sites of the Makapansgat valley. Excursion guide for the Dual Congress of the International Association for the Study of Human Palaeontology and International Association of Human Biologists. Sun City, Republic of South Africa.

Maher, B. A. 1998. Magnetic properties of modern soils and Quaternary loessic paleosols: Paleoclimatic implications. *Palaeogeography, Palaeoclimatology, Palaeoecology* 137:25–54.

Maher, B. A. and R. Thompson. 1995. Paleorainfall reconstructions from pedogenic magnetic susceptibility variations in the Chinese loess and paleosols. Quaternary Research 44:383–391.

Mason, R. J. 1988. *Cave of Hearths Makapansgat, Transvaal. Archaeology Research Unit Occasional Paper* 21. Johannesburg: University of the Witwatersrand Press.

Marean, C.W., Bar-Matthews, M, Bernatchez, J., Fisher, E., Glodberg, P., Herries, A.I.R., Jacobs, Z., Jerardino, A., Karkanas, P., Minichillo, T., Nilssen, P.J., Thompson, E., Watts, I., Williams, H.W., 2007. Early Human use of marine resources and pigment in South Africa during the Middle Pleistocene. *Nature*. 449, 905-908.

Mooney, S. D., C. Geiss and M. A. Smith. 2003. The use of mineral magnetic parameters to characterize archaeological ochres. *Journal of Archaeological Science* 30:511–523.

Morinaga, H., H. Inokuchi, H. Yamashita, A. Ono and T. Inada. 1999. Magnetic detection of heated soils at Palaeolithic sites in Japan. *Geoarchaeology* 14:377–399.

Mullins, C.E. 1977. Magnetic susceptibility of the soil and its significance in soil science — a review. *Journal of Soil Science* 28:223–246.

Ogg, J. G. and A. G. Smith. 2004. The geomagnetic polarity time scale. In F. Gradstein, J. Ogg and A. Smith (eds), *A geologic time scale* 2004, pp 63–86. Cambridge: Cambridge University Press.

Openshaw, S. J., A. Latham, J. Shaw and Z. Xuewen. 1993. Preliminary results on recent palaeomagnetic secular variation recorded in speleothems from Xingwen, Sichuan, China. *Cave Science* 20:93–99.

Peck, J. A., R. R. Green, T. Shanahan, J. W, King, J. T. Overpeck and C. A. Scholz. 2004. A magnetic mineral record of Late Quaternary tropical climate variability from Lake Bosumtwi, Ghana. *Palaeogeography, Palaeoclimatology, Palaeoecology* 215:37–57.

Peters, C., M. J. Church and C. Mitchell. 2001. Investigation of fire ash residues using mineral magnetism. *Archaeological Prospection* 8:227–237.

Peters, C., R. Thompson, A. Harrison and M. J. Church. 2002. Low temperature magnetic characterisation of fire ash residues. *Physics and Chemistry of the Earth* 27:1355–1361.

Roberts, A.P. 2006. High-resolution magnetic analysis of sediment cores: Strengths, limitations and strategies for maximizing the value of long-core magnetic data. *Physics of the Earth and Planetary Interiors* 156:162–178.

Tamrat, E., N. Thouveny, M. Taieb and N. D. Opdyke. 1995. Revised magnetostratigraphy of the Plio-Pleistocene sedimentary sequence of the Olduvai Formation (Tanzania). *Palaeogeography, Palaeoclimate, Paleoecology* 114:273–283.

Thompson, R. and F. Oldfield. 1986. *Environmental magnetism.* London: Allen and Unwin.

Valladas, H., L. Wadley, N. Mercier, L. Froget, C. Tribolo, J. L. Reyss and J. L. Joron. 2005. Thermoluminescence dating on burnt lithics from Middle Stone Age layers at Rose Cottage Cave. *South African Journal of Science* 101:169–174.

Wadley, L. and Z. Jacobs. 2006. Sibudu Cave: Background to the excavations, stratigraphy and dating. *Southern African Humanities* 18:1–26.

Weiner, S., Q. Xu, P. Goldberg, J. Liu and O. Bar-Yosef. 1998. Evidence for the use of fire at Zhoukoudian, China. *Science* 281:251–253.

Williamson, D., A. Jelinowska, C. Kissel, P. Tucholka, E. Gibert, F. Gasse, M. Massault, M. Taieb, E. Van Campo and K. Wieckowski. 1998. Mineral-magnetic proxies of erosion/oxidation cycles in tropical maar-lake sediments (Lake Tritrivakely, Madagascar): Paleoenvironmental implications. *Earth and Planetary Science Letters* 155:205–219.

Zhu, R., Z. An, R. Potts and K. A. Hoffman. 2003. Magnetostratigraphic dating of early humans in China. *Earth-Science Reviews* 61:341–359.

17

The role of the conservator in the preservation of megafaunal bone from the excavations at Cuddie Springs, NSW

Colin Macgregor

Materials Conservation Unit
Australian Museum
6 College Street
Sydney NSW 2010

Abstract

Including a conservator on an excavation team can provide benefits to the archaeologist in addition to the improved preservation of excavated materials. Additional information can also be obtained from the material through conservation processing and cleaning on site. The nature and conditions of the site should be considered when deciding if a conservator is required on a team. The work of the site conservator is illustrated by the functions carried out by the author as conservator on the excavations by Dr Judith Field (1997–2005) at Cuddie Springs, NSW. During these excavations, the conservator carried out the consolidation, lifting, packing and treatment of crushed megafauna bones which contributed to more effective long-term survival of the material.

Introduction

In the Australia/Pacific Region, it is less common for conservators to be actively employed in archaeological excavations than is the case in North America or Europe. This results from a combination of factors: the nature of the materials, the degree of preservation of material excavated in the region, and the availability of trained archaeological conservators. Most archaeological materials (particularly organics and metals) have reached an equilibrium state with their burial environment which is disturbed upon excavation. Taking appropriate measures on site to halt rapid deterioration immediately can be crucial in the prevention of major damage occurring before the finds reach the laboratory.

This paper will outline some of the conservation techniques that should be employed on site and the benefits to the archaeologist. It will also provide a case study showing the application of specific techniques at Cuddie Springs.

The contribution of a conservator to the outcomes of any excavation can be considerably more than simply increasing the prospects of the long-term survival of the excavated material (O'Connor 1996). Broadly the benefits for the excavation may be:

- successful lifting of fragile material for analysis
- increased amount of information accessed from excavated material
- easier post-excavation management
- long-term survival of trace materials for analysis.

Given these benefits, the principal aim of the conservator is to ensure the survival of the material in the best possible condition whilst assisting in unlocking the information that an object has to offer (Foley 1984).

Part 1: Benefits of on-site conservation

Slotting the conservator into the excavation
Is a conservator required at all on site? This question is best answered by assessing the nature of the site. When required, the conservator's role should be defined at the planning stage, in order to ensure that the conservator can deliver satisfactory outcomes and be effectively integrated into the excavation team (Cronyn 1990:10–11).

Planning conservation
The questions to consider when planning the conservation resources for an excavation are what materials may be found and what is their likely state of preservation.

Materials: The period, culture and function of the site should be considered, in order to anticipate the types materials that are likely to be excavated.

Condition: An appraisal of the likely physical nature and chemistry of the site before the excavation commences allows the archaeologist to anticipate the types of material that will survive and the expected condition of them.

The exact details of what will have survived on the site and the quality of preservation is impossible to predict with certainty. There are many variables relating to the post-depositional environment that may have affected the buried materials, such as changes in the moisture content of the sediment.

Examples of types of sites and influence on survival of finds:
- damp-acidic environment: poor survival of metals and bone
- damp-alkaline environment: good survival of metals and bone but little organic material
- waterlogged sites: good survival of organic materials and metals if conditions are anaerobic and not alkaline.

Identifying resources required
Using information about the nature of the site, a conservator can advise on the materials and equipment that may be required on site. These will be used for:
- stabilization and consolidation of finds
- lifting techniques for fragile materials
- storage systems for finds
- transportation from site

On-site conservation strategies
Armed with this information the archaeologist can decide, in consultation with a conservator, whether a trained conservator is required on site. If not, the finds co-ordinator may handle the first-aid for objects and storage (Leigh 1981: 3–7) and call in a conservator if unexpectedly difficult problems arise.

The work carried out by a conservator in the field can assist the archaeologist by both unlocking information about the finds and improving the prospects of the survival of the objects with the minimum of deterioration. Examples include:

Consolidation techniques

Fragile and crushed bones may require strengthening before lifting can be attempted by applying a reversible water-based polymer in situ.

Lifting techniques

Soft waterlogged wood structures such as trackways can be supported by a semi-rigid sheet such as steel plate inserted through the soft matrix (Coles 1980).

Fragmentary materials such as bone or pottery may have to be encapsulated in a plaster or a foam jacket which will form a rigid cradle for the object after lifting.

Halting deterioration and storage

Metals can oxidize and deteriorate rapidly when exposed to the air so transfer to sealed containers with dry silica gel will remove the moisture and halt deterioration (Sease 1987:68).

Cleaning and stabilisation on site

Cleaning coins on site will reveal sufficient details to assist with the dating of contexts and interpretation of the site.

Reconstruction on site

During longer excavation seasons, reconstructions of profiles of pottery can enable the pottery to be drawn and recorded during excavation and facilitate the identification of pottery styles and chronologies (Figure 1).

Figure 1. Reconstruction of pottery in the field.

Packing of finds for transport

Preparing well-padded boxing systems for finds that limit movement and cushion shock and vibration will increase the chances of materials arriving in good condition, particularly from areas with unsealed roads. Creating appropriate micro-climates around unstable finds such as wood or metals will improve their chances of surviving the journey and storage until processed.

Planning the future of material

The long-term storage of the finds should also be considered before the excavation proceeds. Any site which produces a high volume of finds will require suitable storage space. Some materials will need an additional investment of funds to buy suitable storage units. For example, metals will require sealed polyethylene boxes and silica gel to keep the humidity low around the objects, whereas waterlogged wood requires tanks and biocides for storage prior to treatment. The cost of conserving large waterlogged objects is so great that reburial has often been the preferred option in recent years unless the cultural significance of the object can attract the necessary funding (Goodburn-Brown and Hughes 1996).

Part 2: A case study: Lifting and conservation of megafauna bones at Cuddie Springs, NSW

The site of Cuddie Springs is located in north west New South Wales (Field and Dodson 1999). It is a claypan that was formerly a lake, now silted up by drainage from the surrounding areas. It was recognised as a significant source of megafauna material during excavations by the Australian Museum in 1933 (Anderson and Fletcher 1934). The deposits containing bones are 10 metres deep. In 1991 further excavations (Furby et al. 1993) revealed the presence of stone tools in close proximity to the remains of Genyornis and Diprotodon in deposits dated at 28–35,000 years BP. The nature of these deposits makes the site significant to archaeologists interested in the causes of the demise of the megafauna.

The condition of the bones varied enormously between the oldest and most recent levels. In the lower levels the bone consisted of totally fossilised hard black material, whereas the upper archaeological levels contained delicate semi-mineralised dark-brown bones. The archaeological levels were 1.0–1.7 metres below the surface of the clay pan. In these levels, bones were more likely to have been encased in damp clay since deposition. The fine, even-grained silt formed an excellent support for the bones. The pH of the sediment was around 9. This alkalinity caused the organic components in the bone to deteriorate to the point beyond detection. The remaining inorganic structure had absorbed minerals from the surrounding soil, including manganese, resulting in a darkening of the material. Many of the bones were crushed, with multiple fractures but the nature of the clay had held the pieces together.

Preservation on site was generally only possible where the material was kept damp until it had been lifted and transported to the laboratory in Sydney. Rapid drying occurred in the surface layers and the clay shrank and hardened to a concrete consistency. The shrinkage caused physical disruption of the bone and the adhesion between the fragments was lost. Regular application of water using a hand-spray was required to keep the bones damp prior to lifting. Evaporation of moisture was also slowed by placing food-wrap (Glad® wrap) and aluminium foil over the bones, weighted down with miniature sandbags. Over night and during breaks the site was covered with polythene sheeting. This also aided the excavation by keeping the clay surface soft.

Bone material that had been fully exposed and supported by a pedestal of clay sometimes required 'facing-up' to hold the fragmentary surface together during lifting and transport. This was achieved by adhering strips of nylon gossamer tissue across the breaks (Figure 2). The adhesive chosen for the job was Plextol B500, a water-based dispersion of an acrylic co-polymer that has sufficient strength for this work but is also easily removable in the laboratory (Horton-James et al. 1991). On bone that had areas of soft or friable surfaces, especially the terminals of large bones, Primal WS24 was applied as an overall surface consolidant (Howie 1984). WS24 is an acrylic dispersion in water and on bone that is wet the polymer does not penetrate much beyond the surface. It is, however, an effective surface consolidant forming a skin that holds the surface together.

Figure 2. Consolidation of bone prior to lifting.

Lifting and Transport: The fragmentary nature of the bones embedded in a matrix of clay necessitated some form of block-lifting to ensure the pieces stayed together during lifting and transport. The jacket could have been formed from polyurethane foam or plaster of Paris. The first stage of the process was the same for both materials. A layer of food-wrap was spread across the bone. A layer of heavy-duty aluminium foil was then applied over this. Both were then stippled into the contours of the bones using a soft paint brush. The layer of food-wrap was an extra layer of protection to prevent leakage through pinholes in the foil. Polyurethane foam (Jones 1980) space filler, which can be purchased in aerosol cans at hardware stores, was sprayed onto the aluminium foil and allowed to expand and harden. It is not advisable to apply too much foam in any one application as foam trapped at the base will not set for days, causing dimensional movement in the support jacket.

Plaster of Paris and bandages are cheaper alternatives and faster setting materials, although more difficult to apply. In some cases the weight of the plaster makes it a preferred option. For example, the polyurethane foam will expand and push itself out of undercuts and narrow channels whereas plaster 'stays put' after application.

Once the jacket had set, the bone was removed by cutting through the clay beneath the pedestal. A selection of well-sharpened spatulae and knives with thin flexible blades were found to slice through the clay effectively (Figure 3). The jacket was then inverted with the spatula holding the bone in position.

Figure 3. Using a block of polyurethane foam to support the bone as it is lifted and inverted.

Treatment in the laboratory started with the removal of the remaining clay on the bones. This was carried out while the clay was still damp. Larger sections of clay were removed with a scalpel and dental tools. The smaller deposits that hid the fine detail in the surface were removed with a brush, hand-spray and dental aspirator. This drew a slurry of water and clay straight off the surface without allowing the water to penetrate and left a very clean surface (Figure 4).

The nylon gossamer applied on site to hold the fragments together was gently peeled off before consolidation. It was frequently detached from the surface manually. However, acetone was brushed under the edge to release the more stubborn pieces.

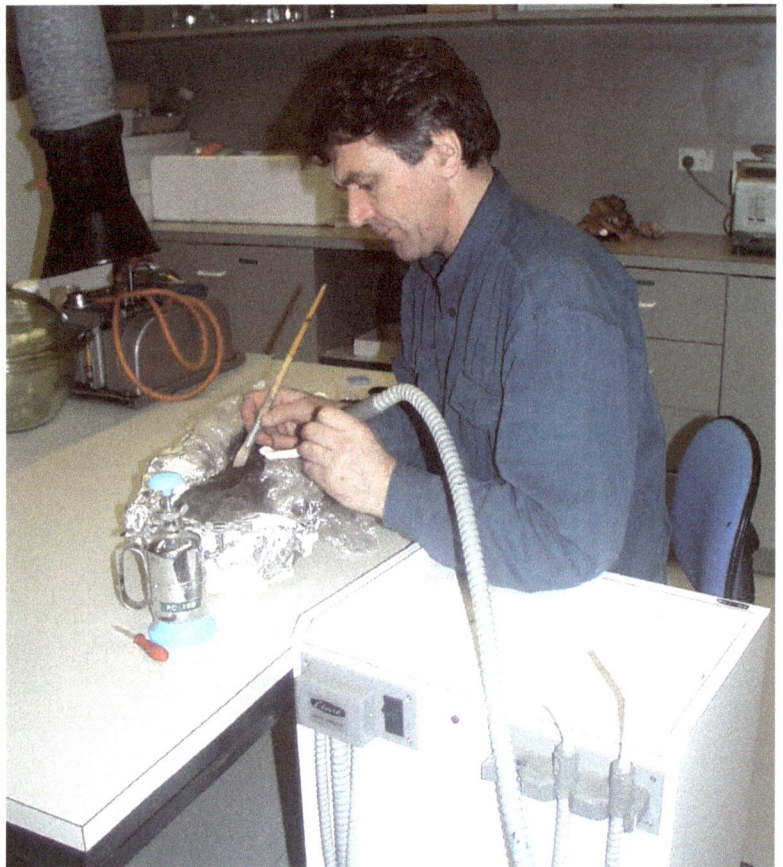

Figure 4. Cleaning bones using vacuum technique with water and dental aspirator.

The decision to consolidate the internal structure of the bone was made after assessing the strength of the remains. It was essential that all samples for analysis or dating were taken before treatment was carried out. Structural consolidation was best carried out with a polymer in an organic solvent solution. The high surface tension of the water made penetration by aqueous solutions problematic. It was therefore necessary to dry the bones slowly before consolidation. This resulted in a temporary weakening of the bones, as the adhesion of the damp clay was lost. Careful handling was crucial at this stage. The polymer chosen for consolidation was Paraloid B72, an acrylic co-polymer, that is recognized as one of the most stable resins available in conservation (Koob 1986). It also remains soluble over long periods, so that it can be removed again in the future, if required.

The three methods available for deep consolidation are:

- immersion in solution under vacuum
- immersion in solution without vacuum
- application of consolidant with pipette

Immersion under vacuum exerts considerable stress on a porous specimen as gas bubbles are drawn out by the reduction in pressure. This is particularly true with cancellous bone that will contain pockets of trapped air. For this reason vacuum was seldom used on the extremely fragile specimens from Cuddie Springs.

Immersion in solution without vacuum may also to lead to collapse of some bones, as the bone and clay matrix softens and loses all structural strength in the solution. Where danger from this exists, specimens should be wrapped in gauze bandage or supported in a cradle of foil before immersion. These can be removed with acetone after drying.

Application by pipette allows the consolidant solution to be drawn into the bone by capillary action. It is preferable to apply as much solution as possible with each application as the dried polymer resin will inhibit subsequent applications. This was the safest option and was used on the majority of the bones treated, although it did not necessarily result in as thorough consolidation as the immersion techniques.

Contoured supports for storage and handling the most fragile bones are made by adapting the polyurethane foam jackets that were applied for lifting on site, or sculpted from closed-cell polyethylene foam. The PE foam lasts a great deal longer than polyurethane foam, as it is more chemically stable.

Summary

In order to achieve the most effective conservation for excavation material and to support the research goals of the project it was necessary to:

- anticipate the likely materials that would be found
- consider the nature of the burial environment
- halt chemical and physical changes of finds immediately after excavation
- ensure treatments did not compromise analytical and dating goals of project
- consider the possible impact of treatment on future analytical techniques
- plan the destination of material for long-term storage.

Acknowledgements

Dr Judith Field and Dr Joe Dortch, University of Sydney, and the Australian Museum for photographs.

References

Anderson, C. and H. O. Fletcher. 1934. The Cuddie Springs bone. *The Australian Museum Magazine* 5:152–158.

Coles, J. 1980. Archaeology in the Somerset Levels 1979. In J. Coles (ed), *Somerset Levels Papers* 6:52–56.

Cronyn, J. 1990. *The Elements of archaeological conservation*. London: Routledge.

Field, J. H., and J. Dodson. 1999. Late Pleistocene megafauna and archaeology from Cuddie Springs, South-eastern Australia. *Proceedings of the Prehistoric Society* 65:275–301.

Foley, K. 1984. The role of the objects conservator in field archaeology. In N. P. Stanley-Price (ed.), *Conservation on archaeological excavations : Proceedings of the ICCROM Conference*, pp. 11–19. Rome: ICCROM.

Furby, J. H., R. Fullagar, J. R. Dodson and I. Prosser. 1993. The Cuddie Springs bone bed revisited 1991. In M. Smith, M. Spriggs and B. Frankhauser (eds), *Sahul in review: Pleistocene archaeology in Australia, New Guinea and Island Melanesia, pp 204–210. Occasional Papers in Prehistory* 24. Canberra: Department of Prehistory, RSPacS, ANU.

Goodburn-Brown, D. and R. Hughes. 1996. A review of some conservation procedures for the reburial of archaeological sites in London. In A. Roy and P. Smith (eds), *Archaeological conservation and its consequences: Proceedings of the IIC Copenhagen Congress 1996*, pp 65–69. London: International Institute for Conservation.

Horton-James, D., S. Walston and S. Zounis. 1991. Evaluation of the stability, appearance and performance of resins for the adhesion of flaking paint on ethnographic objects. Studies in *Conservation* 36:203–221.

Howie, F. 1984. Materials used for conserving fossil specimens since 1930: A review. In N. S. Brommelle, E. M.Pye, P. Smith and G. Thomson (eds), *IIC Preprints of the Contributions to the Paris Congress*, pp 92–97. London: International Institute for Conservation.

Jones, J. 1980. The use of polyurethane foam in lifting large, fragile objects on site. *The Conservator* 4:31–33.

Koob, S. 1986. The use of Paraloid B-72 as an adhesive: Its application for archaeological ceramics and other materials. *Studies in Conservation* 31:7–14.

Leigh, D. 1981. First aid for finds. *A practical guide for archaeologists*. Hertford: Rescue.

O'Connor, S. 1996. Developing a Conservation Strategy in a Rescue Archaeology Environment. In A. Roy, and P. Smith (eds), *Archaeological Conservation and Its Consequences: Proceedings of the IIC Copenhagen Congress 1996*, pp. 133–136. London,IIC

Sease, C. 1987. *A conservation manual for the field archaeologist. Archaeological Research Tools* 4. Los Angeles: Institute of Archaeology, UCLA.